MECHANICS OF ELASTIC STRUCTURES

MECHANICS OF ELASTIC STRUCTURES

J. T. ODEN

Associate Professor of Engineering Mechanics
University of Alabama in Huntsville

McGRAW-HILL BOOK COMPANY

New York St. Louis
San Francisco
Toronto London Sydney

MECHANICS OF ELASTIC STRUCTURES

Library of Congress Catalog Card Number 67-14675

ISBN 07-047599-7

4 5 6 7 8 9 0 KPKP 7 9 8 7 6 5

To my Parents

PREFACE

This book is intended to serve as a text for advanced undergraduate structures students in aeronautical, civil, and mechanical engineering and in engineering mechanics. It is designed for the senior and beginning graduate courses offered by most engineering schools, which attempt to bridge the gap between elementary strength of materials and the more advanced courses in structural analysis and structural mechanics. The book should also be of interest to the structural engineer who wishes to strengthen his background in linear structural mechanics.

With few exceptions, all developments are confined to linearly elastic structures undergoing small deformations. An attempt is made to provide the student with a background in the classical theory of elasticity. This is followed by the development of approximate theories for studying practical one- and two-dimensional structural problems. In every case, the relationships between the simplified and the more general theories are indicated and the limitations of the approximate methods are identified. No attempt is made to cover the entire spectrum of structural problems which may confront the engineer. Rather, emphasis is placed on the systematic solution of a number of representative problems which demonstrate the basic principles of structural analysis, permit a definite physical interpretation of the results, and are within the grasp of students who have a limited background in structural mechanics. Emphasis is given to the physics of structural behavior, and the reader is expected to have no more mathematical background than a knowledge of ordinary differential equations. When additional mathematical tools are needed, they are given adequate coverage in the text.

The book contains nine chapters. It begins with a brief introductory chapter in which the purpose and the scope of the text are established. Here fundamental definitions are presented and the basic assumptions of the linear theory are discussed.

In the second chapter, the general equilibrium, strain-displacement, and compatibility equations for three-dimensional bodies are presented along with the general kinematic and static boundary conditions. Parallel to these developments, the corresponding equations for simplified one-dimensional structural elements are also discussed so that the reader becomes immediately aware of some of the limitations and the ranges of applicability of the elementary theories.

The third chapter is dedicated to the theory of torsion of prismatic bars. Few subjects permit more sound physical arguments in their development and, at the same time, allow a more clear and simple demonstration of the significance of the equations of linear elasticity. Once the elements of Saint-Venant's theory of torsion are established, applications to solid shafts, open sections, tubes, and multicell thin-walled beams are presented.

Chapter 4 deals with the stress analysis of coplanar bars of any shape. Equations are developed from the generalization of the Winkler theory of curved bars. Radial stresses and unsymmetrical bending are also considered. This is followed by a chapter on shearing stresses and shear flow in curved bars, beams of variable depth, monocoque and semimonocoque shells, and multicell tubes. Shear lag in reinforced panels and combined bending and torsion of unrestrained multicell tubes are also discussed.

The sixth chapter begins with the development of general differential equations for the elastic curves of coplanar beams of arbitrary shape. By appropriately modifying these equations, it is shown that a variety of related topics can be studied. These include anticlastic bending, deflections of beam columns and ties, shear deformation in beams, beams on elastic foundations, and the theory of cables. Elastic stability is not treated as a separate topic; instead, it is regarded as one of several important characteristics of structural behavior and is, therefore, included in discussions of the behavior of specific types of structural elements.

Chapter 7 contains a systematic development of the theory of thin-walled members under combined bending and torsion. The Navier hypothesis concerning the conservation of planes during deformation is abandoned, and relatively general formulas are derived for the analysis of such structures. Buckling of thin-walled open sections is discussed, and the governing equations for the analysis of thin-walled closed sections are derived.

Chapters 8 and 9 deal with the work and energy principles of structural mechanics and they comprise approximately one third of the book. In Chap. 8, the principles of virtual displacements and virtual forces are developed largely from physical arguments for systems of particles and rigid bodies. Once the basic ideas have been established, these principles are extended to general three-dimensional bodies. This is followed by applications to the analysis of statically determinate and indeterminate structural systems, wherein the fundamentals of the stiffness and the flexibility methods of structural analysis are presented. Chapter 9 contains a thorough introductory coverage of the energy principles of structural mechanics. The principles of minimum potential energy and minimum complementary energy are derived from the principles of virtual work; from these follow the theorems of Castigliano and Engesser, among others. The ranges of applicability and the relationships among various energy theorems are also discussed.

With the possible exception of Chaps. 1, 2, and 9, each chapter is essentially self-contained. Relatively general discussions are given in the beginning articles of each chapter, and more specialized topics are covered near the end of each chapter. Thus, the instructor can omit some of the special topics at his discretion without loss in continuity. By appropriate selection of topics, the book can be used as a text for courses on aircraft structures, statically indeterminate structures, advanced strength of materials, and energy methods in structural mechanics.

Many example problems are presented throughout the text to demonstrate various aspects of the theory. Also, more than two hundred problems are provided to give the reader an opportunity to test his understanding of the subject matter.

The author wishes to express his sincere gratitude to his wife, Barbara, who not only typed part of the manuscript and helped with the proofreading but who also sacrificed many weekends with her husband so that he could finish this book. In addition, the author thanks Mrs. Patricia Campbell, Miss Nancy Manning, and Mrs. Sandra Atkinson, who typed various portions of the manuscript.

J. T. Oden

CONTENTS

one

INTRODUCTION

1.1 Structural mechanics. We are familiar from our study of elementary mechanics with the concept of particles and bodies. A particle, we recall, is a material dimensionless point having mass, while a body is merely a collection of particles which may be treated as a single object. In fact, it is the province of the science of mechanics to describe and predict the behavior of particles and bodies under the action of forces.

Mechanics is a broad science having many interrelated disciplines, and sharp dividing lines between one branch of mechanics and another are often difficult to define. Frequently, the major subdivisions of mechanics are classified according to the nature of the particles or bodies under consideration. For example, one may study the behavior of fluid particles or solid bodies and in so doing concern himself with the fields of fluid or solid mechanics. Or, one may analyze solid bodies whose deformations do not appreciably influence their behavior and, thus, study rigid-body mechanics. In this volume, the underlying theory of a portion of one of the most interesting and important branches of mechanics is presented, namely, the mechanics of structures.

A structure may be defined as a collection of bodies arranged and supported so that it can resist and transmit loads. Chairs, tables, tree branches, building frames; aircraft, missiles, even the delicate web of a spider fall within the general definition of a structure. The body of knowledge associated with the description and prediction of the behavior of such systems is *structural mechanics*.

1.2 Subclassifications of structural mechanics. To attempt to cover this entire field in a single volume would indeed be a formidable task, if not an impossible one, for structural mechanics, itself, is a subject with many branches. To identify the limited area upon which we intend to focus our attention, we note that structural mechanics is often subclassified according to the material properties or the geometry of the bodies comprised by the system to be studied. The theories of elasticity, plasticity, and viscoelasticity, for example, are each concerned with the study of bodies of various shapes made of materials which possess certain prescribed physical properties.

Furthermore, though structures, in general, may be of any shape, the analytical methods developed to study their behavior often take advantage of geometric characteristics of certain classes of structures. The theories of bars, plates, and shells, for instance, deal with bodies having one or more characteristic dimensions which are small in comparison with the others. Bars, for example, are bodies whose cross-sectional dimensions are small compared with their lengths. A plate is a flat body whose thickness is small compared with its other dimensions. Curved plates are called shells. More generally, a shell is a material surface of relatively small thickness; light bulbs, missile hulls, and parachutes are examples of shells.

The theories of bars, plates, and shells, of course, can be further broken down into a variety of subclassifications. When the relative dimensions of a body do not fit the above definitions, a more general theory—perhaps the theory of elasticity if the body is elastic—must be used to study its behavior.

1.3 The linear theory of structures. Although many of the principles that we are to cover apply to plates, shells, and even bodies of general shape, the majority of their applications are demonstrated through the analysis of bars and bar systems. In the chapters to follow, we propose to establish methods for predicting the behavior of such structures under the action of given applied loads and to identify, when possible, the limitations of our results by comparing them with more general theories. Our primary concern will be with systems at rest, and the majority of applications of the principles which we will establish will deal with the analysis of one- and two-dimensional structures. It is in the analysis of these types of systems that the details of the theory are often more clearly demonstrated.

With few exceptions, our investigations fall into that portion of structural mechanics generally referred to as *the linear theory of structures*. Unless noted otherwise, we assume that all members of the structural system are constructed of linearly elastic and continuously distributed materials which are homogeneous (have the same properties throughout) and isotropic (have the same properties in every direction); and we make the additional important assumption that the displacements of points on any member of the structure are small compared with the dimensions of that member.

The latter assumption is justifiable for a great many structural systems; the deformation of most structural systems under normal working loads can seldom be detected with the unaided eye. No structural material, however, is perfectly homogeneous and isotropic. Under a microscope a piece of mild steel appears to be an inhomogeneous mass of individual crystals each being far from isotropic. A look at a specimen of concrete is even more discouraging. Yet, the equations of the linear theory can predict quite accurately the stresses and deformations of many steel and even concrete structures. This is primarily because there are millions of crystals or component parts oriented at random within each specimen. Each part is so small compared with the dimensions of the body that the behavior of the specimen represents the statistical behavior of its many parts. The elastic properties that we measure in the laboratory, then, are essentially averages of the properties of each component part of the specimen. Obviously, if the dimensions of these parts are not small compared with those of the body and if they are not oriented at random, the assumptions of homogeneity and isotropy are no longer valid.

The behavior of linear structures, as the name implies, can be described with linear equations. We will find that often (but not always) linear structures may be assumed to obey *the principle of superposition*, which simply states that the combined effect of a number of loads acting on a structure is equal to the sum of the effects of each load applied separately. This is one of the most important principles of the linear theory of structures, and many of the methods developed to study such systems are based, in part, on the assumption that this principle is valid. In some instances, however, superposition is not valid, at least without some definite limitations. These cases are pointed out during the course of our study.

two

STRUCTURAL BEHAVIOR

2.1 General. Our principal course has been set as one of determining the behavior of structural systems, particularly linear systems, at rest. Behavior, by definition, is the way a system responds to a stimulus; and in structural systems stimuli may be present in many forms. For example, such things as systems of static or dynamic loads, errors in the fabrication and assembly of members of the structure, the displacement of structural supports, and changes in temperature or material properties of portions of the structure are structural stimuli.

In the theory of structures we assume that the stimuli are known and that they can be represented with sufficient accuracy by mathematical formulas. These mathematical abstractions are often symbolic and ideal; their correspondence with actual stimuli has a significant influence on the accuracy with which we are able to predict the response of a structure.

The response of structural systems, for our purposes, is characterized in two ways: first by the distribution of stresses and stress resultants developed in the system and second by the nature of strains and displacements at all points in the system. When, for a given stimulus, these two features of structural response are determined, the behavior of the structure is completely defined.

2.2 Forces and stresses in structural systems. Our study of the first characteristic of structural behavior is essentially based on the concept of stress and on the special cases of Newton's laws of motion pertaining to bodies at rest, namely, the familiar laws of statics:

$$\Sigma F_x = 0 \qquad \Sigma F_y = 0 \qquad \Sigma F_z = 0 \tag{2.1}$$

and

$$\Sigma M_x = 0 \qquad \Sigma M_y = 0 \qquad \Sigma M_z = 0 \tag{2.2}$$

In words, the sum of forces in any three mutually perpendicular directions and the sum of moments about lines in these directions must vanish for equilibrium to exist. With these equations we impose certain restrictions on the nature of the *external* and the *internal* forces and moments which exist in a system.

The *external* forces which act on a deformable body may be in the form of known applied loads or support reactions. In general, external forces are independent of the deformation and the material properties of the

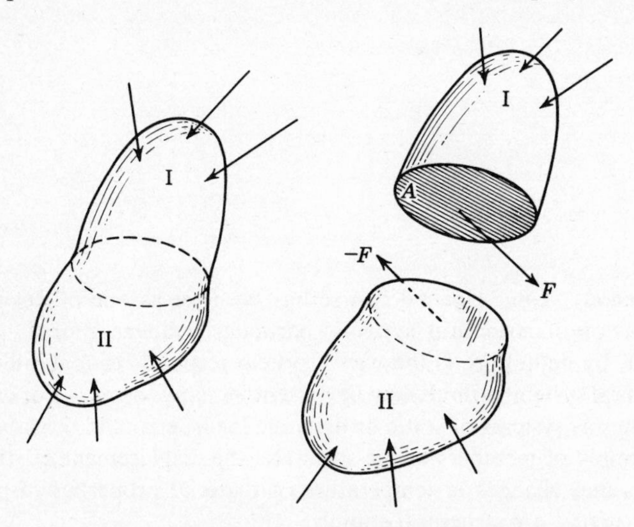

FIGURE 2.1 Internal force F acting on a plane section of area A within a three-dimensional body.

body, though they may depend upon the distribution of the mass of the body. We refer to forces produced by effects external to the region occupied by a body as external forces, even though these forces may actually be developed within the boundaries of the body. For example, the source of the effects of the earth's gravitational field on a body is external to the body, but it produces forces at each point within the body in proportion to the mass density at the point. Thus, the unit weight of a body is regarded as an external force distributed throughout the volume of the body. In the general case of a three-dimensional deformable body, we separate external forces into two categories: we classify them as either *body forces*, which are given in terms of force per unit volume and which act within the boundaries of the body, or we classify them as *surface forces*, which are given in terms of force per unit surface area and which act on the boundary surfaces of the body. Body forces may represent effects such as the weight of the body, centrifugal forces, etc. Forces such as distributed support reactions or those due to hydrostatic pressure may be treated as surface forces. To represent the distribution of external forces mathematically, we resolve them into components acting parallel to the axes of a right-handed cartesian coordinate system, x, y, z. Thus, we denote the x, y, z components of the body force per unit volume by X_b, Y_b, Z_b, respectively, and the x, y, z components of the surface force per unit area by X_s, Y_s, Z_s, respectively. External forces usually represent a stimulus of the structural behavior of the body.

A body responds to the application of external forces by deforming and by developing *internal* forces. Consider, for example, the three-dimensional body in Fig. 2.1, which is in equilibrium under the action of a general system of external forces. We can imagine that the body is separated into two parts, I and II, by slicing it along a plane section of area A which cuts the boundaries along the dashed line indicated in the figure. Since both portions of the body must be in equilibrium, forces must be developed on the contact areas A of each free body in order to balance the external forces acting on these bodies. According to Newton's third law, if F is the force on area A of body I, an equal and opposite force $-F$ acts on area A of body II. Clearly, the precise nature of F depends upon the section we select to separate the two free bodies. Since the area A on which F acts lies within the boundaries of the body after portions I and II are reconnected, we refer to F as an internal force. We assume that the internal force F results from forces which are continuously distributed over A.

The familiar notion of stress is a measure of the internal force per unit area within a body. In general, if ΔA is an element of plane area within a body and if, as we assumed earlier, the material of the body is distributed continuously within its boundaries, we may reduce ΔA continuously to a point P. Further, if ΔF is an increment of internal force normal to ΔA, we define *the component of unit normal stress at the point P in the direction of* ΔF *by*

$$\sigma = \lim_{\Delta A \to 0} \frac{\Delta F}{\Delta A} \tag{2.3}$$

If ΔF is parallel rather than normal to ΔA, we refer to the limit of the ratio in Eq. (2.3) as τ, *the component of unit shearing stress at P in the direction of* ΔF. Clearly, the specific values of σ and τ which are obtained from these definitions depend upon the orientation of the plane area ΔA through P. These components are indicated in Fig. 2.2.

To describe the *state of stress at a point* within a body, we refer the components of normal and shearing stress to the *xyz* coordinate axes and, following a standard means of identification, give normal stresses a single subscript to stand for the directions in which they act and shearing stresses two subscripts, the first indicating the direction of a normal to the plane on which the stress occurs and the second indicating the direction of the stress. Thus, σ_x is a normal stress acting in the x direction and τ_{xy} is shearing stress on a plane normal to the x axis acting in the y direction (see Fig. 2.3). Since the material is continuous, we assume that the components of stress are continuous functions of the coordinates x, y, and z. From this it follows that the state of stress at any point is defined by, at most, nine components of stress: σ_x, σ_y, σ_z, τ_{xy}, τ_{yx}, τ_{xz}, τ_{zx}, τ_{yz}, and τ_{zy}.

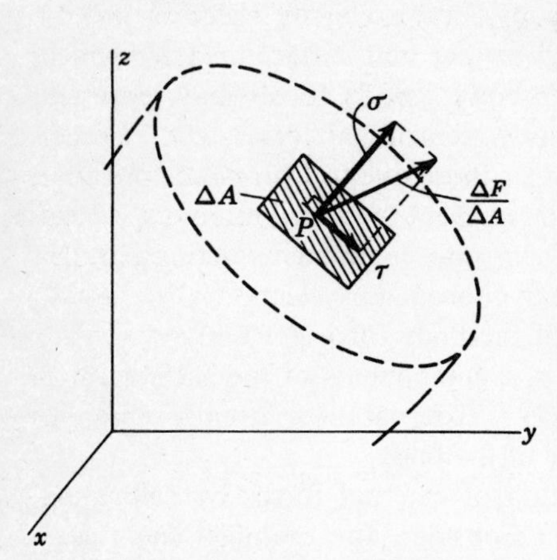

FIGURE 2.2 Normal and shearing stress components at point P.

2.3 Equilibrium conditions in three-dimensional bodies. Let us now examine the equilibrium requirements on external and internal forces at a point P given by the coordinates (x,y,z) in the deformed body. To do this, we consider a point P' close to P which has the coordinates $(x + \Delta x, y + \Delta y, z + \Delta z)$ and note that all stress components acquire a small change from P to P'. In so doing we define a differential element of volume $\Delta x\, \Delta y\, \Delta z$ in equilibrium under the action of the stress system indicated in Fig. 2.3*a*. For clarity, only the forces acting in the x direction on the element are shown in Fig. 2.3*b*. Note that no surface forces are present since the element is within the boundaries of the body. Thus, summing forces in the x direction, we find

$$\frac{\partial \sigma_x}{\partial x} \Delta x\, \Delta y\, \Delta z + \frac{\partial \tau_{yx}}{\partial y} \Delta x\, \Delta y\, \Delta z + \frac{\partial \tau_{zx}}{\partial z} \Delta z\, \Delta x\, \Delta y + X_b\, \Delta x\, \Delta y\, \Delta z = 0$$

where, again, X_b is the x component of the body force at P.

Dividing by $\Delta x\, \Delta y\, \Delta z$ and taking the limit as Δx, Δy, and Δz approach zero, we arrive at the equilibrium condition

$$\frac{\partial \sigma_x}{\partial x} + \frac{\partial \tau_{yx}}{\partial y} + \frac{\partial \tau_{zx}}{\partial z} + X_b = 0$$

By continuing in this manner, we obtain the equilibrium conditions for forces in the y and z directions. In summary, the differential equations of equilibrium in terms of stresses and body forces are

$$\begin{aligned}\frac{\partial \sigma_x}{\partial x} + \frac{\partial \tau_{yx}}{\partial y} + \frac{\partial \tau_{zx}}{\partial z} + X_b = 0 \\ \frac{\partial \tau_{xy}}{\partial x} + \frac{\partial \sigma_y}{\partial y} + \frac{\partial \tau_{zy}}{\partial z} + Y_b = 0 \\ \frac{\partial \tau_{xz}}{\partial x} + \frac{\partial \tau_{yz}}{\partial y} + \frac{\partial \sigma_z}{\partial z} + Z_b = 0\end{aligned} \tag{2.4}$$

We have used only three of the six independent equations of statics. Three additional equilibrium conditions are furnished by applying Eqs. (2.2). For example, summing moments about the x axis and taking the limit as Δx, Δy, and Δz approach zero gives

$$\tau_{yz} - \tau_{zy} = 0$$

Similarly, summing moments about the y and z axes yields two equations involving the remaining components of shearing stress. From these, we find that the moments of forces on the element are in equilibrium provided

$$\tau_{xy} = \tau_{yx} \qquad \tau_{xz} = \tau_{zx} \qquad \tau_{yz} = \tau_{zy} \tag{2.5}$$

Thus, the state of stress at P is defined by only six independent components of stress: σ_x, σ_y, σ_z, τ_{xy}, τ_{xz}, and τ_{yz}.

On an element of surface area ΔS we denote the stresses due to the external forces by the components X_s, Y_s, Z_s and refer to them as surface forces. The conditions of equilibrium at a point on the boundary of the body are in terms of stresses and surface forces, and they are often termed *static boundary conditions.* To evaluate these conditions, let us consider the tetrahedral element shown in Fig. 2.4. The outside face of the tetrahedron is an element of surface area ΔS on which we construct a normal having direction cosines l, m, and n. The three remaining faces are within the body and hence are subjected to internal forces in terms of the components of stress. The interior faces are of area $l\,\Delta S$, $m\,\Delta S$, and $n\,\Delta S$, and the volume of the element is $2\sqrt{lmn/3}\,\Delta S^{3/2}$. Thus, summing forces in the x direction, we find

$$X_b 2\sqrt{\frac{lmn}{3}}\,\Delta S^{3/2} + X_s\,\Delta S - \sigma_x l\,\Delta S - \tau_{xy} m\,\Delta S - \tau_{xz} n\,\Delta S = 0$$

Dividing this result by ΔS and taking the limit as ΔS approaches zero, we find that the body force term vanishes and that the resulting static boundary condition is

$$X_s - \sigma_x l - \tau_{xy} m - \tau_{xz} n = 0$$

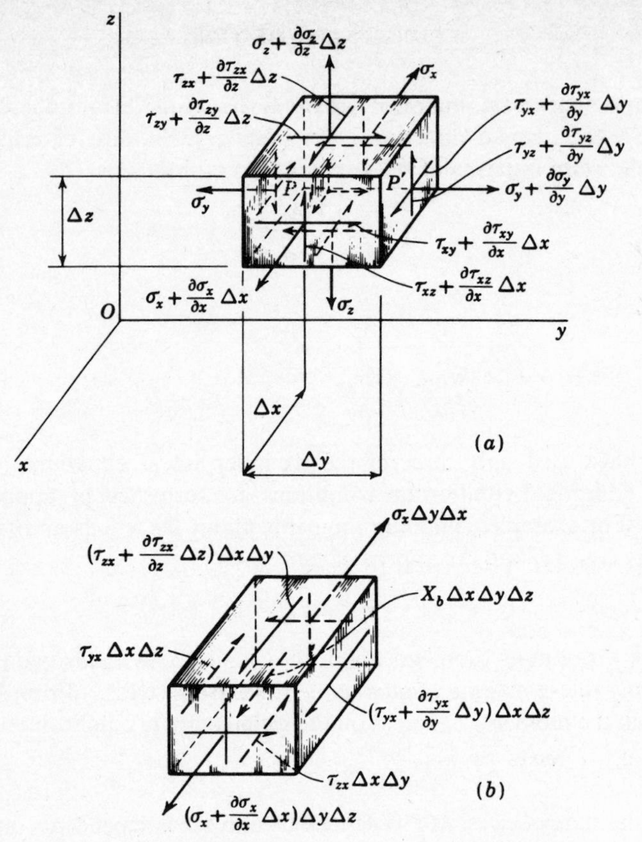

FIGURE 2.3 (*a*) General state of stress and its variation from a point P to P'; (*b*) internal and external forces in the x direction acting on a typical volume element.

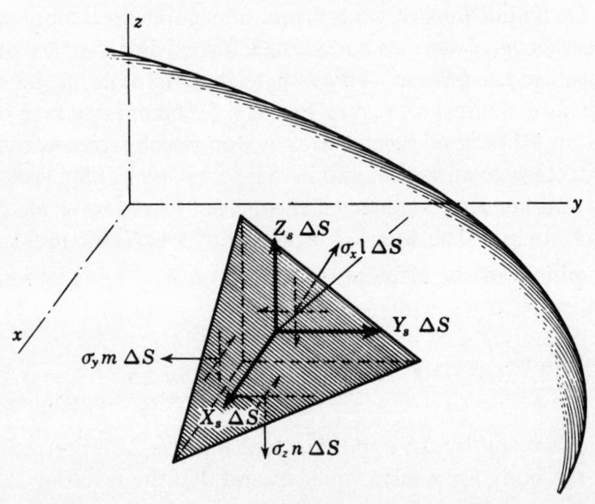

FIGURE 2.4 Forces acting on a boundary element of surface area ΔS.

By proceeding in this same manner, we obtain two other conditions involving forces and stresses in the y and z directions. Finally, we have for the three independent static boundary conditions

$$
\begin{aligned}
X_s &= \sigma_x l + \tau_{xy} m + \tau_{xz} n \\
Y_s &= \tau_{xy} l + \sigma_y m + \tau_{yz} n \\
Z_s &= \tau_{xz} l + \tau_{yz} m + \sigma_z n
\end{aligned}
\tag{2.6}
$$

2.4 Stress resultants in bars. Equations (2.4), (2.5), and (2.6) are the general equations of equilibrium for a three-dimensional body. In a study of bars and bar systems we shall seldom have the occasion to use them in their general form. We introduce these relationships only because we shall often find it beneficial to relate simplified developments to the more general case in order to evaluate their limitations. In fact, the equilibrium conditions to be satisfied by internal and external forces in bar systems are considerably simpler than those of the general three-dimensional body.

All bars, of course, are in reality three-dimensional bodies; but because of their relative dimensions it is usually possible to predict their behavior by analyzing *mathematical models* of actual bars. By a mathematical model, we mean an idealization of the actual bar which can be studied mathematically and which displays with sufficient accuracy enough characteristics of the actual bar to allow us to predict its behavior. The mathematical model of a bar is essentially a curve representing the bar's geometric axis and the set of cross-sectional planes normal to this axis. External forces are then represented by systems of forces and moments applied at points along this curve, and internal forces are defined in terms of stresses developed on the cross-sectional planes.

The nature of this mathematical model makes it more convenient to express the equilibrium conditions of bar elements in terms of stress resultants rather than the stress components themselves. Stress resultants, as their name implies, are the internal forces and moments which result from integrating a given stress distribution over the area of a cross-sectional plane. Consider, for example, the bar of general shape shown in Fig. 2.5*b*. On a typical plane section normal to the bar's geometric axis we establish an orthogonal coordinate system; two axes y and z lie in the plane of the section and the third axis x is normal to the section. Under the most general loading, the stress on an element of cross-sectional area ΔA is given by only three independent components: σ_x, the normal stress, and τ_{xy} and τ_{xz}, the shearing stresses on the section. The other normal stress components, of course, may exist; but owing to the geometry and loading they are often negligible.[1] When this stress distribution is

[1] Such stresses are investigated in Chap. 5.

integrated over the cross section it yields six[2] independent *stress resultants:* two *shearing forces* V_y and V_z acting parallel to the section, a *normal force* N_x, two *bending moments* M_y and M_z acting about the y and z axes, and a *torque* or *twisting moment* M_x acting about the x axis (Fig. 2.5*b*). Thus,

$$\begin{aligned} N_x &= \int \sigma_x \, dA & M_x &= \int (\tau_{xz} y - \tau_{xy} z) \, dA \\ V_y &= \int \tau_{xy} \, dA & M_y &= \int \sigma_x z \, dA \\ V_z &= \int \tau_{xz} \, dA & M_z &= \int \sigma_x y \, dA \end{aligned} \tag{2.7}$$

In correspondence with the stress components in Eqs. (2.4), these stress resultants must be in equilibrium with the external forces within the boundaries of the bar.

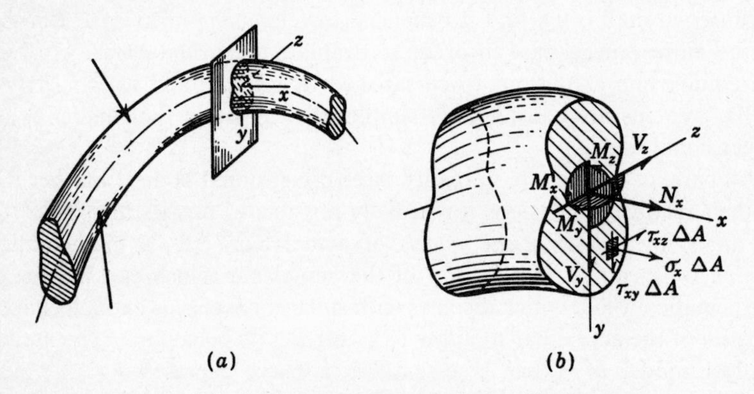

FIGURE 2.5 (*a*) A bar of general shape; (*b*) the stress resultants on a typical cross section of the bar.

The precise form of the differential equations of equilibrium for bars depends upon the loading and the geometry of the bar and, hence, varies from case to case. The special case of a straight prismatic bar loaded only in the xy plane provides a simple example.[3] Referring to Fig. 2.6, we isolate an element between cross sections a distance x and $x + \Delta x$ from the origin and apply the equations of statics. By then allowing Δx to approach zero, we find that the stress resultants shown are in equilibrium at x provided

$$\frac{dV_y}{dx} = -p_y \qquad \text{and} \qquad \frac{dM_z}{dx} = V_y$$

where p_y is the y component of the external force per unit length, in analogy with Y_b in Eqs. (2.4).

[2] It is shown in Chap. 7 that in some cases it is possible to develop more than six stress resultants.

[3] A more general form of the equilibrium conditions for bars is given in Art. 4.2.

Similarly, stress resultants in bars must satisfy static boundary conditions analogous to Eqs. (2.6) involving stresses. The boundaries of a bar are simply its end points, and here surface forces reduce to end moments and forces. For example, if end A of the bar in Fig. 2.6 is subjected only to a force P in the y direction, the static boundary conditions at A are

$$(V_y)_{x=0} = -P$$

and $$V_z = N_x = M_x = M_y = M_z = 0 \qquad \text{at } x = 0$$

In this case P is equivalent to the integral of Y_s over the surface area normal to the x axis at A.

This practice of replacing the surface forces at the ends of bars by statically equivalent forces and moments obviously leads to some error in the true stress distribution at the boundaries of the bar. The validity of our mathematical model of the bar's behavior is based on the premise that these errors are insignificant so far as the overall behavior of the bar is concerned and that they are confined to small regions near the ends of the bar. Fortunately, this premise is supported by a fundamental principle of solid mechanics known as *Saint-Venant's principle*.[4] This principle states that two different but statically equivalent force systems acting on a small portion of the surface of an elastic body produce the same stress distribution at distances large in comparison with the linear dimensions of the portion of the surface on which these forces act. For the bar in Fig. 2.6, for example, these linear dimensions are those of the cross section at end A. Saint-Venant's principle is applicable to bars of solid cross section; however, in the case of thin-walled bars under general loading, applications of the principle may lead to significant errors.[5]

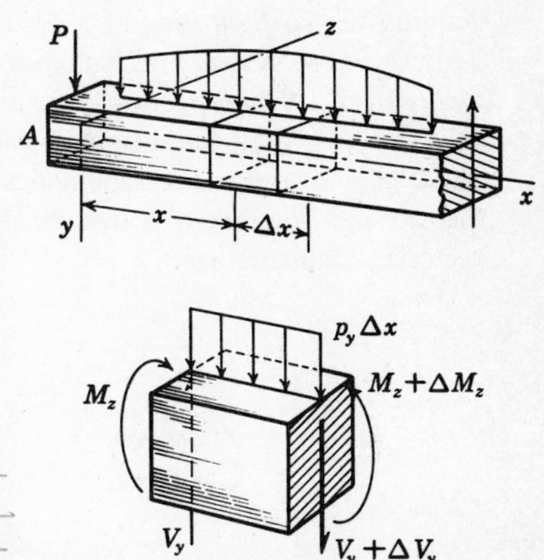

FIGURE 2.6 Straight prismatic bar loaded in the xy plane.

It is important to note that the conditions of equilibrium seldom provide us with enough information to completely determine the stresses in a body. Structures in which the distribution of stresses or stress resultants cannot

[4] This principle was first presented in 1855 by Barré Saint-Venant.

[5] See Chap. 7.

be completely determined by statics are said to be *statically indeterminate*, and the additional stress resultants developed which are not needed for equilibrium are called *statical redundants*. To solve statically indeterminate stress problems, we must supplement the equations of statics with additional conditions of deformation. We refer to the least number of additional deformation conditions needed to determine the stresses as the *degree of statical indeterminacy* of the body. Thus, without mentioning specific stress-strain or stress resultant-displacement relations, we see that the two characteristics of structural response are necessarily related and that seldom can one be defined without considering the other.

2.5 Displacements and strains in structural systems. Our study of the second feature of structural response closely parallels the first. We begin by defining the *configuration* of a system as the simultaneous position of all particles in the system and a change in the configuration of a system as a *displacement*. When portions of a system are displaced relative to one another we say that the original system has *deformed*. Our present objective is to examine the kinematics of deformation in a general way so that we can identify a convenient set of variables which completely characterizes deformation.

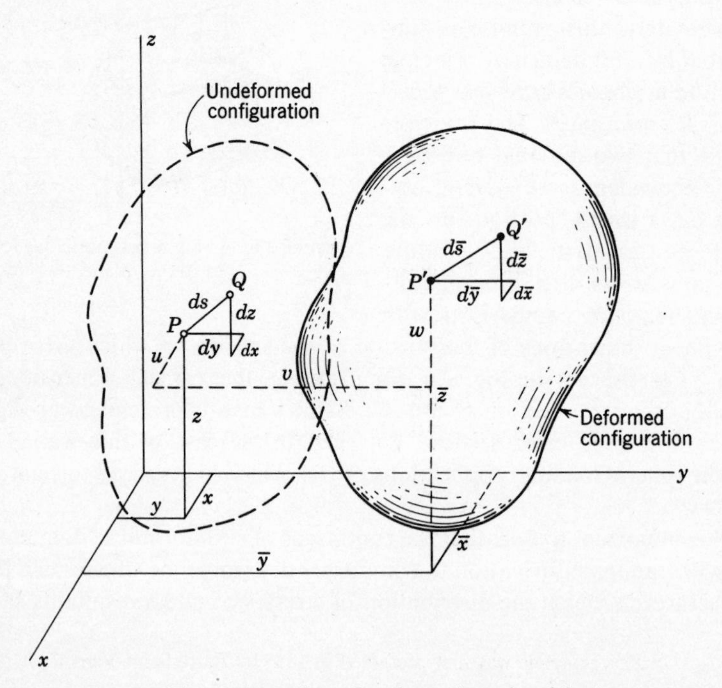

FIGURE 2.7 Deformation of a three-dimensional body.

To this end, consider two points P and Q in the initial (undeformed) configuration of a three-dimensional body which are the end points of a differential line element ds (Fig. 2.7). Before deformation, the coordinates of the point P are (x,y,z) and the coordinates of Q are $(x + dx, y + dy, z + dz)$, where dx, dy, and dz are the projections of ds on the fixed x, y, z coordinate axes. According to the familiar theorem of Pythagoras, the square of the length of this line element is given by

$$ds^2 = dx^2 + dy^2 + dz^2 \tag{2.8}$$

Now suppose that the body is deformed so that the particles at P and Q are displaced from their original positions to points P' and Q' whose coordinates are $(\bar{x},\bar{y},\bar{z})$ and $(\bar{x} + d\bar{x}, \bar{y} + d\bar{y}, \bar{z} + d\bar{z})$, respectively. Thus, a material point with coordinates (x,y,z) in the undeformed body has coordinates $(\bar{x},\bar{y},\bar{z})$ in the deformed body. In general, the line element ds is stretched and rotated during the deformation so that the displaced points are connected by a line of length $d\bar{s}$ instead of ds. It follows that the square of the length of the line element in the deformed body is given by

$$d\bar{s}^2 = d\bar{x}^2 + d\bar{y}^2 + d\bar{z}^2 \tag{2.9}$$

Since the term *displacement* refers to a change in the configuration of a system, the functions

$$u = \bar{x} - x \qquad v = \bar{y} - y \qquad w = \bar{z} - z \tag{2.10}$$

define the displacements of points in the x, y, and z directions, respectively. These functions are called the *components of displacement*, and we assume that they are continuous functions of the fixed coordinates x, y, and z.

Solving Eqs. (2.10) for $\bar{x}$, $\bar{y}$, and $\bar{z}$ and substituting the results into Eq. (2.9), we get

$$d\bar{s}^2 = dx^2 + dy^2 + dz^2 + 2du\,dx + 2dv\,dy + 2dw\,dz + du^2 + dv^2 + dw^2$$

or, in view of Eq. (2.8),

$$d\bar{s}^2 - ds^2 = 2(du\,dx + dv\,dy + dw\,dz) + du^2 + dv^2 + dw^2 \tag{2.11}$$

From the chain rule of partial differentiation, we recall that

$$\begin{aligned} du &= \frac{\partial u}{\partial x}dx + \frac{\partial u}{\partial y}dy + \frac{\partial u}{\partial z}dz \\ dv &= \frac{\partial v}{\partial x}dx + \frac{\partial v}{\partial y}dy + \frac{\partial v}{\partial z}dz \\ dw &= \frac{\partial w}{\partial x}dx + \frac{\partial w}{\partial y}dy + \frac{\partial w}{\partial z}dz \end{aligned} \tag{2.12}$$

Thus, substituting Eqs. (2.12) into Eq. (2.11) and simplifying, we obtain

$$
\begin{aligned}
d\bar{s}^2 - ds^2 = {} & 2\left\{\frac{\partial u}{\partial x} + \frac{1}{2}\left[\left(\frac{\partial u}{\partial x}\right)^2 + \left(\frac{dv}{\partial x}\right)^2 + \left(\frac{\partial w}{\partial x}\right)^2\right]\right\} dx^2 \\
& + 2\left\{\frac{\partial v}{\partial y} + \frac{1}{2}\left[\left(\frac{\partial u}{\partial y}\right)^2 + \left(\frac{\partial v}{\partial y}\right)^2 + \left(\frac{\partial w}{\partial y}\right)^2\right]\right\} dy^2 \\
& + 2\left\{\frac{\partial w}{\partial z} + \frac{1}{2}\left[\left(\frac{\partial u}{\partial z}\right)^2 + \left(\frac{\partial v}{\partial z}\right)^2 + \left(\frac{\partial w}{\partial z}\right)^2\right]\right\} dz^2 \\
& + 2\left(\frac{\partial v}{\partial x} + \frac{\partial u}{\partial y} + \frac{\partial u}{\partial x}\frac{\partial u}{\partial y} + \frac{\partial v}{\partial x}\frac{\partial v}{\partial y} + \frac{\partial w}{\partial x}\frac{\partial w}{\partial y}\right) dx\,dy \\
& + 2\left(\frac{\partial w}{\partial x} + \frac{\partial u}{\partial z} + \frac{\partial u}{\partial x}\frac{\partial u}{\partial z} + \frac{\partial v}{\partial x}\frac{\partial v}{\partial z} + \frac{\partial w}{\partial x}\frac{\partial w}{\partial z}\right) dx\,dz \\
& + 2\left(\frac{\partial w}{\partial y} + \frac{\partial v}{\partial z} + \frac{\partial u}{\partial y}\frac{\partial u}{\partial z} + \frac{\partial v}{\partial y}\frac{\partial v}{\partial z} + \frac{\partial w}{\partial y}\frac{\partial w}{\partial z}\right) dy\,dz
\end{aligned}
\tag{2.13}
$$

The quantity $d\bar{s}^2 - ds^2$ thus provides a convenient and natural means for describing the deformation of the body. For example, if

$$d\bar{s}^2 - ds^2 = 0 \tag{2.14}$$

no relative displacement occurs between particles originally at points P and Q. Thus, if the components of displacement are such that their first partial derivatives vanish, Eq. (2.14) is satisfied and we say that the body has undergone a *rigid-body* motion. On the other hand, if $d\bar{s}^2 - ds^2$ is other than zero, it follows that the original line elements dx, dy, and dz have changed in length and in their orientation relative to one another during the motion of the body. If such is the case, we say that fibers in the body have been *strained* and that points in the body are in a *state of strain.* Since the character of this strained state is indicated by the dimensionless coefficients of the terms dx^2, dy^2, dz^2, $dx\,dy$, $dx\,dz$, and $dy\,dz$ on the right side of Eq. (2.13), it is logical to take these quantities as *strain measures.* These coefficients completely characterize the deformation of the body.

With this in mind, we adopt a standard notation and rewrite Eq. (2.13) in the form

$$
\begin{aligned}
d\bar{s}^2 - ds^2 = {} & 2\epsilon_x\,dx^2 + 2\epsilon_y\,dy^2 + 2\epsilon_z\,dz^2 \\
& + 2\gamma_{xy}\,dx\,dy + 2\gamma_{xz}\,dx\,dz + 2\gamma_{yz}\,dy\,dz
\end{aligned}
\tag{2.15}
$$

where

$$
\begin{aligned}
\epsilon_x &= \frac{\partial u}{\partial x} + \frac{1}{2}\left[\left(\frac{\partial u}{\partial x}\right)^2 + \left(\frac{\partial v}{\partial x}\right)^2 + \left(\frac{\partial w}{\partial x}\right)^2\right] \\
\epsilon_y &= \frac{\partial v}{\partial y} + \frac{1}{2}\left[\left(\frac{\partial u}{\partial y}\right)^2 + \left(\frac{\partial v}{\partial y}\right)^2 + \left(\frac{\partial w}{\partial y}\right)^2\right] \\
\epsilon_z &= \frac{\partial w}{\partial z} + \frac{1}{2}\left[\left(\frac{\partial u}{\partial z}\right)^2 + \left(\frac{\partial v}{\partial z}\right)^2 + \left(\frac{\partial w}{\partial z}\right)^2\right]
\end{aligned}
\tag{2.16a}
$$

and
$$\gamma_{xy} = \frac{\partial v}{\partial x} + \frac{\partial u}{\partial y} + \frac{\partial u}{\partial x}\frac{\partial u}{\partial y} + \frac{\partial v}{\partial x}\frac{\partial v}{\partial y} + \frac{\partial w}{\partial x}\frac{\partial w}{\partial y}$$

$$\gamma_{xz} = \frac{\partial w}{\partial x} + \frac{\partial u}{\partial z} + \frac{\partial u}{\partial x}\frac{\partial u}{\partial z} + \frac{\partial v}{\partial x}\frac{\partial v}{\partial z} + \frac{\partial w}{\partial x}\frac{\partial w}{\partial z} \tag{2.16b}$$

$$\gamma_{yz} = \frac{\partial w}{\partial y} + \frac{\partial v}{\partial z} + \frac{\partial u}{\partial y}\frac{\partial u}{\partial z} + \frac{\partial v}{\partial y}\frac{\partial v}{\partial z} + \frac{\partial w}{\partial y}\frac{\partial w}{\partial z}$$

The quantities ϵ_x, ϵ_y, ϵ_z, γ_{xy}, γ_{xz}, and γ_{yz} are called the *components of strain* and are said to define the state of strain at points in the deformed body. We refer to Eqs. (2.16) as the *strain-displacement relations*. If the strain components are specified, these relations provide a system of highly nonlinear partial differential equations in the unknown components of displacement.

Hereafter, we assume that u, v, and w are continuous, single-valued functions of x, y, z and that they are small compared with characteristic dimensions of the body. In addition, we assume that squares and products of the first partial derivatives of u, v, and w are negligible compared with the derivatives themselves. Thus, we treat the strains and the derivatives of the displacements as infinitesimal quantities, and in so doing we considerably simplify the analysis of deformable bodies. If the displacements and their derivatives are small, the strains of fibers in one plane are not influenced by out-of-plane displacements, the geometry of the undeformed body can be used when writing the equilibrium conditions, and the strain-displacement formulas in Eqs. (2.16) reduce to the linear relations

$$\epsilon_x = \frac{\partial u}{\partial x} \qquad \epsilon_y = \frac{\partial v}{\partial y} \qquad \epsilon_z = \frac{\partial w}{\partial z} \tag{2.17a}$$

and
$$\gamma_{xy} = \frac{\partial v}{\partial x} + \frac{\partial u}{\partial y} \qquad \gamma_{xz} = \frac{\partial w}{\partial x} + \frac{\partial u}{\partial z} \qquad \gamma_{yz} = \frac{\partial w}{\partial y} + \frac{\partial v}{\partial z} \tag{2.17b}$$

When we write the strain-displacement relations in the form given in Eqs. (2.17), the strain components can be given simple physical interpretations. The component ϵ_x, for example, clearly represents the rate of change of the displacement u with respect to x. In simpler terms, ϵ_x is the change in length per unit length of fibers originally parallel to the x axis. Similar interpretations apply to ϵ_y and ϵ_z. It is for these reasons that the strain components ϵ_x, ϵ_y, and ϵ_z are referred to as *extensional* or *longitudinal* strains.

The remaining strain components γ_{xy}, γ_{xz}, and γ_{yz} are called *shearing strains*. When defined as in Eqs. (2.17*b*), these quantities represent the changes in the original right angles between line elements dx, dy, and dz

due to deformation. Consider, for example, the distortion of line elements dx and dy in the xy plane, as is indicated in Fig. 2.8. After deformation, these line elements have rotated through angles γ_1 and γ_2 relative to their original orientation, as shown in the figure. Since the derivatives

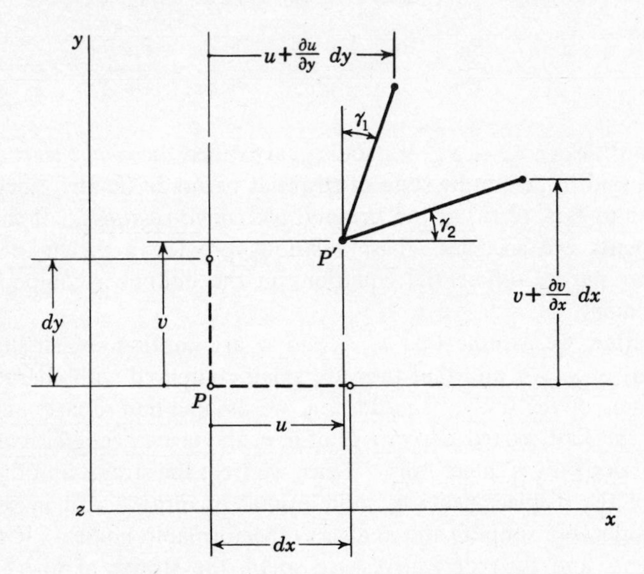

FIGURE 2.8 Displacements and rotations of line elements in the xy plane.

of the displacements are small, the sines of these angles are approximately equal to the angles themselves. Hence

$$\sin \gamma_1 \approx \gamma_1 = \frac{\partial u}{\partial y} \qquad \text{and} \qquad \sin \gamma_2 \approx \gamma_2 = \frac{\partial v}{\partial x}$$

The total decrease in the right angle is equal to the component of shearing strain γ_{xy}:

$$\gamma_{xy} = \gamma_1 + \gamma_2 = \frac{\partial u}{\partial y} + \frac{\partial v}{\partial x}$$

Similar interpretations apply to γ_{xz} and γ_{yz}.

2.6 Deformation conditions in deformable bodies. Just as stresses must satisfy conditions of equilibrium, the components of strain and displacement must satisfy certain conditions of deformation at all points in a body. Perhaps the most obvious condition to be satisfied by displacements is that they must be consistent with constraints imposed at the boundaries.

At a "fixed" boundary point, for example, the functions u, v, and w must be zero. Since descriptions of the motion and deformation of a body need not involve forces and are thus based on purely kinematic considerations, we refer to conditions that we impose on the displacements at points on the boundary as *kinematic boundary conditions.* In analogy with static conditions, when the displacements cannot be completely determined through kinematic considerations, we say that the body is *kinematically indeterminate*, and we must then supplement the equations of deformation with a number of equilibrium conditions equal to the *degree of kinematic indeterminacy* of the body.

Elements within a body may deform in any manner so long as their deformations are consistent with those of adjacent elements. This means that the displacements u, v, and w must not only satisfy certain kinematic boundary conditions but must also be such that no gaps or discontinuities occur in deforming the body. Hence, they must satisfy certain continuity conditions throughout the body and, in view of our previous assumptions, they must be single-valued.

To examine this point further, we temporarily confine our attention to the two-dimensional case of *plane strain* in which particles originally in a plane remain in the same plane after deformation. We do this only for simplicity, since the same line of reasoning applies to the three-dimensional case. In the two-dimensional case, the strain-displacement relations are

$$\epsilon_x = \frac{\partial u}{\partial x} \qquad \epsilon_y = \frac{\partial v}{\partial y} \qquad \gamma_{xy} = \frac{\partial v}{\partial x} + \frac{\partial u}{\partial y} \tag{2.18}$$

Now if the displacement components u and v are given as continuous functions of x and y, we can use Eqs. (2.18) to evaluate the strains at every point in the body. Strains so computed are automatically consistent with the displacements, and the continuity conditions are automatically satisfied throughout the body. On the other hand, if the strain components are specified instead of the displacements, it becomes necessary to regard Eqs. (2.18) as a system of three simultaneous partial differential equations in the unknowns u and v. This system of equations is "overdetermined," since there are three equations and only two unknowns. It follows that we cannot determine u and v uniquely by arbitrarily specifying three strain components. Therefore, we must establish additional conditions to be imposed on the strain components in order to ensure the existence of single-valued continuous solutions u and v to Eqs. (2.18). These conditions are known as the *compatibility conditions*.

To interpret these conditions in another way, consider two points A and B of a deformed body in a state of plane strain. Suppose that the components of displacement and their derivatives are known at point A and that we wish to find the components of displacement of B by summing up (integrating) the changes in u and v along a path from A to B. For example, if u_A is the value of u at point A and if u is a continuous single-valued function, we should be able to calculate u_B from the relation

$$\int_A^B du = u_B - u_A$$

or
$$u_B = u_A + \int_A^B du \tag{2.19}$$

We assume here that du is known in terms of the strains. Equation (2.19), of course, follows directly from the fundamental theorem of integral calculus; it is valid so long as du is an exact differential. However, if the strains are arbitrarily specified, the displacement of B may not be uniquely determined and Eq. (2.19) may yield different values of u_B for different paths of integration. Therefore, *the strain components must be such that the integral in Eq.* (2.19) *is independent of the path of integration from A to B.*[6]

Noting that

$$du = \frac{\partial u}{\partial x} dx + \frac{\partial u}{\partial y} dy = \epsilon_x \, dx + \left(\gamma_{xy} - \frac{\partial v}{\partial x}\right) dy$$

we rewrite Eq. (2.19) in the form

$$u_B = u_A + \int_A^B \epsilon_x \, d(x - x_B) + \int_A^B \left(\gamma_{xy} - \frac{\partial v}{\partial x}\right) d(y - y_B)$$

where (x_B, y_B) are the coordinates of point B and we have made use of the fact that $d(x - x_B) = dx$ and $d(y - y_B) = dy$. Assuming that the strains are continuously differentiable at least through their second partial derivatives, we integrate once by parts and obtain

$$u_B = \left[u_A + (x_B - x_A)(\epsilon_x)_A + (y_B - y_A)\left(\gamma_{xy} - \frac{\partial v}{\partial x}\right)_A\right] - \int_A^B [F(x,y)\, dx + G(x,y)\, dy] \tag{2.20}$$

where
$$F(x,y) = (x - x_B)\frac{\partial \epsilon_x}{\partial x} + (y - y_B)\frac{\partial \epsilon_x}{\partial y} \tag{2.21a}$$

$$G(x,y) = (x - x_B)\frac{\partial \epsilon_x}{\partial y} + (y - y_B)\left(\frac{\partial \gamma_{xy}}{\partial y} - \frac{\partial \epsilon_y}{\partial x}\right) \tag{2.21b}$$

and the notation $(\ )_A$ indicates that the function is evaluated at point A. Since the term in brackets in Eq. (2.20) involves known constants, the problem reduces to one of determining conditions for which the integral

[6] A similar interpretation and derivation is found in Ref. 58, pp. 25–29.

in this equation is independent of the path of integration from A to B. From our previous discussion, we recall that this requirement is equivalent to the condition that the integrand be an exact differential. A necessary and sufficient condition for this integrand to be an exact differential[7] is that

$$\frac{\partial F}{\partial y} - \frac{\partial G}{\partial x} = 0 \tag{2.22}$$

Thus, substituting Eqs. (2.21) into Eq. (2.22) and simplifying, we get

$$(y - y_B)\left(\frac{\partial^2 \epsilon_y}{\partial x^2} + \frac{\partial^2 \epsilon_x}{\partial y^2} - \frac{\partial^2 \gamma_{xy}}{\partial x\, \partial y}\right) = 0$$

Since this must be satisfied for all values of y, it follows that

$$\frac{\partial^2 \epsilon_y}{\partial x^2} + \frac{\partial^2 \epsilon_x}{\partial y^2} = \frac{\partial^2 \gamma_{xy}}{\partial x\, \partial y} \tag{2.23}$$

Equation (2.23) is the compatibility condition for two-dimensional (plane) problems. For obvious reasons, this condition is also referred to as an *integrability condition* for Eqs. (2.18). Since this equation can also be obtained by eliminating u and v from Eqs. (2.18) by differentiation, it is both a necessary and a sufficient condition for the existence of continuous single-valued displacements. Satisfaction of the compatibility conditions becomes an essential part of problems in which the strain components are specified.

In the three-dimensional case, instead of Eq. (2.23) we have six compatibility conditions:

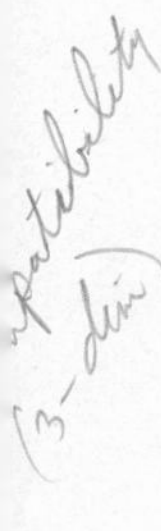

$$\begin{aligned}
&\frac{\partial^2 \epsilon_x}{\partial y^2} + \frac{\partial^2 \epsilon_y}{\partial x^2} = \frac{\partial^2 \gamma_{xy}}{\partial x\, \partial y} \qquad && 2\frac{\partial^2 \epsilon_x}{\partial y\, \partial z} = \frac{\partial}{\partial x}\left(-\frac{\partial \gamma_{yz}}{\partial x} + \frac{\partial \gamma_{xz}}{\partial y} + \frac{\partial \gamma_{xy}}{\partial z}\right)\\
&\frac{\partial^2 \epsilon_y}{\partial z^2} + \frac{\partial^2 \epsilon_z}{\partial y^2} = \frac{\partial^2 \gamma_{yz}}{\partial y\, \partial z} && 2\frac{\partial^2 \epsilon_y}{\partial z\, \partial x} = \frac{\partial}{\partial y}\left(\frac{\partial \gamma_{yz}}{\partial x} - \frac{\partial \gamma_{xz}}{\partial y} + \frac{\partial \gamma_{xy}}{\partial z}\right)\\
&\frac{\partial^2 \epsilon_z}{\partial x^2} + \frac{\partial^2 \epsilon_x}{\partial z^2} = \frac{\partial^2 \gamma_{xz}}{\partial x\, \partial z} && 2\frac{\partial^2 \epsilon_z}{\partial x\, \partial y} = \frac{\partial}{\partial z}\left(\frac{\partial \gamma_{yz}}{\partial x} + \frac{\partial \gamma_{xz}}{\partial y} - \frac{\partial \gamma_{xy}}{\partial z}\right)
\end{aligned} \tag{2.24}$$

We can obtain these equations in a manner similar to that used in the two-dimensional case, or we can obtain them directly from Eqs. (2.17) by eliminating u, v, and w by appropriate differentiations.

Thus, the components of stress must satisfy equilibrium conditions at all points within the body, whereas the components of strain must satisfy compatibility conditions at all points. Similarly, the stresses must satisfy static boundary conditions, whereas strains and displacements must satisfy

[7] This follows from Green's lemma; see, for example, Ref. 59, pp. 391–395.

kinematic boundary conditions at points on the boundary surfaces of the body. Thus, the structural problem involves at least two independent types of conditions—the static conditions, which involve only kinetic variables (e.g., forces, stresses), and the kinematic conditions, which involve only kinematic variables (e.g., displacements, strains). Since both types of conditions can be formulated independently, they do not in themselves determine uniquely a solution to the structural problem. A dependency must be established between the kinetic and the kinematic variables, and the solutions to the static and kinematic conditions must also be consistent with this third set of conditions. These relations between the kinetic and the kinematic variables depend upon the constitution (physical properties) of the material and are thus called *constitutive equations*. For a given problem in which the body forces are prescribed within the body and the surface forces or displacements are prescribed at all points on the boundaries, those distributions of stresses, strains, and displacements which simultaneously satisfy both the equilibrium conditions and the deformation conditions at all points within the body and at all points on its boundaries and which are also consistent with the constitutive equations of the material constitute a unique solution to the problem.

2.7 Deformation of bars. In a study of bar deformations we shall seldom have the occasion to use all of the strain-displacement relations or the conditions of compatibility in their general form. Again, we introduce them here primarily to serve as a basis for evaluating the limitations of more specialized developments. In fact, just as we find it more convenient to deal with stress resultants than stresses in the study of bars, we also find that the behavior of bars is more conveniently described in terms of displacements than strains.

For this reason, in analyzing bars we express the deformations of bar elements in terms of linear displacements, angular displacements, or combinations of both. We refer to linear displacements normal to the bar's axis as *deflections*, and we refer to angular displacements as *rotations*. Rotations transverse to the axis of a bar are called *slopes*, and those about the axis are called *twists*. Again, the straight bar in Fig. 2.9 provides a simple example: u, v, and w are deflections and θ_x, θ_y, and θ_z denote rotations of line elements about the x, y, and z axes. The twist θ_x of a line on a cross section is more frequently denoted ϕ. We are usually able to define the rotations in terms of derivatives of the displacements.

Consistent with these definitions and with the mathematical model of the bar, we refer to the locus of points as the *deflection curve of the bar*. Deflection curves of bars which are capable of returning to their undeformed configuration upon the removal of external effects are called *elastic curves.* For the bar in Fig. 2.9, we might possibly define the shape of the deflection curve through equations of the form

$$u = f(x) \qquad v = g(x) \qquad w = h(x)$$

where f, g, and h are continuous functions.

We shall always find it possible, of course, to relate bar displacements to the components of strain. The elastic curve of a bar, for example, is the accumulative result of the infinitesimal strains of its longitudinal elements. Furthermore, in correspondence with strains in the three-dimensional body, bar displacements must satisfy compatibility conditions as

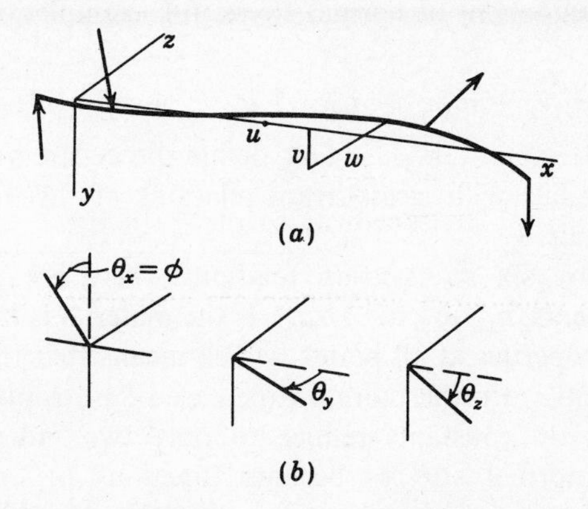

FIGURE 2.9 Deformation of a straight bar. (*a*) Bar deflections; (*b*) slopes.

well as kinematic boundary conditions. The deflection of each bar element, for example, must be consistent with that of adjacent elements in order that the deflection curve be a continuous function of the coordinates. The precise form of the compatibility conditions for bar elements depends upon the geometry of the bar, the nature of the loading, and the assumptions made in arriving at the mathematical model; hence, it varies from case to case. The kinematic boundary conditions for bars are simply conditions imposed upon the displacements and rotations of the end points. The familiar term *fixed-end bar*, for example, refers to a bar whose end displacements must satisfy the kinematic conditions

$$u = v = w = \theta_x = \theta_y = \theta_z = 0$$

Thus, it appears that we may interpret the concepts of the ends of bars being free, hinged, fixed, etc., as convenient physical analogies used to depict static or kinematic boundary conditions, or combinations of both.

2.8 Stress-strain relations. Thus far, all the equations that we have developed are completely independent of the relationship between stress and strain. They are applicable to any type of continuous body undergoing small displacements. However, we noted earlier that to predict the behavior of a structure it is also necessary to know the components of stress as functions of the components of strain and vice versa. It is through the stress-strain relations (constitutive equations) that the material properties of the body enter the problem.

With few exceptions, in the developments to follow we assume that structures under consideration are made of elastic materials that obey *Hooke's*[8] *law.* In its general form, Hooke's law states that the six components of stress may be expressed as linear functions of the six components of strain. A component of normal stress, for example, obeys equations of the form

$$\sigma_x = a_{11}\epsilon_x + a_{12}\epsilon_y + a_{13}\epsilon_z + a_{14}\gamma_{xy} + a_{15}\gamma_{yz} + a_{16}\gamma_{zx}$$

where the coefficients $a_{11}, a_{12}, \ldots, a_{16}$ define the elastic properties of the material. Materials with stress-strain relations of this form are called *Hookean* materials.

Since there are six stress-strain relations, there are, in general, 36 material coefficients, $a_{11}, a_{12}, \ldots, a_{66}$. If the material is homogeneous, it has the same properties at all points, which means that these coefficients must be constants. Furthermore, in the case of isotropic homogeneous materials, these 36 constants reduce to only two independent elastic constants, the normal stresses become functions of only extensional strains, and each shearing stress becomes a function of only its corresponding shearing strain.

The two constants, of course, must be determined experimentally; and, to this end, it is common practice to attempt to subject a specimen of a material assumed to be homogeneous and isotropic to a single component of stress and to measure the corresponding component of strain. Ideally, we obtain relations of the form

$$\sigma_x = E\epsilon_x \qquad \text{and} \qquad \tau_{xy} = G\gamma_{xy} \tag{2.25}$$

in which E is *Young's modulus of elasticity*[9] and G is the *modulus of rigidity* (or the shear modulus) of the material. When more than one component of normal stress is present, however, the stress-strain relations become more complicated. This is because a normal strain component such as ϵ_x is accompanied by proportional lateral contractions of magnitude $-\nu\epsilon_x$, where ν is *Poisson's ratio*,[10] also a constant for the material. These three constants, E, G, and ν, are related according to the equation

$$G = \frac{E}{2(1 + \nu)} \tag{2.26}$$

[8] Named after Robert Hooke (1635–1703), who established the law in 1678.

[9] Named after Thomas Young (1773–1829).

[10] Discovered in 1830 by S. D. Poisson (1781–1840).

It follows[11] that when all components of stress and strain are present in a homogeneous, isotropic Hookean body, they are related as follows:

$$\begin{aligned}
\epsilon_x &= \frac{1}{E}[\sigma_x - \nu(\sigma_y + \sigma_z)] & \gamma_{xy} &= \frac{\tau_{xy}}{G} \\
\epsilon_y &= \frac{1}{E}[\sigma_y - \nu(\sigma_x + \sigma_z)] & \gamma_{yz} &= \frac{\tau_{yz}}{G} \\
\epsilon_z &= \frac{1}{E}[\sigma_z - \nu(\sigma_x + \sigma_y)] & \gamma_{xz} &= \frac{\tau_{xz}}{G}
\end{aligned} \tag{2.27}$$

Further, if we solve these equations for the stresses we find

$$\begin{aligned}
\sigma_x &= \frac{\nu E(\epsilon_x + \epsilon_y + \epsilon_z)}{(1+\nu)(1-2\nu)} + \frac{E}{1+\nu}\epsilon_x & \tau_{xy} &= G\gamma_{xy} \\
\sigma_y &= \frac{\nu E(\epsilon_x + \epsilon_y + \epsilon_z)}{(1+\nu)(1-2\nu)} + \frac{E}{1+\nu}\epsilon_y & \tau_{yz} &= G\gamma_{yz} \\
\sigma_z &= \frac{\nu E(\epsilon_x + \epsilon_y + \epsilon_z)}{(1+\nu)(1-2\nu)} + \frac{E}{1+\nu}\epsilon_z & \tau_{xz} &= G\gamma_{xz}
\end{aligned} \tag{2.28}$$

2.9 Static and elastic stability. In concluding our qualitative examination of structural behavior, it is fitting that we mention the conditions of *static* and *elastic stability* of structural systems. In contrast to statically indeterminate structures, when the number of stress resultants is *less* than the number of independent equations of statics for any portion of a structure, the structure is *statically unstable*.[12] Statically unstable structures cannot resist any general system of applied loads without first undergoing large rigid-body displacements. Static stability is not a function of the material properties of a structure; it is solely dependent upon the geometry of the structure and the boundary conditions. For this reason, this phenomenon is also referred to as *geometric instability* or *improper constraint*.

The *elastic stability*[13] of a structure, on the other hand, is dependent upon the geometry and the material properties of the structure as well as the nature of the applied loading. The term *elastic stability* pertains to the phenomena of structural bodies possessing more than one possible equilibrium configuration for certain loads. In such cases the changes in geometry due to deformation enter the equilibrium considerations and there is no longer a unique solution to a given problem.

[11] See Ref. 72, pp. 25–31, or Ref. 58, pp. 65–68.

[12] See, for example, Ref. 39, pp. 7–17.

[13] Elastic stability is discussed in Arts. 9.5 to 9.7. See also Arts. 6.9 and 7.10.

PROBLEMS

2.1. Show that the equilibrium equations in polar coordinates for a two-dimensional element are

$$\frac{\partial \sigma_r}{\partial r} + \frac{1}{r}\frac{\partial \tau_{r\alpha}}{\partial \alpha} + \frac{\sigma_r - \sigma_\alpha}{r} + F_r = 0$$

$$\frac{1}{r}\frac{\partial \sigma_\alpha}{\partial \alpha} + \frac{\partial \tau_{r\alpha}}{\partial r} + 2\frac{\tau_{r\alpha}}{r} + F_\alpha = 0$$

where F_r and F_α are the body forces and σ_r, σ_α, and $\tau_{r\alpha}$ are the stress components indicated in the figure below.

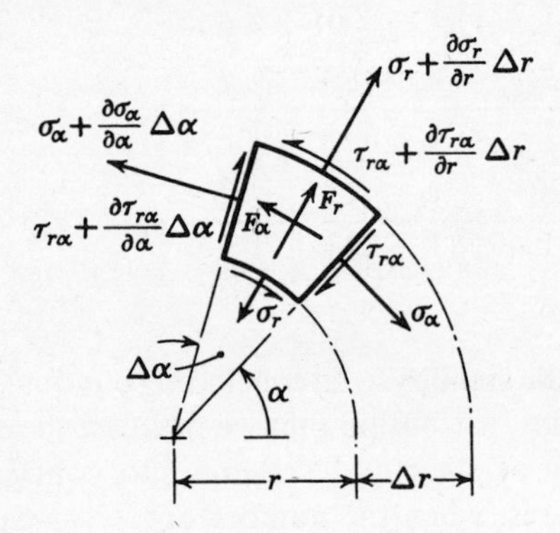

2.2. When the particles within an elastic body are continuously magnetized, a magnetic field produces *body moments* D_x, D_y, and D_z which have dimensions of moment per unit volume. Show that in such cases Eqs. (2.5) are no longer valid.

2.3. Show that the equilibrium equations in terms of the displacements are

$$(\lambda + G)\frac{\partial e}{\partial x} + G\,\nabla^2 u + X_b = 0$$

$$(\lambda + G)\frac{\partial e}{\partial y} + G\,\nabla^2 v + Y_b = 0$$

$$(\lambda + G)\frac{\partial e}{\partial z} + G\,\nabla^2 w + Z_b = 0$$

for an isotropic Hookean material. In these equations $\lambda = \nu E/(1 + \nu) \times (1 - 2\nu)$, $e = \partial u/\partial x + \partial v/\partial y + \partial w/\partial z$, and ∇^2 is the Laplacian operator: $\nabla^2 = \partial^2/\partial x^2 + \partial^2/\partial y^2 + \partial^2/\partial z^2$. Also express the static boundary conditions in terms of the displacements.

2.4. If u and v are the radial and the tangential components of displacement for the two-dimensional element in Prob. 2.1, show that the strain-displacement relations in polar coordinates are

$$\epsilon_r = \frac{\partial u}{\partial r}$$

$$\epsilon_\alpha = \frac{u}{r} + \frac{1}{r}\frac{\partial v}{\partial \alpha}$$

$$\gamma_{r\alpha} = \frac{1}{r}\frac{\partial u}{\partial \alpha} + \frac{\partial v}{\partial r} - \frac{v}{r}$$

2.5. Using the results of Prob. 2.4, show that a compatibility equation in polar coordinates is

$$\frac{\partial^2 \epsilon_\alpha}{\partial r^2} + \frac{\partial^2 \epsilon_r}{r^2\, \partial \alpha^2} + \frac{2}{r}\frac{\partial \epsilon_\alpha}{\partial r} - \frac{1}{r}\frac{\partial \epsilon_r}{\partial r} = \frac{\partial^2 \gamma_{r\alpha}}{r\, \partial r\, \partial \alpha} + \frac{1}{r^2}\frac{\partial \gamma_{r\alpha}}{\partial \alpha}$$

Discuss the significance of this equation.

2.6. Use v, the y component of displacement, instead of u to derive the compatibility condition [Eq. (2.23)] for the case of plane strain.

2.7. Derive Eqs. (2.24) by eliminating u, v, and w from Eqs. (2.17) through appropriate differentiations.

2.8. Verify that

$$\begin{aligned}
\epsilon_x &= Ky^2 + Cxy \\
\epsilon_y &= K(x^2 + y^2) + Cy \\
\gamma_{xy} &= Cxy \\
\epsilon_z &= \gamma_{xz} = \gamma_{yz} = 0
\end{aligned}$$

is a possible state of strain provided $C = 4K$.

2.9. Is the following a possible state of strain?

$$\begin{aligned}
\epsilon_x &= K(x^2 + y^2) + C(yz + z^2) \\
\epsilon_y &= Ky^2 + C(x + y + z) \\
\gamma_{xy} &= 4Kxy + 2Cxz \\
\epsilon_z &= C(y^2 + z^2) \\
\gamma_{xz} &= 2Cxz \\
\gamma_{yz} &= 2Cyz
\end{aligned}$$

where C and K are small constants.

2.10. Write the stress-strain relations for a homogeneous isotropic linearly elastic body in a state of *plane stress* ($\sigma_z = \tau_{xz} = \tau_{yz} = 0$).

2.11. Repeat Prob. 2.10 for the case of *plane strain* ($\epsilon_z = \gamma_{xz} = \gamma_{yz} = 0$).

2.12. Show that, in general, not all of the compatibility conditions are satisfied in the case of plane stress of an isotropic Hookean body.

2.13. The Airy stress function $\Psi(x,y)$ is defined as follows:

$$\sigma_x = \frac{\partial^2 \Psi}{\partial y^2} \qquad \sigma_y = \frac{\partial^2 \Psi}{\partial x^2} \qquad \tau_{xy} = -\frac{\partial^2 \Psi}{\partial x\, \partial y}$$

Show that any continuously differentiable function Ψ automatically satisfies the equilibrium equations of the plane stress problem provided no body forces are present. Then show that Ψ also leads to compatible strains provided

$$\frac{\partial^4 \Psi}{\partial x^4} + 2\frac{\partial^4 \Psi}{\partial x^2\, \partial y^2} + \frac{\partial^4 \Psi}{\partial y^4} = 0$$

2.14. The state of stress at a point P in a two-dimensional body is given by the stress components σ_x, σ_y, and τ_{xy} when referred to the xy coordinate system. Let x', y' be another orthogonal coordinate system at P and σ'_x, σ'_y, and τ'_{xy} be the stress components referred to this system. Show that

$$\sigma'_x = l_1{}^2\sigma_x + m_1{}^2\sigma_y + 2l_1m_1\tau_{xy}$$
$$\sigma'_y = l_2{}^2\sigma_x + m_2{}^2\sigma_y + 2l_2m_2\tau_{xy}$$
$$\tau'_{xy} = l_1l_2\sigma_x + m_1m_2\sigma_y + (l_1m_2 + m_1l_2)\tau_{xy}$$

where l_1, m_1 are the direction cosines of x' and l_2, m_2 are the direction cosines of y' with respect to the xy coordinate system.

2.15. Extend the results of Prob. 2.14 to the three-dimensional case. Denote the direction cosines of x', y', and z' with respect to x, y, and z by (l_1,m_1,n_1), (l_2,m_2,n_2), and (l_3,m_3,n_3), respectively.

2.16. Let α be the angle between x and x' in Prob. 2.14. Show that σ'_x is a maximum and a minimum when

$$\alpha = \tfrac{1}{2} \tan^{-1} \frac{2\tau_{xy}}{\sigma_x - \sigma_y}$$

That is, find the direction of the *principal stresses*.

2.17. Using the results of Prob. 2.16, show that

$$\sigma_1 = \tfrac{1}{2}(\sigma_x + \sigma_y) + \sqrt{\left(\frac{\sigma_x - \sigma_y}{2}\right)^2 + \tau_{xy}{}^2}$$

$$\sigma_2 = \tfrac{1}{2}(\sigma_x + \sigma_y) - \sqrt{\left(\frac{\sigma_x - \sigma_y}{2}\right)^2 + \tau_{xy}{}^2}$$

where σ_1 and σ_2 are the principal stresses.

2.18. Assuming that x and y are principal directions of stress, express σ'_x, σ'_y, and τ'_{xy} of Prob. 2.14 in terms of the principal stresses and sines and cosines of α. Show that the resulting equations represent *Mohr's circle* in two dimensions.

2.19. If ϵ'_x, ϵ'_y, and γ'_{xy} are the components of strain referred to the x', y' axes of a two-dimensional body, use the notation of Prob. 2.14 and show that

$$\epsilon'_x = l_1{}^2\epsilon_x + m_1{}^2\epsilon_y + l_1m_1\gamma_{xy}$$

$$\epsilon'_y = l_2{}^2\epsilon_x + m_2{}^2\epsilon_y + l_2m_2\gamma_{xy}$$

$$\gamma'_{xy} = 2l_1l_2\epsilon_x + 2m_1m_2\epsilon_y + (l_1m_2 + m_1l_2)\gamma_{xy}$$

2.20. Derive the equations for the principal strains in terms of the angle α for a two-dimensional body.

2.21. Derive Eq. (2.26).

2.22. An element in an elastic body in a state of plane stress is subjected to the following stresses: $\sigma_x = 10{,}000$ psi, $\sigma_y = -6{,}000$ psi, and $\tau_{xy} = 6{,}000$ psi. Use the results of Prob. 2.17 to evaluate the principal stresses.

2.23. The components of stress at a point in a deformable body are as follows: $\sigma_x = 12{,}000$ psi, $\sigma_y = -8{,}000$ psi, $\sigma_z = -4{,}000$ psi, $\tau_{xy} = 3{,}000$ psi, $\tau_{xz} = 0$, and $\tau_{yz} = 2{,}000$ psi. Determine the principal stresses and their directions with respect to the x, y, z axes.

2.24. For the states of stress given below, compute the principal stresses and the orientation of the axes along which they act.

(*a*) $\sigma_x = 4{,}000$ psi, $\tau_{xy} = -4{,}000$ psi, $\sigma_y = \sigma_z = \tau_{xz} = \tau_{yx} = 0$

(*b*) $\tau_{xy} = 12{,}000$ psi, $\sigma_x = \sigma_y = \sigma_z = \tau_{xz} = \tau_{yz} = 0$

(*c*) $\sigma_x = -6{,}000$ psi, $\tau_{xz} = 8{,}000$ psi, $\sigma_y = \sigma_z = \tau_{xy} = \tau_{yz} = 0$

(*d*) $\sigma_x = 800$ psi, $\sigma_y = -800$ psi, $\sigma_z = 100$ psi, $\tau_{xy} = 400$ psi, $\tau_{xz} = \tau_{yz} = 0$

2.25. A plane which intersects the axes along which the principal stresses are directed is called the *octahedral plane* and the shearing stress τ_0 on this plane is called the *octahedral shearing stress.* Show that

$$\tau_0 = \tfrac{1}{3}\sqrt{(\sigma_1 - \sigma_2)^2 + (\sigma_2 - \sigma_3)^2 + (\sigma_3 - \sigma_1)^2}$$

where σ_1, σ_2, and σ_3 are the principal stresses.

2.26. Show that the maximum and minimum shearing stress developed at a point in a body which is in a state of plane stress in the xy plane is given by

$$\tau_{\max,\min} = \pm\sqrt{\left(\frac{\sigma_x - \sigma_y}{2}\right)^2 + \tau_{xy}{}^2}$$

2.27. Show that the maximum and minimum shearing strain developed at a point in a body which is in a state of plane strain in the xy plane is given by

$$\gamma_{\max,\min} = \pm\sqrt{(\epsilon_x - \epsilon_y)^2 + \gamma_{xy}{}^2}$$

2.28. Compute the maximum shearing stress at a point in the elastic body described in Prob. 2.22 (see Prob. 2.26).

2.29. Determine the principal strains and the associated principal directions for the following states of strain:

(*a*) $\epsilon_x = 0.0008$, $\epsilon_y = 0.0004$, $\gamma_{xy} = 0.0002$, $\epsilon_z = \gamma_{xz} = \gamma_{yz} = 0$

(*b*) $\gamma_{xz} = 0.0005$, $\gamma_{xz} = 0.0004$, $\epsilon_x = \epsilon_y = \epsilon_z = \gamma_{xy} = 0$

(*c*) $\epsilon_x = -0.0012$, $\epsilon_y = 0.0008$, $\gamma_{xy} = 0.0010$, $\epsilon_z = \gamma_{xz} = \gamma_{yz} = 0$

(*d*) $\gamma_{xy} = 0.0024$, $\epsilon_x = \epsilon_y = \epsilon_z = \gamma_{xz} = \gamma_{yz} = 0$

2.30. Starting with the definitions of E and ν, derive Eqs. (2.27).

three

TORSION OF PRISMATIC BARS

3.1 General. When a straight prismatic bar is subjected solely to moments about its longitudinal axis, it twists about that axis, shearing stresses are developed, and a single stress resultant, the twisting moment M_x, is developed on each transverse cross section. If no cross sections are restrained from deforming under this loading, no normal stresses are developed and the result is a special but interesting example of structural behavior.

3.2 Circular bars. The straight prismatic bar of circular cross section is the most common structural element subjected primarily to torsion. Such bars are used as shafts in most engines and, in fact, serve as the basic test specimens for determining the shearing-stress–shearing-strain relations for most metals.

Developing relationships between the angle of twist, the shearing stresses, and the twisting moment on a circular section is quite simple because of the perfect symmetry of both the structure and the loading. Because of symmetry, we may argue that plane cross sections normal to the axis of the bar remain plane during deformation and that these sections remain undistorted in their own plane. From these observations it follows that a radial line on the cross section oriented any angle α with respect to some reference remains a straight line during deformation. Consequently, the shearing strain $\gamma_{x\alpha}$ varies linearly with r, the radial distance from the unstrained axis of the shaft (see Fig. 3.1). With this relatively simple condition of deformation, the analysis of a circular shaft in torsion reduces to a statically determinate problem.

Consider, for example, the typical segment of the circular bar in pure torsion shown in Fig. 3.1. From the geometry of the deformed element (indicated by dashed lines), we find

$$\gamma_{x\alpha}\,\Delta x = \frac{\tau_{x\alpha}}{G}\,\Delta x = r\,\Delta\phi$$

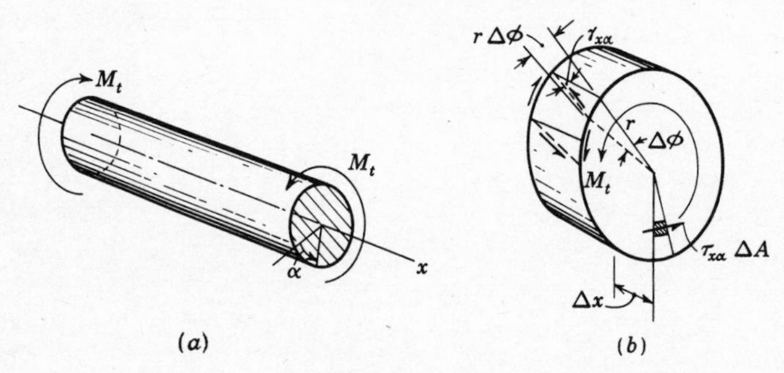

FIGURE 3.1 (*a*) A circular shaft in pure torsion; (*b*) a typical transverse element of the shaft.

where $\Delta\phi$ is an increment in ϕ, the total angle of twist of any radial line on the section. It follows that

$$\tau_{x\alpha} = \lim_{\Delta x\to 0} Gr\frac{\Delta\phi}{\Delta x} = Gr\frac{d\phi}{dx}$$

Here the rate of change of twist $d\phi/dx$ is a constant.

The twisting moment resulting from $\tau_{x\alpha}$ is

$$M_x = \iint_A r(\tau_{x\alpha}\,dA) = G\frac{d\phi}{dx}\iint_A r^2\,dA = G\frac{d\phi}{dx}I_p$$

in which A is the area and I_p is the polar moment of inertia of the cross section. From simple statics, we see that M_x is equal in magnitude to the end moment M_t.

By rearranging terms and introducing the notation

$$\theta = \frac{d\phi}{dx}$$

into the above relationships, we arrive at the equations of the elementary theory of torsion:

$$\tau_{x\alpha} = \frac{M_t r}{I_p} \tag{3.1}$$

$$\theta = \frac{M_t}{GI_p} \tag{3.2}$$

3.3 Torsional behavior of noncircular bars. In the case of torsion of prismatic bars not circular in cross section, we lose the arguments of symmetry and, along with them, the simplicity of the elementary theory. The argument that plane cross sections remain plane during deformation, for example, is now no longer valid.

To prove this, we need only examine the behavior of a number of elements of the prismatic bar shown in Fig. 3.2*a*. For simplicity, elements A and D are shown located on right-angled corners so that no shearing stress can possibly exist on these elements; any component of shearing stress on element A, for example, would require stresses to be developed on the outside surfaces of the bar which, of course, are impossible. Consequently, these elements undergo no shearing strain, which means that their corners remain right angles after deformation. Elements B and C, however, can develop shearing stresses so long as they are parallel to the boundary line $\overline{ad}$. Again, this is because shearing stresses directed normal to any boundary lead to impossible stresses on the outside surface of the bar (Fig. 3.2*b*). It follows that when the sides of elements B and C acquire a shearing strain γ elements A and D must undergo rigid-body rotations of the same amount for the sake of continuity of the material. Hence, the outside corners of elements A and B are displaced out of the plane of the cross section, as indicated in Fig. 3.2*c*. Finally, if we increase the number of elements indefinitely and allow their dimensions to approach zero, the deformed configuration of the boundary line $\overline{ad}$ becomes a smooth curve; the shearing stress at the boundaries is directed parallel to the boundary curve and is zero at points a and d.

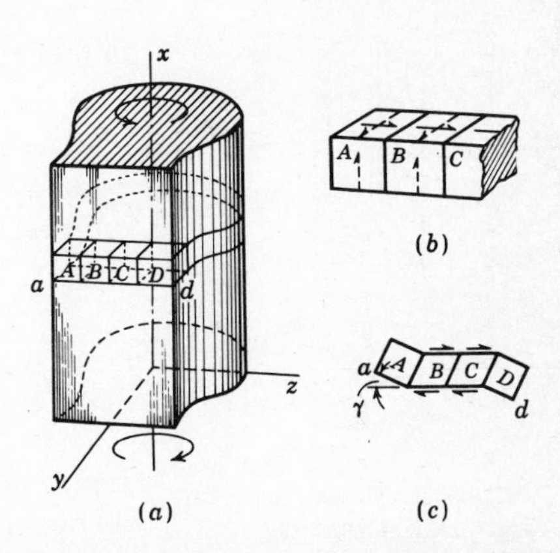

FIGURE 3.2 (*a*) A noncircular prismatic bar in pure torsion; (*b*) elements of the bar with impossible (dashed line) and possible (solid line) shearing stresses; (*c*) deformation of bar elements.

Out-of-plane displacements such as those discussed above are called *warping displacements*. We assume that no external constraints exist which could prevent any cross-sectional plane from *warping*. Thus, no external forces exist in the x direction.

Other than the fact that cross sections warp, perhaps one of the most obvious characteristics of the bar's behavior is the absence of normal stresses. No external forces or bending moments are present, and no end constraints exist; therefore, the only stress components needed to provide the equilibrium of any transverse segment are shearing stresses in the cross-sectional planes. Furthermore, of the three components of shearing stress, only τ_{xy} and τ_{xz} can result in a twisting moment. The component τ_{yz} is zero, which we may also verify by examining plane areas parallel to the bar's axis, such as those shown in Fig. 3.3. If any shearing stresses were developed on these planes, they would result in forces and moments which not only offer no aid in resisting M_t but which also, in general, cannot be in equilibrium. Thus, we conclude that for the noncircular prism in torsion

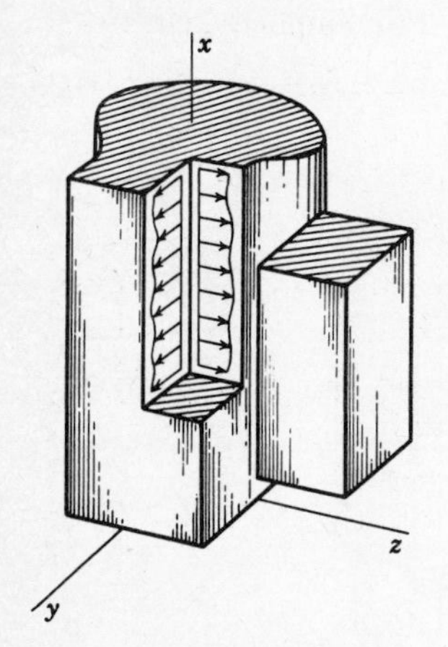

FIGURE 3.3 Nonexistent shearing stresses τ_{yz}.

$$\sigma_x = \sigma_y = \sigma_z = \tau_{yz} = 0 \qquad (3.3)$$

Referring now to Eqs. (2.4) and noting that no body forces are present, we arrive at the differential equations of equilibrium:

$$\frac{\partial \tau_{xy}}{\partial y} + \frac{\partial \tau_{xz}}{\partial z} = 0 \qquad (3.4)$$

and

$$\frac{\partial \tau_{xy}}{\partial x} = \frac{\partial \tau_{xz}}{\partial x} = 0 \qquad (3.5)$$

Equations (3.5) show that the shearing stresses do not vary with x and, hence, have the same distribution on each cross section. Only Eq. (3.4) is useful as an equilibrium condition for stresses on the cross-sectional planes, and it involves two unknowns. Thus, we have a statically indeterminate problem, and we are forced to turn to considerations of strains and displacements for additional information.

Since the material is assumed to be homogenous, isotropic, and linearly elastic, we may introduce Eqs. (3.3) into Eqs. (2.27) to obtain the components of strain. We find

$$\epsilon_x = \epsilon_y = \epsilon_z = \gamma_{yz} = 0 \qquad (3.6)$$

Therefore, according to Eqs. (2.17),

$$\frac{\partial u}{\partial x} = \frac{\partial v}{\partial y} = \frac{\partial w}{\partial z} = 0$$

which means that

$$u = f(y, z) \qquad v = g(x, z) \qquad w = h(x, y) \qquad (3.7)$$

where f, g, and h are continuous functions yet to be determined.

The fact that γ_{yz} is zero is extremely important in our development. From this observation we conclude that cross sections do not distort in their own planes. In other words, the angle between any two lines on a cross section is not changed during the deformation of the bar. This means that deformations of elements, such as that indicated in Fig. 3.4*a*, do *not* exist and that during deformation a point in the *yz* plane merely rotates about a *center of twist*,[1] which we take as coincident with the x axis. This makes it possible for us to define the in-plane displacements of any point on the cross section in terms of the angle of twist ϕ of a straight line on the section drawn from the x axis to the point, as shown in Fig. 3.4*b*.

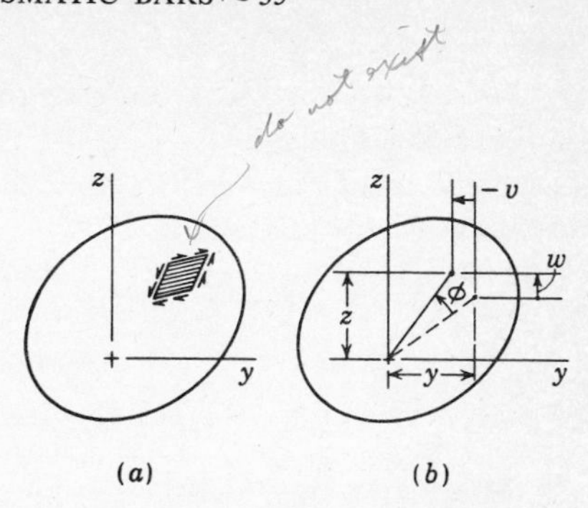

FIGURE 3.4 Geometry of deformation. Distortions such as that shown in (*a*) do not exist since γ_{yz} is zero. Hence, lines rotate as shown in (*b*).

Upon examining the geometry of Fig. 3.4*b*, we find

$$v = -\phi z \qquad \text{and} \qquad w = \phi y \tag{3.8}$$

We will now show that $d\phi/dx$ is constant. Referring to Eqs. (2.27) and (3.5), we find

$$\frac{\partial \tau_{xy}}{\partial x} = G\frac{\partial \gamma_{xy}}{\partial x} = G\frac{\partial}{\partial x}\left(\frac{\partial u}{\partial y} + \frac{\partial v}{\partial x}\right) = 0$$

or, by virtue of Eqs. (3.8),

$$\frac{\partial}{\partial y}\left(\frac{\partial u}{\partial x}\right) - z\frac{d^2\phi}{dx^2} = 0$$

The first term in this equation is zero owing to Eqs. (3.6). Thus

$$\frac{d^2\phi}{dx^2} = 0$$

and

$$\frac{d\phi}{dx} = \theta = \text{constant} \tag{3.9}$$

It follows that the twist relative to the section $x = 0$ is θx and that Eqs. (3.8) may be expressed in terms of θ, x, y, and z. In summary, we express the components of displacement in the form

$$\begin{aligned} u &= f(y,z) \\ v &= -\theta xz \\ w &= \theta xy \end{aligned} \tag{3.10}$$

[1] A more detailed discussion of the center of twist of thin-walled sections under general loading is given in Art. 7.4.

Thus, once θ and $f(y,z)$ are known, the complete displacement pattern can be evaluated.

Finally, from Eqs. (2.17) and (2.28), we relate the shearing stresses and strains to the displacements by

$$\tau_{xy} = G\gamma_{xy} = G\left(\frac{\partial u}{\partial y} + \frac{\partial v}{\partial x}\right)$$

$$\tau_{xz} = G\gamma_{xz} = G\left(\frac{\partial u}{\partial z} + \frac{\partial w}{\partial x}\right)$$

or, introducing Eqs. (3.10),

$$\begin{aligned} \tau_{xy} &= G\left(\frac{\partial u}{\partial y} - \theta z\right) \\ \tau_{xz} &= G\left(\frac{\partial u}{\partial z} + \theta y\right) \end{aligned} \tag{3.11}$$

Differentiating the first of these equations with respect to y and the second with respect to z and substituting the result into Eq. (3.4) yields

$$\frac{\partial^2 u}{\partial y^2} + \frac{\partial^2 u}{\partial z^2} = 0 \tag{3.12}$$

which is the governing partial differential equation for the warping displacement. This relationship is called *Laplace's*[2] *equation for the warping function*, and any solution to Laplace's equation is called a *harmonic function*. Thus, u is a harmonic function.

The torsion problem now reduces to one of determining the four unknowns τ_{xy}, τ_{xz}, u, and θ. To solve this problem we have three relationships—Eq. (3.4), the equilibrium condition, and the two kinematic conditions in Eqs. (3.11)—which we have written in terms of the stresses. Equation (3.12) is not independent since it was obtained from Eqs. (3.4) and (3.11). The fourth relationship necessary to solve the problem is the simple static condition that τ_{xy} and τ_{xz} must result in a twisting moment of magnitude M_t on each cross section.

3.4 Saint-Venant's stress function. The torsion problem may be reduced to one of determining a single unknown by using a scheme first presented by Saint-Venant in 1855. He assumed the existence of a continuously differentiable function $\Phi(y,z)$, now called *Saint-Venant's stress function*, which has the property

$$\tau_{xy} = \frac{\partial \Phi}{\partial z}$$

and

$$\tau_{xz} = -\frac{\partial \Phi}{\partial y} \tag{3.13}$$

[2] Named after the great French mathematician Pierre Simon Laplace (1749–1827).

When we introduce these definitions into Eq. (3.4) we obtain

$$\frac{\partial^2\Phi}{\partial y\,\partial z} - \frac{\partial^2\Phi}{\partial z\,\partial y} = 0$$

which is satisfied by any function continuous through its second derivative. Thus, any such continuous function will automatically satisfy Eq. (3.4) and, therefore, lead to shearing stresses which are in equilibrium. The correct solution to the torsion problem, however, must be a state of stress providing not only equilibrium but also compatible strains and displacements. Thus, out of the infinite number of functions Φ which satisfy Eq. (3.4) we must choose those which also satisfy a condition of compatibility.

To arrive at this condition, we introduce Eqs. (3.13) into Eqs. (3.11) and differentiate the first with respect to z and the second with respect to y:

$$\frac{\partial}{\partial z}\left(\frac{\partial\Phi}{\partial z}\right) = G\frac{\partial}{\partial z}\left(\frac{\partial u}{\partial y} - \theta z\right) \quad \text{and} \quad \frac{\partial}{\partial y}\left(-\frac{\partial\Phi}{\partial y}\right) = G\frac{\partial}{\partial y}\left(\frac{\partial u}{\partial z} + \theta y\right)$$

Recalling that u is also continuously differentiable, we subtract the second equation from the first and find

$$\frac{\partial^2\Phi}{\partial y^2} + \frac{\partial^2\Phi}{\partial z^2} = -2G\theta \tag{3.14}$$

This is the *equation of compatibility* for the problem of torsion of prismatic bars. Any function Φ continuous through its second derivatives which satisfies Eq. (3.14) now automatically provides both equilibrium and compatibility. Any partial differential equation of this form is also called *Poisson's equation*.[3]

We may visualize Φ as being a curved surface spread over the cross section of the bar. According to its definition in Eqs. (3.13), the slope of the surface in the z direction is the stress in the y direction, and its slope in the y direction is the negative of the stress in the z direction. In fact, if n is any direction oriented α with respect to the y axis, as shown in Fig. 3.5, the stress directed normal to n is clearly

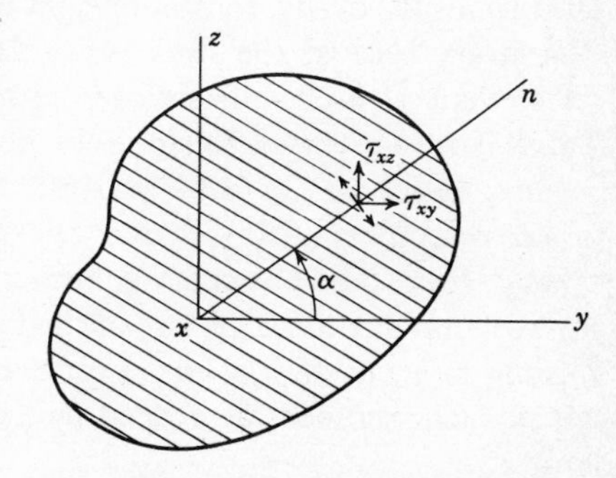

FIGURE 3.5 An arbitrary direction n on the cross section.

$$-\tau_{xz}\cos\alpha + \tau_{xy}\sin\alpha$$

[3] Named after Siméon-Denis Poisson (1781–1840), who also discovered Poisson's ratio.

The slope of Φ in the n direction is, by definition,

$$\frac{d\Phi}{dn} = \frac{\partial\Phi}{\partial y}\frac{dy}{dn} + \frac{\partial\Phi}{\partial z}\frac{dz}{dn}$$

or, since $dy/dn = \cos\alpha$ and $dz/dn = \sin\alpha$,

$$\frac{d\Phi}{dn} = -\tau_{xz}\cos\alpha + \tau_{xy}\sin\alpha \tag{3.15}$$

Hence, *the slope of the Φ surface in any direction is equal to the shearing stress in the perpendicular direction.*

Furthermore, since we proved earlier that no shearing-stress components can act normal to the boundary of the cross section, the slope of Φ parallel to the boundary must be zero. This is possible *only if Φ is a constant along the boundary*. We may also verify that this is true by noting that the slope of Φ parallel to the boundary curve s is

$$\frac{d\Phi}{ds} = \frac{\partial\Phi}{\partial y}\frac{dy}{ds} + \frac{\partial\Phi}{\partial z}\frac{dz}{ds} = -\tau_{xz}(-n) + \tau_{xy}m$$

where m and n are the direction cosines of a normal to the curve. Since no surface forces are present, the right side of this equation is zero by virtue of the first condition in Eqs. (2.6). Thus, Φ must satisfy the boundary condition

$$\frac{d\Phi}{ds} = 0 \tag{3.16}$$

The magnitude of this constant height of the Φ surface along the boundary is arbitrary because the stresses are defined in terms of derivatives of Φ rather than Φ itself. Therefore, without loss in generality, we assume that Φ is zero everywhere along the boundary of the cross section.

Any solution to Eq. (3.14) provides both equilibrium and compatibility for cross sections of any shape. To ensure that it also leads to stresses which satisfy static boundary conditions at the ends of the bar (or on any cross section due to the nature of the loading) we must also relate Φ to the twisting moment developed on each section. This is a matter of simple statics. The moment developed by the shearing stresses about the x axis must be

$$M_x = M_t = \iint_A (-\tau_{xy}z + \tau_{xz}y)\,dy\,dz$$

where the integration is carried out over the entire area of the cross section. Introducing Φ, this equation becomes

$$M_t = -\iint_A \left(\frac{\partial\Phi}{\partial z}z + \frac{\partial\Phi}{\partial y}y\right)dy\,dz$$

We now write the integral in the form

$$-\int\left(\int_A^B \frac{d\Phi}{dz} z\, dz\right) dy - \int\left(\int_C^D \frac{d\Phi}{dy} y\, dy\right) dz$$

where the limits A and B stand for boundary points along a line $y =$ constant, and C and D stand for boundary points along some line $z =$ constant. Integrating by parts, we find

$$M_t = -\int(\Phi_B z_B - \Phi_A z_A - \int\Phi\, dz)\, dy - \int(\Phi_D y_D - \Phi_C y_C - \int\Phi\, dy)\, dz$$

Now Φ_A, Φ_B, Φ_C, and Φ_D denote values of Φ at the boundary points which, according to our earlier discussion, are zero. Hence, the terms within the parentheses vanish and we have

$$M_t = 2\iint_A \Phi\, dy\, dz \tag{3.17}$$

This final result states that *the total twisting moment on any section is equal to twice the volume under the surface* Φ.

We would now like to obtain the twisting moment in terms of three independent quantities as we did for the circular bar: G, the modulus of rigidity, which depends upon the material; θ, the angle of twist per unit length; and a constant J, which depends upon the geometry of the cross section. In other words, for every cross section there exists a constant J such that

$$M_t = GJ\theta \tag{3.18}$$

J is called the *torsional* constant of the bar. The product GJ is called the *torsional stiffness* of the bar. The formula for J follows directly from Eqs. (3.17) and (3.18):

$$J = \frac{2}{G\theta}\iint_A \Phi\, dy\, dz \tag{3.19}$$

Comparing Eq. (3.18) with Eq. (3.2), we see that for the very special case of circular cross section, J is the polar moment of inertia of the cross section.

3.5 The membrane analogy. It is of interest to note that Eq. (3.14), which is the governing equation in our theory of torsion, closely resembles the equilibrium equation describing small deflections of a flat membrane subjected to an internal pressure p. If $\bar{w}$ is the deflection of the membrane and T is the constant tension per unit length in the membrane, it is easily shown that an element is in equilibrium provided

$$\frac{\partial^2 \bar{w}}{\partial y^2} + \frac{\partial^2 \bar{w}}{\partial z^2} = -\frac{p}{T} \tag{3.20}$$

Comparing this with Eq. (3.14), we see that $\bar{w}$ is analogous to Φ and that T/p is analogous to $2G\theta$. Furthermore, owing to Eq. (3.17), the volume under the membrane is proportional to the twisting moment developed on a bar in torsion of the same shape of the membrane.

This analogy was first discovered by Ludwig Prandtl in 1903 and is known as *Prandtl's membrane analogy*. Prandtl took full advantage of the analogy and devised clever experiments with membranes. By measuring the volumes under membranes formed by a soap film subjected to a known pressure, he was able to evaluate torsional constants. By obtaining the contour lines of the membranes he determined stress distributions. A number of other analogies to the torsion problem have also been proposed.[4]

For our present purposes, the chief advantage of the membrane analogy is as an aid to the visualization of the Φ surface. Contour lines on a surface, such as those on a map of mountainous terrain, indicate the intensity of the slope of the surface, which, in turn, indicates the intensity of the shearing stress. Thus, at points on the Φ surface where contour lines are grouped closely together, we expect high concentrations of stress.

3.6 Bars of solid section. In general, to analyze the torsional behavior of bars of any shape, it is necessary to obtain an exact solution to Eq. (3.14). This often requires a knowledge of partial differential equations, and we shall not attempt to solve such problems here. The correct stress function for a number of simple shapes (the circle, the ellipse, the equilateral triangle, etc.), however, can be obtained by examining the equations for the boundary curves.

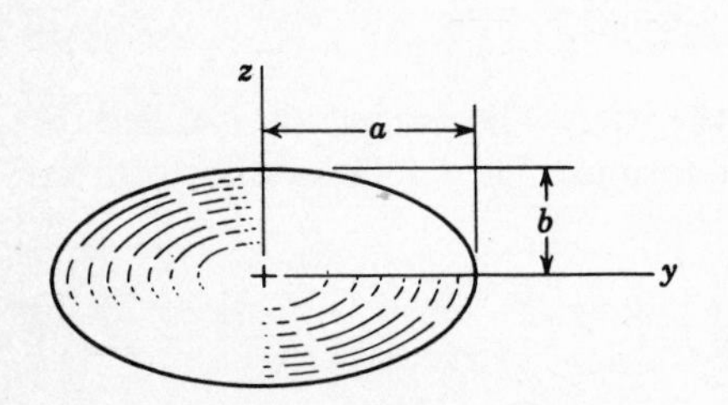

FIGURE 3.6 Bar of elliptical cross section.

To demonstrate this procedure, let us examine the stress function for the bar of elliptical cross section shown in Fig. 3.6. By assuming a solution to Eq. (3.14) of the form

$$\Phi = C\left(\frac{y^2}{a^2} + \frac{z^2}{b^2} - 1\right) \tag{a}$$

where C is some constant, we automatically satisfy the conditions that Φ be zero along the boundary and that its slopes parallel to the boundary be zero. This is because the function inside the parentheses is the equation

[4] See, for example, Ref. 27, pp. 700–827.

of the elliptical boundary curve. We now must check to see if the assumed Φ satisfies Eq. (3.14). Substituting this function into (3.14) gives

$$C\left(\frac{2}{a^2} + \frac{2}{b^2}\right) = -2G\theta$$

Thus, Eq. (*a*) is the correct solution if we set

$$C = -\frac{a^2b^2G\theta}{a^2 + b^2} \tag{b}$$

Introducing Φ into Eq. (3.17), we find

$$M_t = -\frac{2a^2b^2G\theta}{a^2 + b^2}\int_{-b}^{b}\int_{-\frac{a}{b}\sqrt{b^2-z^2}}^{\frac{a}{b}\sqrt{b^2-z^2}}\left(\frac{y^2}{a^2} + \frac{z^2}{b^2} - 1\right) dy\, dz$$

or, after integrating,

$$M_t = G\theta\,\frac{\pi a^3b^3}{a^2 + b^2} \tag{c}$$

Thus, from Eq. (3.18),

$$J = \frac{\pi a^3b^3}{a^2 + b^2} \tag{3.21}$$

and

$$C = -\frac{a^2b^2M_t}{J(a^2 + b^2)} = -\frac{M_t}{\pi ab}$$

Finally, we evaluate the shearing stresses by direct application of Eqs. (3.13):

$$\begin{aligned}\tau_{xy} &= \frac{\partial\Phi}{\partial z} = \frac{2C}{b^2}z = -\frac{2M_tz}{\pi ab^3}\\ \tau_{xz} &= -\frac{\partial\Phi}{\partial y} = -\frac{2C}{a^2}y = \frac{2M_ty}{\pi a^3b}\end{aligned} \tag{3.22}$$

If $b < a$, the maximum stress is τ_{xy} at $z = \pm b$:

$$\tau_{\max} = \frac{2M_t}{\pi ab^2} \tag{3.23}$$

Noting that the resultant force on a unit area at any point on the cross section is

$$({\tau_{xy}}^2 + {\tau_{xz}}^2)^{\frac{1}{2}} = \frac{2M_t}{\pi ab}\left(\frac{y^2}{a^4} + \frac{z^2}{b^4}\right)^{\frac{1}{2}}$$

we see that the contour lines of constant stress on the Φ surface form a family of ellipses, as indicated by the dashed lines in Fig. 3.6.

We may now determine the warping displacement u by introducing the above results into either of Eqs. (3.11):

$$\frac{\partial u}{\partial y} = \frac{\tau_{xy}}{G} + \theta z = \frac{M_t}{G}\left(\frac{-2}{\pi ab^3} + \frac{1}{J}\right)z$$

Integrating and simplifying the result, we find

$$u = \frac{M_t}{\pi ab^3}(b^2 - a^2)yz + g(y)$$

where $g(y)$ is some function of y. By using the second relation in Eqs. (3.11), we find

$$u = \frac{M_t}{\pi ab^3}(b^2 - a^2)yz + h(z)$$

which can be equal to the first result only if $g(y)$ and $h(z)$ are zero. Thus,

$$u = \frac{M_t}{\pi ab^3}(b^2 - a^2)yz \tag{3.24}$$

This displacement pattern is in the form of a hyperbolic paraboloid. The lines of $y = 0$ and $z = 0$ do not displace and the deformed shape is antisymmetrical with respect to the y and z axes.

The procedure illustrated above can be applied to a number of solid sections with relatively simple cross-sectional shapes. In such cases, a stress function in the form of a polynomial is assumed and the coefficients of each term are adjusted so that the function satisfies Eq. (3.14) as well as the boundary conditions. If this proves to be impossible, it is necessary, as mentioned earlier, to attack Eq. (3.14) directly and obtain an exact solution or to use some approximate or empirical method. In this regard, it is worthy to note that Saint-Venant also presented an approximate expression for the torsional constant of any solid section:

$$J \approx \frac{0.025A^4}{I_p} \tag{3.25}$$

where A is the cross-sectional area and I_p is the polar moment of inertia. This equation yields accurate values of J except for elongated sections, that is, for sections having one dimension which is much larger than the rest. We investigate the torsional behavior of such narrow thin-walled sections in the following article.

3.7 Thin-walled open sections. A thin-walled section is said to be *open* when the locus of points defining the center line of the walls is not a closed curve. Channels, angles, I beams, and wide-flange sections are among many common structural shapes characterized by combinations of thin-walled rectangular elements; a variety of different thin-walled curved sections are used in aircraft and missile structures. The basic characteristic of these sections is that the thicknesses of the component elements are small compared with the other dimensions.

The key to the torsional analysis of this type of structure lies in the study of the simplest thin-walled open section, the narrow rectangle. The surface Φ for such a shape is of the form indicated in Fig. 3.7. Obviously, it must be symmetrical with respect to both the y and z axes, and it is very flat. Contour lines are grouped closely together at the boundaries parallel to the y axis, but Φ is essentially constant in the y direction. We may assume, therefore, that Φ is solely a function of z and does not vary with y. Then Eq. (3.14) reduces to the ordinary differential equation

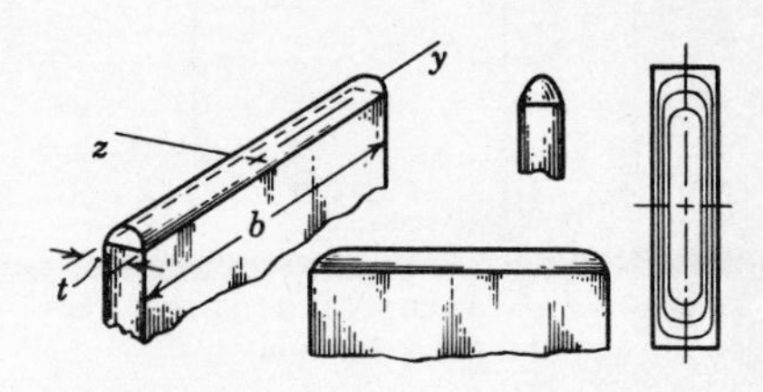

FIGURE 3.7 The surface Φ for a narrow rectangular section.

$$\frac{d^2\Phi}{dz^2} = -2G\theta$$

which, on integrating twice, gives

$$\Phi = -G\theta z^2 + C_1 z + C_2$$

where C_1 and C_2 are constants of integration.

Since Φ is zero at $z = \pm t/2$, we find that C_1 is zero and C_2 is $G\theta t^2/4$. Hence,

$$\Phi = -G\theta\left(z^2 - \frac{t^2}{4}\right)$$

From Eqs. (3.13),

$$\tau_{xz} = -\frac{\partial\Phi}{\partial y} = 0 \tag{3.26}$$

and

$$\tau_{xy} = \frac{\partial\Phi}{\partial z} = -2G\theta z \tag{3.27}$$

Thus, the shearing stress varies linearly over the thickness and is zero along the center line of the rectangle.

Introducing Φ into Eq. (3.19), we find

$$J = -\int_{-b/2}^{b/2}\int_{-t/2}^{t/2}\left(2z^2 - \frac{t^2}{2}\right) dy\, dz$$

or

$$J = \frac{bt^3}{3} \tag{3.28}$$

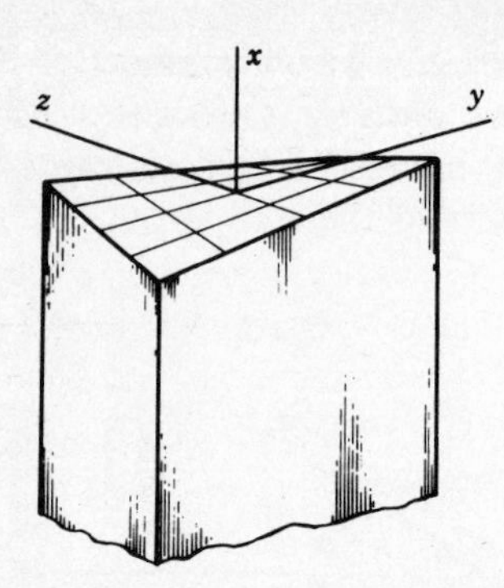

FIGURE 3.8 Warping of a narrow rectangular section.

Referring to Eq. (3.18), we see that Eq. (3.27) can also be written

$$\tau_{xy} = -\frac{2M_t}{J}z \tag{3.29}$$

The maximum stress occurs at $z = \pm t/2$:

$$\tau_{\max} = \mp\frac{M_t t}{J} = \mp\frac{3M_t}{bt^2} \tag{3.30}$$

The warping u is now easily obtained from Eqs. (3.11):

$$\frac{\partial u}{\partial z} = -\theta y$$

$$u = -\theta yz + h(z)$$

Comparing this with the result obtained using the first of Eqs. (3.11) shows that $h(z)$ is zero. Thus, replacing θ with M_t/GJ, we have

$$u = -\frac{M_t}{GJ}yz \tag{3.31}$$

which is the hyperbolic-paraboloid surface shown in Fig. 3.8.

The above formulas are applicable only if t is much smaller than b. When this is not the case, it may be shown that[5]

$$J \approx \frac{bt^3}{3}\left(1 - \frac{192}{\pi^5 b}\tanh\frac{\pi b}{2t}\right)$$

$$\tau_{\max} = \frac{M_t t}{J}\left[1 - \frac{8}{\pi^2}\sum_{n=0}^{\infty}\frac{\operatorname{sech}(2n+1)\pi b/t}{(2n+1)^2}\right] \tag{3.32}$$

These more general equations can be written in the simplified form

$$J = c_1 bt^3 \qquad \tau_{\max} = \frac{c_2 M_t}{bt^2} \tag{3.33}$$

where c_1 and c_2 are the constants recorded in Table 3.1. The maximum stress occurs at points closest to the center of the section. We see that the simplified formulas give reasonably accurate results for $b/t > 10$.

Table 3.1 Torsional constants c_1 and c_2 for a rectangle

b/t	∞	10	5	3	2.5	2.0	1.5	1.2	1.0
c_1	0.333	0.312	0.291	0.263	0.249	0.229	0.196	0.166	0.141
c_2	3.00	3.20	3.44	3.74	3.88	4.06	4.33	4.57	4.80

[5] See Ref. 61, p. 276.

Torsional constants and maximum shearing stresses can now be found for many more complex thin-walled open sections by using the results obtained for the narrow rectangle. This is possible because M_t and J are directly proportional to the volume under the Φ surface. This volume, for a section composed of several thin narrow elements, is clearly equal to the sum of the volumes of each element (neglecting a small error at corners or points of intersection of the elements). Thus, we may obtain the torsional constant J for such sections by simply adding the J's for each element calculated by using Eq. (3.28). For example, the torsional constant for the section shown in Fig. 3.9*a* is

$$J = \frac{b_1 t_1^3}{3} + \frac{b_2 t_2^3}{3} + \frac{b_3 t_3^3}{3} + \frac{b_4 t_4^3}{3}$$

Torsional constants for some other thin-walled open sections are also given in Fig. 3.9. Note that a smooth curvature of an element does not

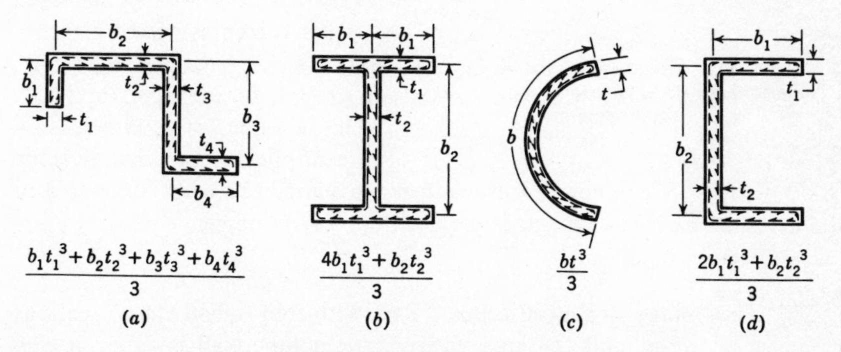

FIGURE 3.9 Typical thin-walled open sections and their corresponding torsional constants J.

alter the results. In the general case of a section with n elements,

$$J = \tfrac{1}{3} \sum_{i=1}^{n} b_i t_i^3 \tag{3.34}$$

The maximum shearing stress on any element i is still given by Eq. (3.30), except that the percentage of the twisting moment developed on this element is $J_i M_t / J$. Thus, for element i,

$$(\tau_{\max})_i = \frac{M_t (J_i/J)(t_i)_{\max}}{J_i} = \frac{M_t (t_i)_{\max}}{J} \tag{3.35}$$

The maximum stress on the entire section is

$$\tau_{\max} = \frac{M_t t_{\max}}{J} \tag{3.36}$$

where $t_{\max}$ is the maximum thickness and J is given by Eq. (3.34). As before, the stress varies linearly over the thickness.

An inspection of the contours of the Φ surface at reentrant corners, such as that shown in Fig. 3.10, indicates that high stress concentrations occur which are not accounted for by the above approximate formulas. Because of this, such corners are rounded by *fillets*, in practice, to allow a smooth "flow" of stress from one element to another. To evaluate the maximum stress at the fillet τ_f we use the empirical formula presented by Trefftz in 1922:

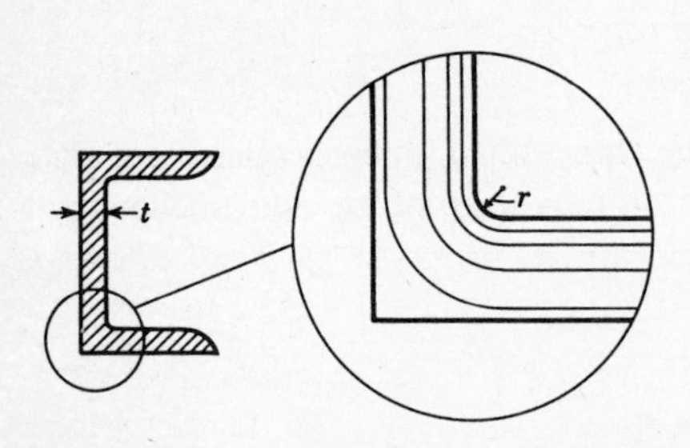

FIGURE 3.10 Contours of the Φ surface indicate high stress concentrations at reentrant corners.

$$\tau_f = \tau_{\max} 1.74 \left(\frac{t}{r}\right)^{\frac{1}{3}} \tag{3.37}$$

where r is the radius of the fillet, t is the thickness of the element, and $\tau_{\max}$ is given by Eq. (3.36). If $t/r = 1$, for example, Eq. (3.37) yields a value for τ_f which is 74 percent greater than that given by Eq. (3.36). These concentrations become less severe when the angle of intersection of the elements is greater than a right angle.

3.8 Thin-walled single-cell tubes. Bars with thin-walled closed sections are called *tubes*, and the area enclosed by a tube wall is called a *cell*. The boundary of *single-cell* tubes encloses only one cell and that of a *multicell* tube, more than one.

The formulas that we developed earlier for the analysis of open sections cannot be used when the sections are closed. This is fundamentally due to the way that the member must develop shearing stresses to resist applied torques. For example, on the open circular segment shown in Fig. 3.11*a*, there is developed a linear stress distribution across its thickness which results in a "loop" of forces, as is indicated. The hollow circular shaft shown in Fig. 3.11*b* is definitely not a thin-walled section, but it is obvious that the twisting moment is resisted in an entirely different manner. According to Eq. (3.1), the stress is still linear; but, in this case, it varies linearly with the radius r. If we keep the outside diameter of the circular shaft constant and increase the inside diameter until the wall thickness is very small compared with other dimensions (Fig. 3.11*c*), the

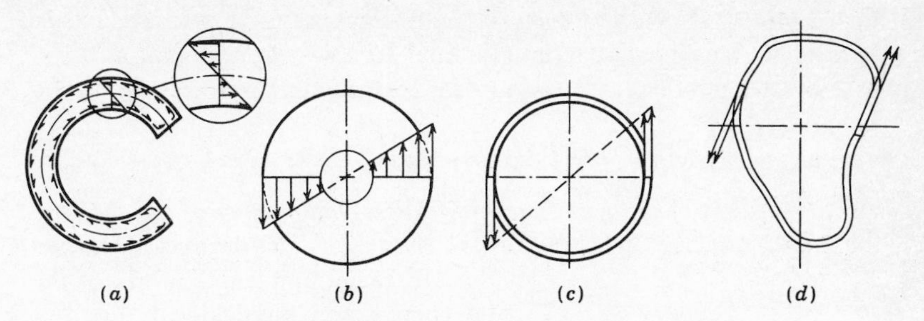

FIGURE 3.11 Distribution of torsional shearing stresses in open (*a*) and closed (*b–d*) sections.

exact shearing stress is still given by Eq. (3.1); but we note one important consequence: the stress has "no room to vary" and is practically uniform across the thickness. We have, of course, examined the very special case of the circular section, but an inspection of the shape of the stress function surface for other sections implies that the stress is as shown in Fig. 3.11*d*. This important observation considerably simplifies our analysis of such structures.

Let us now examine the segment of a single-cell tube shown in Fig. 3.12 and, in particular, a portion of the tube wall between points 1 and 2. The shearing stress is uniform over the thickness of the wall and is directed tangent to the boundary curve s describing the center line of the wall. Stresses normal to s are negligible because of the small wall thickness.

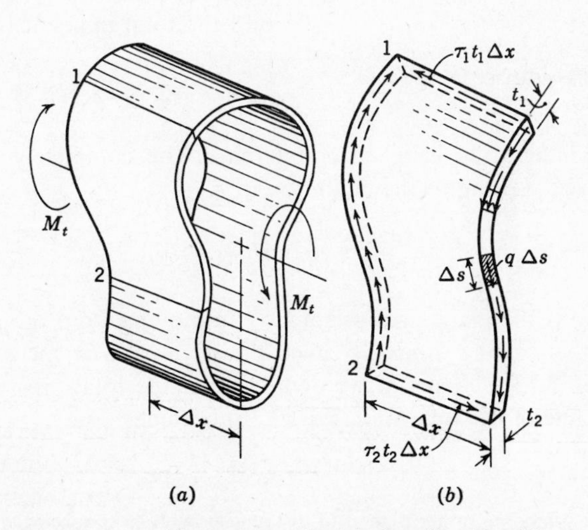

FIGURE 3.12 (*a*) Segment of a single-cell tube; (*b*) a slice taken from the segment.

If τ_1, τ_2, and t_1, t_2 denote the stresses and thicknesses at points 1 and 2, we see from the figure that longitudinal forces are developed at these points of magnitude $\tau_1 t_1 \Delta x$ and $\tau_2 t_2 \Delta x$. Summing forces in the x direction, we get

$$\tau_1 t_1 = \tau_2 t_2$$

Hence, *regardless of the variation in thickness, the product of the shearing stress and the thickness at every point is a constant.* This product represents a force per unit length of arc of the curve s, and, physically, its distribution resembles the flow of some substance along the tube wall. For this reason, we refer to this product as the *shear flow* on the section. Thus, if q is the shear flow, the shearing stress at point i is simply q/t_i and the force developed on an infinitesimal element of arc ds is $q\,ds$.

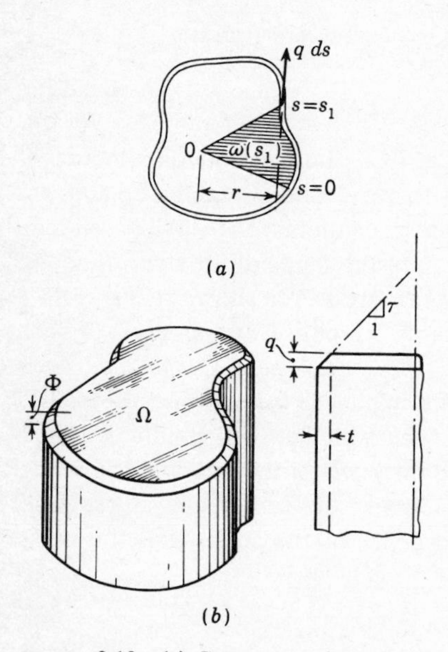

FIGURE 3.13 (*a*) Geometry of a closed tube; (*b*) Φ surface for the tube.

To evaluate the total moment developed by the shear flow, we denote by r the perpendicular distance from some point 0 on the cross-sectional plane to a tangent drawn to any point on the boundary curve (Fig. 3.13*a*). The moment of the force $q\,ds$ about 0 is simply $rq\,ds$ and the total moment is

$$M_t = \oint rq\,ds = q \oint r\,ds \tag{3.38}$$

where the integration is taken counterclockwise completely around the closed curve. Owing to the definition of r,

$$\int_0^{s_1} r\,ds = 2\omega(s_1) \tag{3.39}$$

where $\omega(s_1)$ is the plane area swept by r in moving from the point $s = 0$ to some other point s_1. $\omega(s)$ is called the *sectorial area*;[6] it is the shaded area in Fig. 3.13*a*. Thus, if we denote by Ω the total area enclosed by the center line of the tube wall, Eq. (3.38) becomes

$$M_t = 2\Omega q \tag{3.40}$$

[6] We use the symbol ω rather than A to distinguish between the sectorial area and the cross-sectional area $A = \oint t\,ds$.

The stress at any point is clearly

$$\tau = \frac{M_t}{2\Omega t} \tag{3.41}$$

which, in contrast with the thin-walled open section, acquires a maximum value at the point of smallest thickness.

We obtain this same result by examining the Φ surface for the tube shown in Fig. 3.13*b*. Its slope must be zero everywhere except over the wall thickness, where it is a constant. Its height is $\tau t = q$ at every point, so that the volume under the surface is simply Ωq for t very small. Twice this quantity, according to Eq. (3.17), is M_t.

We now proceed to the evaluation of the rate of twist of the tube. Owing to the relatively small wall thickness and the fact that τ is uniform over t, we are able to describe the deformation of the thin-walled tubes in terms of only two components of displacement, the warping u in the x direction and a tangential component η in the s direction. The projections of η on the y and z axes are v and w, respectively. Since we deduced earlier that cross sections do not distort in their own planes, points on the tube wall rotate θx about some point in the yz plane, as shown in Fig. 3.14; from the geometry of this figure, we see that

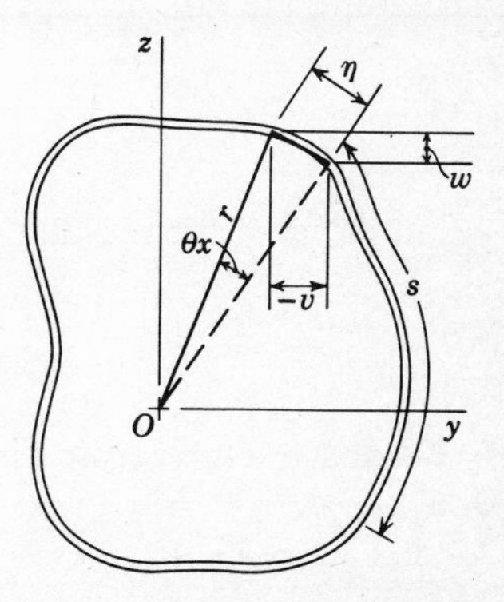

FIGURE 3.14 In-plane displacements of a point due to twisting of the section.

$$\eta = r\theta x \tag{3.42}$$

Now, according to the strain-displacement relations in Eqs. (2.17), an element of tube wall undergoes a shearing strain given by

$$\gamma_{xs} = \frac{\partial u}{\partial s} + \frac{\partial \eta}{\partial x} \tag{3.43}$$

since u and η are orthogonal. Introducing Eq. (3.42) into this relation, we find

$$\gamma_{xs} = \frac{\tau_{xs}}{G} = \frac{\partial u}{\partial s} + r\theta$$

which can also be written

$$\frac{\partial u}{\partial s} = \frac{\tau_{xs}}{G} - r\theta \tag{3.44}$$

Recalling that $\partial u/\partial x$ is zero, we multiply both sides of Eq. (3.44) by ds and integrate around the total periphery S of the tube:

$$\oint du = \frac{q}{G}\oint \frac{ds}{t} - \theta \oint r\, ds$$

The integral on the left side of this equation is clearly zero because it equals the difference in u at $s = S$ and $s = 0$, which, since the tube is closed, are identical. Furthermore, from our previous discussion we recognize that the last integral on the right side of the equation is 2Ω. Therefore, the rate of twist of the section is given by

$$\theta = \frac{q}{2\Omega G}\oint \frac{ds}{t} \tag{3.45}$$

or, in terms of the twisting moment,

$$\theta = \frac{M_t}{4\Omega^2 G}\oint \frac{ds}{t} \tag{3.46}$$

Comparing this result with Eq. (3.18), we see that

$$J = \frac{4\Omega^2}{\oint ds/t} \tag{3.47}$$

or, if t is constant,

$$J = \frac{4\Omega^2 t}{S} \tag{3.48}$$

For hybrid sections composed of a closed cell plus open "fin" elements, such as that shown in Fig. 3.15, Eqs. (3.29) and (3.41) are still applicable; stresses in the fins are given by Eq. (3.29), and those in the closed tube are given by Eq. (3.41). In the case of n fins,

$$J = \frac{4\Omega^2}{\oint ds/t} + \frac{1}{3}\sum_{i=1}^{n} b_i t_i^3 \tag{3.49}$$

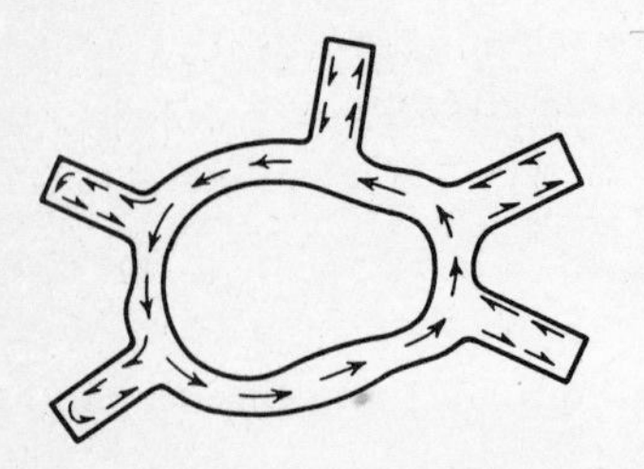

FIGURE 3.15 Closed tube with fins.

The above formulas are applicable so long as t is small compared with the other cross-sectional dimensions. We obtain some indication of the range of validity of Eqs. (3.47) and (3.48) by comparing them with the exact formula for the hollow circular shaft, namely,

$$J = I_p = \frac{\pi}{2}(R_o^4 - R_i^4)$$

where R_o and R_i are the outside and inside radii, respectively. In this case $t = R_o - R_i$, $S = 2\pi(R_o + R_i)/2$, and $\Omega = \pi[(R_o + R_i)/2]^2$, so that Eq. (3.48) gives

$$J = 2\pi t\left(\frac{R_o + R_i}{2}\right)^3$$

Hence,

$$\frac{J}{I_p} = \frac{1}{2}\left(1 + \frac{2\lambda}{1 + \lambda^2}\right) \tag{3.50}$$

where

$$\lambda = \frac{R_i}{R_o}$$

Similar calculations show that

$$\frac{\tau_t}{\tau_0} = \frac{1 + \lambda^2}{1 + \lambda} \tag{3.51}$$

where τ_t and τ_0 denote the maximum shearing stresses in the tube and the hollow shaft, respectively.

Graphs of Eqs. (3.50) and (3.51) are shown in Fig. 3.16. We see that in the limiting case of the solid section, Eq. (3.48) gives only half of the exact value. As t becomes smaller, however, this error rapidly decreases and we find that Eq. (3.48) is in error only 10 percent for values of t/R_o as high as 0.5. We are again reminded that J is equal to the polar moment of inertia only for the special case of a circular cross section. Equation (3.41), on the other hand, is in error no more than 20 percent for all values of λ; but it yields values less than 10 percent in error only for $t/R_o < 0.24$ (or for $t/R_o > 0.875$).

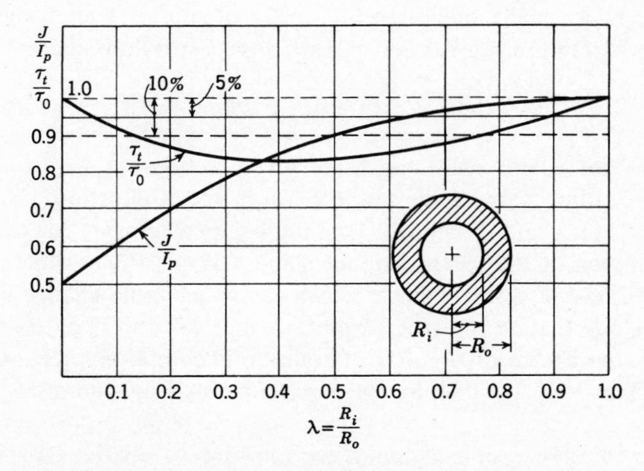

FIGURE 3.16 Comparison of the exact and approximate torsional constants and maximum stresses for a hollow shaft.

3.9 Warping of thin-walled sections. We now evaluate the warping displacement u of a thin-walled member in pure torsion. To accomplish this, we refer to Eq. (3.44), multiply both sides by ds, and integrate from the origin $s = 0$ to some other point s on the tube wall. We get

$$u - u_0 = \frac{1}{G}\int \tau_{xs}\, ds - \theta \int r\, ds$$

where u_0 is the displacement of the point $s = 0$ in the x direction. Comparing the second integral on the right side of this equation with Eq. (3.39), we see that it is equal to twice $\omega(s)$, the sectorial area. Thus,

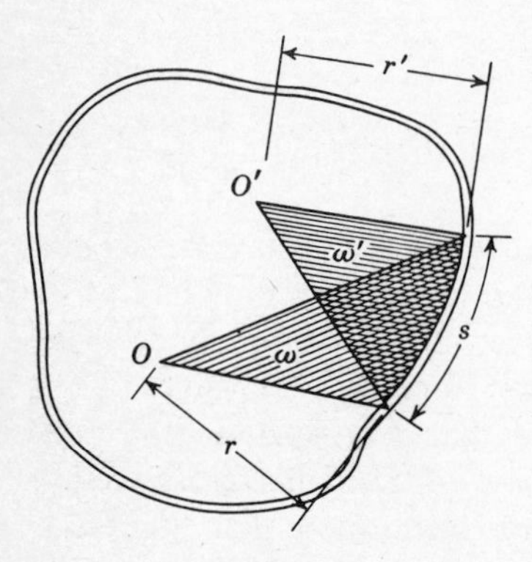

FIGURE 3.17 Thin-walled tube in pure torsion.

$$u = \frac{1}{G}\int \tau_{xs}\, ds - 2\theta\omega(s) + u_0 \qquad (3.52)$$

As before, the radius r in the integral expression for $2\omega(s)$ is measured from some point O on the cross section to a tangent to the center line of the tube wall at any point s. It is interesting to examine the consequences of choosing a center of rotation other than O. Suppose, for example, the radius r is measured from O' rather than O, as is indicated in Fig. (3.17). The new radius r' sweeps through a different sectorial area ω', and u and u_0 change to u' and u_0', respectively. The integral in Eq. (3.52), of course, is unchanged if we measure s from the same origin in each case. It follows that

$$u' = u + (u_0' - u_0) + 2\theta(\omega - \omega')$$

Since u_0 and u_0' are constants for a given section, $u_0' - u_0$ is simply a rigid-body translation of the point $s = 0$ in the x direction. Similarly, the term $2\theta(\omega - \omega')$ is merely a longitudinal displacement due to a rigid-body rotation of the cross-sectional plane. Therefore, u' differs from u only by *rigid-body displacements* which can result in no change in stress. We conclude that in the case of pure torsion with no restraints against warping, the choice of the axis of rotation is completely arbitrary; any longitudinal axis parallel to the centroidal axis of the tube can be used.

With this observation in mind, we introduce a relative warping displacement defined by

$$\bar{u} = u - u_0$$

and thereby refer warping displacements to the plane containing the origin of the s coordinate. Hence, the displacement of any point relative to this origin is

$$\bar{u} = \frac{1}{G}\int \tau_{xs}\, ds - 2\theta\omega \tag{3.53}$$

The development of Eq. (3.53) is based on the general strain-displacement relation given in Eq. (3.43). Thus, Eq. (3.53) is applicable to both closed and open sections. However, in the case of an open section it is necessary to pay closer attention to the definition of $\bar{u}$. The shearing stress τ_{xs} in the open section varies linearly throughout the wall thickness and is zero at the center line of the wall. This means that s cannot be measured along the center line; else the integral is zero. If, on the other hand, we use the maximum value of the stress at a point for τ_{xs} and let s be measured along paths on the outside and then the inside periphery to two points on opposite sides of the wall, the integral would represent the *difference* between the warping displacements of these two points—a very small quantity for thin-walled sections. It appears that for open sections the integral in Eq. (3.53) is negligible in comparison with the term $2\theta\omega$. Physically, it is easy to see that this is true by noting that the open section is many times more flexible than the closed section. In fact, the torsional stiffness of a closed section is often several thousand times greater than that of an open section of the same dimensions. Thus, the shearing strain γ_{xs} in Eq. (3.43) is negligible, and the warping displacements $\bar{u}$ are due almost exclusively to the twist of the section. It follows that Eq. (3.53) becomes for the open section

$$\bar{u} = -2\theta\omega \tag{3.54}$$

In the case of closed sections, γ_{xs} is no longer negligible, $\tau_{xs} = q/t$, and Eq. (3.53) becomes

$$\bar{u} = \frac{q}{G}\int \frac{ds}{t} - 2\theta\omega \tag{3.55}$$

or, in terms of the twisting moment,

$$\bar{u} = \frac{M_t}{G}\left(\frac{1}{2\Omega}\int \frac{ds}{t} - \frac{2}{J}\,\omega\right) \tag{3.56}$$

Note that $\bar{u}$ is a function of s and is zero when the integration is taken completely around the closed curve.

3.10 Multicell thin-walled tubes. Torsion of a single-cell tube free to warp is a statically determinate problem—the shear flow and, hence, the shear stress are independent of the elastic properties of the material. If an additional cell is introduced, however, there exists an additional, independent shear flow associated with this cell, which cannot be determined by statics. The multicell tube in pure torsion, then, is a statically indeterminate problem. The degree of indeterminacy of a tube having n cells is $n - 1$, and $n - 1$ conditions of deformation must be used to supplement the single equation of equilibrium that is available.

Let us consider the n-celled tube of general shape shown in Fig. 3.18*a*. The n independent shear flows are taken as unknowns. Referring to Eq. (3.40), we write the equation of equilibrium,

$$M_t = 2\sum_{j=1}^{n} q_j\Omega_j \tag{3.57}$$

where q_j is the shear flow in cell j and Ω_j is the area of cell j. Note that the sum $\Sigma q_j\Omega_j$ represents the volume under the Φ surface.

Now let us examine the three typical cells i, j, and k shown in Fig. 3.18*b*. If we temporarily assume that only cell j is effective, according to Eq. (3.45) the rate of twist of this cell is

$$\theta_j = \frac{q_j}{2G\Omega_j}\oint_{s_j}\frac{ds}{t} \tag{3.58}$$

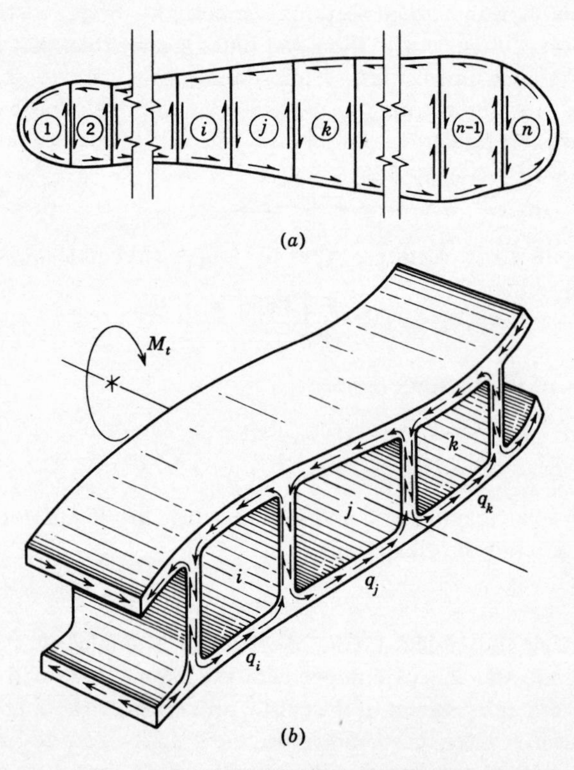

FIGURE 3.18 Multicell tube in torsion.

where s_j indicates integration around cell j. If shear flows q_i and q_k are now introduced in cells i and k, the flows in web s_{ji} common to cell i and j and in web s_{jk} common to cells j and k are reduced to $q_j - q_i$ and $q_j - q_k$, respectively, as is shown in the figure. Thus, the rate of twist of cell j becomes

$$\theta = \frac{1}{2G\Omega_j}\left(q_j \oint_{s_j} \frac{ds}{t} - q_i \int_{s_{ji}} \frac{ds}{t} - q_k \int_{s_{jk}} \frac{ds}{t}\right) \tag{3.59}$$

where, as before, the subscripts on the integrals indicate the path over which the integration is to be performed.

We found earlier that although cross sections warp, they do not distort in their own plane. This means that the entire cross section and each cell rotate at the same rate of twist θ and that, therefore,

$$\theta_1 = \theta_2 = \cdots = \theta_j = \cdots = \theta_n = \theta \tag{3.60}$$

It follows that for the general case in which cell j is bounded by m cells rather than two,

$$\theta = \frac{1}{2G\Omega_j}\left[q_j \oint_{s_j} \frac{ds}{t} - \sum_{r=1}^{m}\left(q_r \int_{s_{jr}} \frac{ds}{t}\right)\right] \tag{3.61}$$

Equations (3.60) and (3.61) are called *equations of consistent deformation.* The shear flows satisfying Eq. (3.61), for example, will be those which result in a twist of cell j that is *compatible* (that is, equal) to that of cells i and k. We may write one such equation for each of the n cells, and the resulting n linearly independent equations plus Eq. (3.57) provide $n + 1$ equations in the $n + 1$ unknowns, $q_1, q_2, \ldots, q_n$, and θ.

The integral coefficients of the shear flows in Eq. (3.59) are called *flexibilities.* Physically, the quantity

$$\frac{\oint_{s_j} ds/t}{2G\Omega_j}$$

is θ due to a unit shear flow in cell j ($q_j = 1, q_i = q_k = 0$). Similarly,

$$\frac{\oint_{s_{ji}} ds/t}{2G\Omega_j}$$

is the rate of twist of cell j due to a unit flow along the web s_{ji} ($q_i = 1$, $q_j = q_k = 0$), etc. However, we obtain a more meaningful physical interpretation by multiplying both sides of Eq. (3.59) or (3.61) by $2\Omega_j$ and comparing the resulting coefficients with Eq. (3.55). The equation of consistent deformation for cell j becomes

$$\delta_{ji}q_i + \delta_{jj}q_j + \delta_{jk}q_k - 2\Omega_j\theta = 0 \tag{3.62}$$

where

$$\delta_{ji} = -\frac{1}{G}\int_{s_{ji}} \frac{ds}{t} \qquad \delta_{jk} = -\frac{1}{G}\int_{s_{jk}} \frac{ds}{t} \tag{3.63}$$

and

$$\delta_{jj} = \frac{1}{G}\oint_{s_j} \frac{ds}{t} \tag{3.64}$$

These coefficients are called *warping flexibilities*. To interpret these quantities physically, we assume that a longitudinal cut has been introduced in the bottom wall of each cell so that the tube is temporarily open. If a torque is applied, each section will twist at a rate θ and, according to Eq. (3.54), there will occur a relative warping of opposite faces of the slit in cell j of magnitude $-2\Omega_j\theta$, regardless of the location of the slit. If constant shear flows are introduced in cells i, j, and k, they will produce a relative warping displacement at the slit in cell j of $q_i\delta_{ji}$, $q_j\delta_{jj}$, and $q_k\delta_{jk}$, respectively. Thus, Eq. (3.62) merely states that the magnitudes of these flows must be such that the total relative warping is zero (that is, that deformations of opposite faces of the slit are compatible). It follows that δ_{ji} and δ_{jk} are the relative warping displacements of opposite sides of the slit in cell j due to unit shear flows in cells i and k, respectively. Similarly, δ_{jj} is the relative warping at this same imaginary cut due to a unit shear flow in cell j.

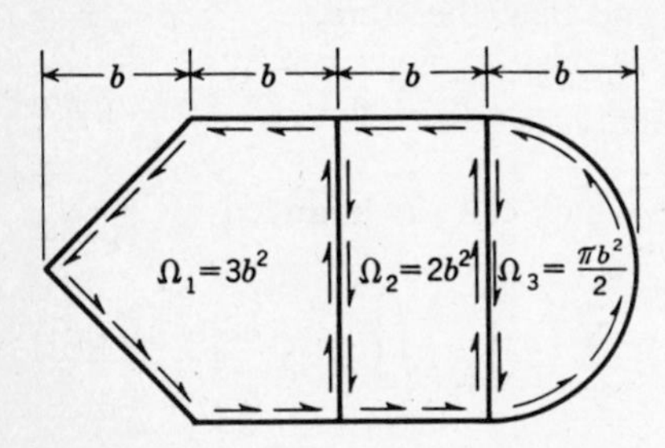

FIGURE 3.19 Three-celled tube in torsion.

Let us consider as a simple example the three-celled tube shown in Fig. 3.19. The tube is subjected to a torque M_t, and the skin thickness t is constant. By using Eqs. (3.63) and (3.64) we obtain

$$\begin{aligned} \delta_{11} &= \frac{2b(2+\sqrt{2})}{Gt} \\ \delta_{12} = \delta_{21} &= -\frac{2b}{Gt} \\ \delta_{22} &= \frac{6b}{Gt} \\ \delta_{23} = \delta_{32} &= -\frac{2b}{Gt} \\ \delta_{33} &= \frac{b(\pi+2)}{Gt} \end{aligned} \tag{a}$$

Introducing these into Eq. (3.62) for each cell and simplifying, we find

$$
\begin{aligned}
2b(2+\sqrt{2})q_1 - 2bq_2 - 2Gt(3b^2)\theta &= 0 \\
-2bq_1 + 6bq_2 - 2bq_3 - 2Gt(2b^2)\theta &= 0 \\
-2bq_2 + 6(2+\pi)q_3 - 2Gt\frac{\pi b^2}{2}\theta &= 0
\end{aligned}
\tag{b}
$$

Solving these equations, we find

$$q_1 = 1.316Gtb\theta \qquad q_2 = 1.502Gtb\theta \qquad q_3 = 1.194Gtb\theta \tag{c}$$

Referring now to Eq. (3.57), the condition of equilibrium of moments about the tube's axis is satisfied if

$$M_t = b^2(6q_1 + 4q_2 + \pi q_3) \tag{d}$$

Substituting Eqs. (*c*) into Eq. (*d*), we find for the rate of twist

$$\theta = 0.0566\frac{M_t}{Gtb^3} \tag{e}$$

Finally, introducing this value into Eqs. (*c*) gives

$$q_1 = 0.0746\frac{M_t}{b^2} \qquad q_2 = 0.0851\frac{M_t}{b^2} \qquad q_3 = 0.0676\frac{M_t}{b^2} \tag{f}$$

The evaluation of torsional constant J for multicell tubes j is also of interest. We obtain J by direct substitution of Eq. (3.57) into Eq. (3.18):

$$J = \frac{2}{G\theta}\sum_{j=1}^{n} q_j\Omega_j \tag{3.65}$$

For the tube in Fig. 3.19 we find from Eq. (*e*) that

$$J = 17.6540tb^3 \tag{g}$$

PROBLEMS

3.1. Show that in the torsion problem, the stresses τ_{xy} and τ_{xz} are harmonic functions.

3.2. Verify the elementary formulas, Eqs. (3.1) and (3.2), for a circular section of radius a by examining the stress function

$$\Phi = C(y^2 + z^2 - a^2)$$

where C is a constant.

3.3. Show that

$$\Phi = C[\sqrt{3}a(y^2 + z^2) - 2y^3 + 6yz^2 + A]$$

where A, C, and a are constants, is a permissible Saint-Venant stress function and evaluate C in terms of G and θ.

3.4. Show that Φ of Prob. 3.3 is the stress function for the equilateral triangle shown below. Find A so that Φ satisfies the boundary conditions.

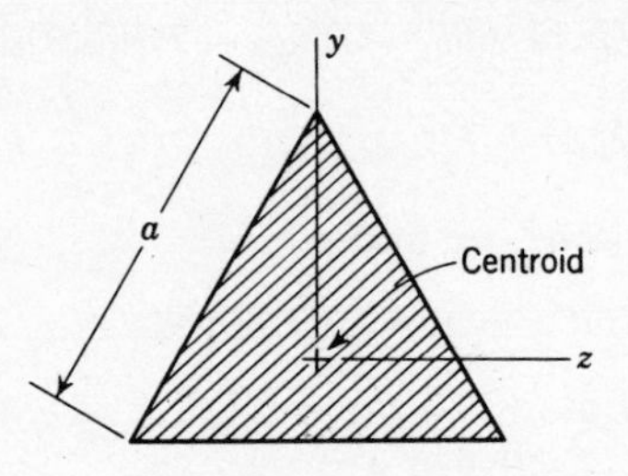

3.5. Using the results of Probs. 3.3 and 3.4, evaluate the stresses and the torsional constant J for an equilateral triangle.

3.6. Compare the torsional constants calculated using Eq. (3.25) with those for the rectangular section given by Eq. (3.33) and Table 3.1.

3.7. (*a*) Evaluate the torsional constant and the maximum stress for the closed thin-walled tube shown.

(*b*) A longitudinal cut is made along the dashed line indicated so that the section becomes open. Calculate J and τ_{max} and compare them with those for the closed section.

(*c*) If the allowable stress is 10,000 psi, compute the maximum torque that each of the sections in (*a*) and (*b*) can carry safely.

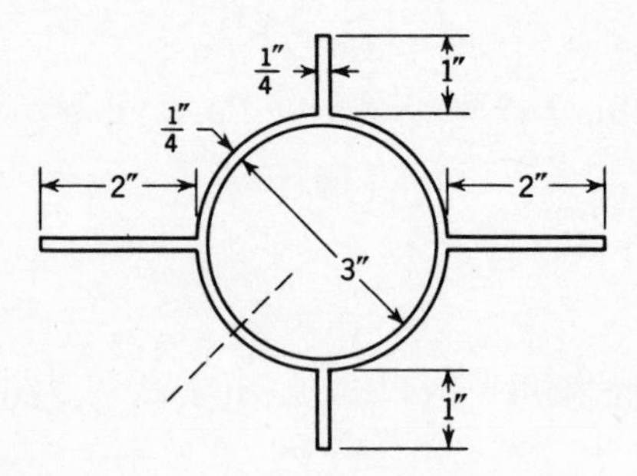

3.8. Evaluate J and τ_{max} for the section shown.

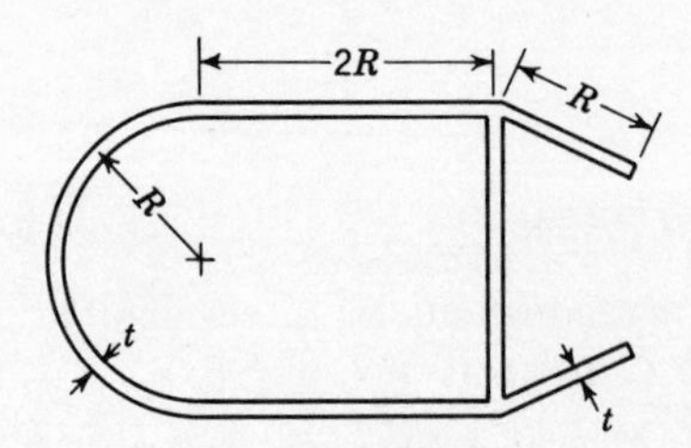

3.9. Evaluate J and τ_{max} for the section shown.

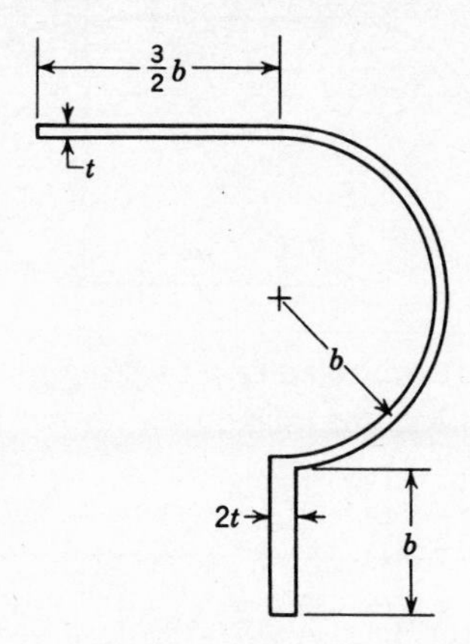

3.10. An elliptical thin-walled tube is shown below.

(*a*) Evaluate J and τ_{max}.

(*b*) Evaluate J and τ_{max} assuming that the section is opened by introducing a small slit at the point indicated by the dashed line.

(*c*) Compare the results of parts (*a*) and (*b*) and with J and τ_{max} for a solid section with the same outside dimensions.

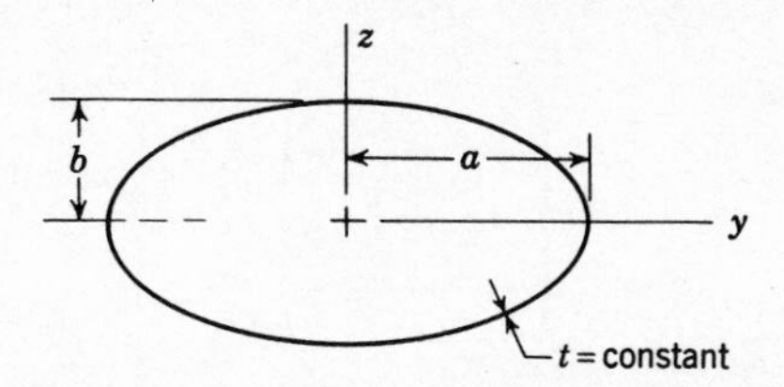

3.11. The hollow section shown is designed to withstand a maximum shearing stress of 4,000 psi. The wall thickness is a constant.

(*a*) Find the largest twisting moment that can be safely carried by the section and the angle of twist developed in a 100-in. length.

(*b*) Repeat part (*a*) assuming that both cells are closed. Compare the two solutions.

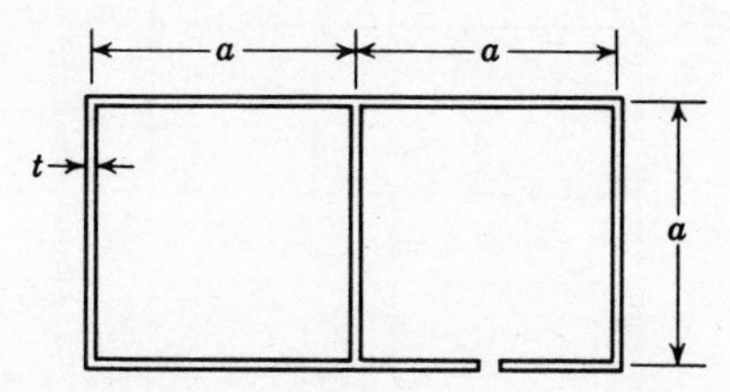

3.12. Compute the shear flows, the rate of twist, and the torsional constant for the three-celled tube shown.

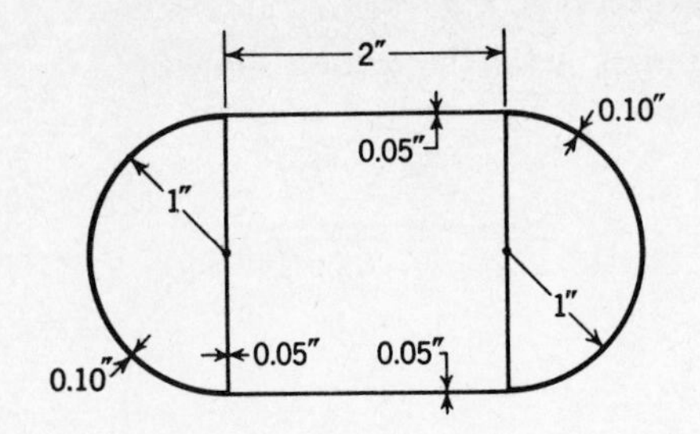

3.13. The tube shown is subjected to a twisting moment of 100,000 in.-lb. Compute the shear flows, the maximum stress, and the rate of twist. Take $G = 12 \times 10^6$ psi.

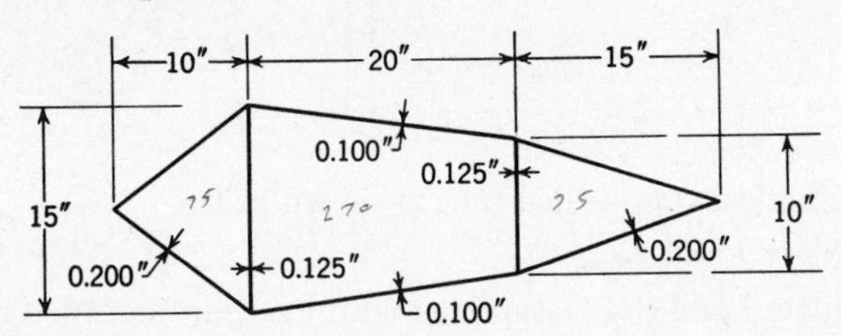

3.14. Compute the shear flows and the rate of twist developed in the tube shown in terms of G and the applied torque T. The wall thickness is constant.

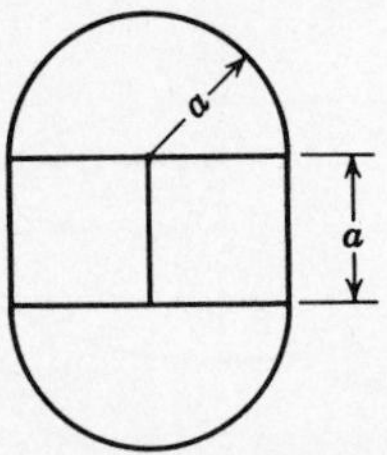

3.15. Compute the shear flows and the rate of twist developed in the thin-walled tube shown in terms of G and the applied torque T. The wall thickness is constant.

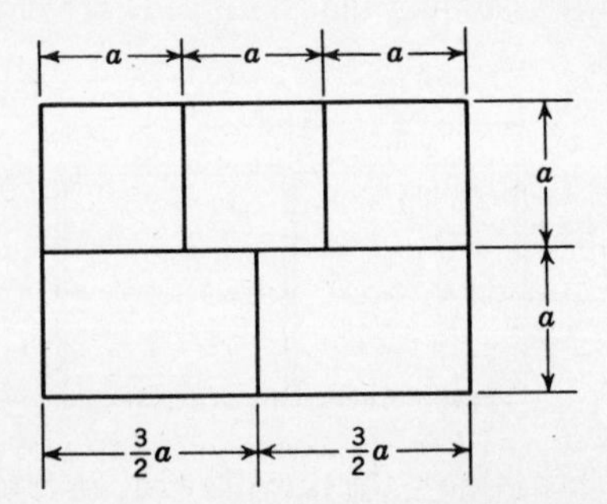

3.16. Assuming that both cells of the thin-walled section in Prob. 3.11 are closed, compute the maximum stress in the interior (center) web.

four

STRESSES AND STRESS RESULTANTS IN BARS

4.1 Introduction. In the preceding chapter we studied the behavior of bars subjected to only one of the six stress resultants. We now turn our attention to the remaining five for the purpose of examining relationships between stresses and stress resultants developed in the bending and stretching of bars. Once these relationships have been established, the problem of evaluating the stress distribution on any cross section reduces to one of determining the stress resultants developed at every point on the axis of the bar.

We begin our study by considering the curved coplanar bar of constant cross section shown in Fig. 4.1*a*. We denote the radius of curvature at any point by R and define the geometry of the bar by establishing two orthogonal coordinate systems: one fixed coordinate system $(x,\bar{y},z)$, with its origin located on some convenient cross section, as shown, and another curvilinear system (s,y,z) in which s is the arc length measured along the geometric axis, y is a radial coordinate directed toward the center of curvature of this axis, and z is directed normal to the plane of the bar. We denote the y and z components of the external forces per unit length

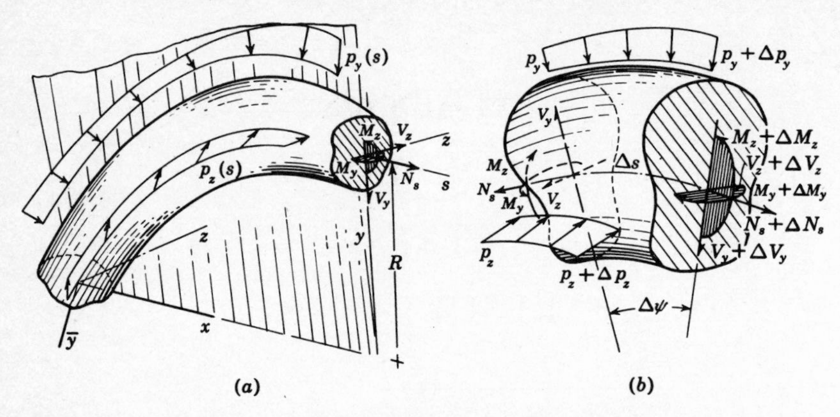

FIGURE 4.1 (*a*) Curved coplanar bar; (*b*) stress resultants developed on a typical bar element.

arc by $p_y(s)$ and $p_z(s)$, respectively, and we assume that the twisting of any cross section is zero or negligible.[1]

Referring to Eqs. (2.7), we find that in this case five stress resultants can be developed at a point on the bar's axis: N_s, V_y, V_z, M_y, and M_z, as shown in Fig. 4.1*b*. Positive stress resultants are referred to the cross section nearest to the origin when a cut is made at some point s on the bar's axis. Positive forces act in the direction of increasing s, y, and z, and positive moments produce tension in the positive y and z quadrants of the section. These quantities result from stresses σ_s, τ_{sy}, and τ_{sz} and are assumed to be known functions of s determined, perhaps, in terms of end reactions and applied loads by simple statics.

4.2 The equilibrium equations. Since we are to describe the behavior of the bar in terms of these internal forces and moments rather than stresses, we must reformulate the equilibrium conditions in terms of stress resultants. To do this, we proceed as in the three-dimensional case by first considering the stress resultant state at some point P on the bar's axis, whose location in the $x\bar{y}$ plane is given by the coordinate s. The portion of the bar between P and another point P' as $s + \Delta s$, arbitrarily close to P, represents an element of the bar. In general, the stress resultants acquire incremental changes of ΔN_s, ΔV_y, ΔV_z, ΔM_y, and ΔM_z in the interval Δs between P and P', as is indicated in Fig. 4.1*b*. The load intensities as well as the radius of curvature also acquire increments Δp_y, Δp_z, and ΔR; but as Δs approaches zero the influence of these changes on the equilibrium equations vanishes and, for clarity, we do not consider them.

[1] In order that the torque be zero on the cross section, the resultant of the shearing forces must pass through the shear center of the cross section. See Art. 5.5. Combined bending and torsion is discussed in Chap. 7.

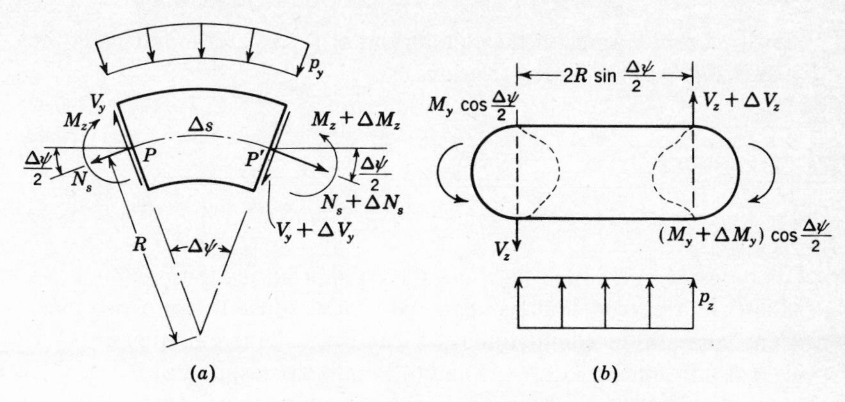

FIGURE 4.2 Statics of a typical curved bar element.

Two views of a typical bar element are shown in Fig. 4.2. In Fig. 4.2*a* we see only those forces and moments affecting equilibrium in the $x\bar{y}$ plane. Denoting by $\Delta\psi$ the angle between the cross-sectional planes at P and P' ($\Delta\psi = \Delta s/R$), we find on summing forces in the vertical direction

$$V_y \cos \frac{\Delta\psi}{2} - (V_y + \Delta V_y) \cos \frac{\Delta\psi}{2} - p_y \Delta s - N_s \sin \frac{\Delta\psi}{2} - (N_s + \Delta N_s) \sin \frac{\Delta\psi}{2} = 0$$

As Δs approaches zero, $\cos(\Delta\psi/2)$ approaches unity and $\sin(\Delta\psi/2)$ approaches $\Delta\psi/2 = \Delta s/2R$. Thus, dividing the above equation by Δs and taking the limit as Δs approaches zero, we get

$$\frac{dV_y}{ds} = -p_y - \frac{N_s}{R} \tag{4.1}$$

Similarly, summing forces in the horizontal direction gives

$$(N_s + \Delta N_s - N_s) \cos \frac{\Delta\psi}{2} - (V_y + V_y + \Delta V_y) \sin \frac{\Delta\psi}{2} = 0$$

or, in the limit,

$$\frac{dN_s}{ds} = \frac{V_y}{R} \tag{4.2}$$

Referring to Fig. 4.2*a*, we now sum moments about point P':

$$M_z + \Delta M_z - M_z + p_y \Delta s\, R \sin \frac{\Delta\psi}{2} - V_y \cos \frac{\Delta\psi}{2} 2R \sin \frac{\Delta\psi}{2} + N_s \sin \frac{\Delta\psi}{2} 2R \sin \frac{\Delta\psi}{2} = 0$$

which, as Δs approaches zero, reduces to

$$\frac{dM_z}{ds} = V_y \tag{4.3}$$

Similar considerations of the equilibrium of forces and moments in the xz plane (Fig. 4.2*b*) yield the relations

$$\frac{dV_z}{ds} = -p_z \tag{4.4}$$

and

$$\frac{dM_y}{ds} = V_z \tag{4.5}$$

Equations (4.1) through (4.5) are the equilibrium conditions on stress resultants in a curved coplanar bar. We may reduce these to two independent equations by eliminating N_s, V_y, and V_z. Differentiating Eq. (4.3) twice and introducing Eqs. (4.1) and (4.2) into the result gives

$$\frac{d^3M_z}{ds^3} + \frac{R'}{R}\frac{d^2M_z}{ds^2} + \frac{1}{R^2}\frac{dM_z}{ds} + \frac{dp_y}{ds} + \frac{R'}{R}p_y = 0 \tag{4.6}$$

where $R' = dR/ds$. If R is constant, this equation becomes

$$\frac{d^3M_z}{ds^3} + \frac{1}{R^2}\frac{dM_z}{ds} + \frac{dp_y}{ds} = 0 \tag{4.7}$$

Similarly, introducing Eq. (4.4) into the derivative of Eq. (4.5), we find

$$\frac{d^2M_y}{ds^2} + p_z = 0 \tag{4.8}$$

The above conditions amount to a one-dimensional equivalent of the general equilibrium conditions in Eqs. (2.4). This equivalence, however, is not apparent in their present form; and we must now establish relationships between the stress resultants and the components of stress.

4.3 Normal stresses in curved bars. As mentioned earlier, the stress resultants are statically equivalent to a stress distribution σ_s, τ_{sy}, and τ_{sz} on each cross-sectional plane (see Fig. 4.3); of the remaining components of stress σ_z and τ_{yz} need only be large enough to provide local resistance to the transverse loading. We justifiably neglect their influence on the cross-sectional stresses. The radial stress σ_y, however, may be of the same order of magnitude as σ_s due to the initial curvature of the bar, and, according to Eqs. (2.27), it alters the longitudinal strain in the s direction by an amount $-\nu\sigma_y/E$. Nevertheless,

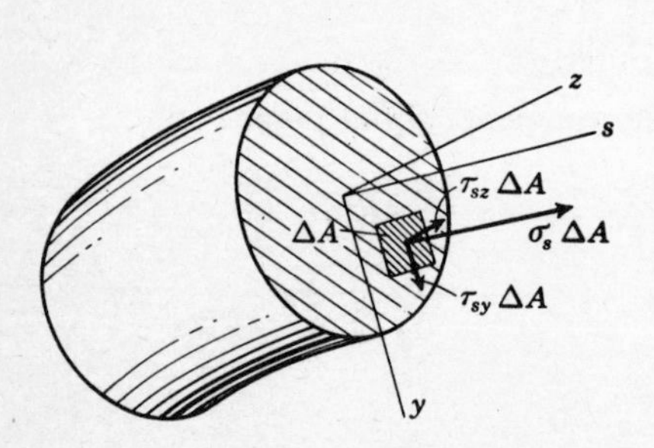

FIGURE 4.3 Stresses at a point on the cross section of a curved bar.

we shall find that for common cross-sectional shapes, σ_s and σ_y acquire maximum values at different points on the cross section so that their net influence on one another is very small. For this reason, we temporarily neglect the effects of σ_y and concentrate on evaluating σ_s.

According to Eqs. (2.7), the normal stress distribution on a cross section is statically equivalent to the normal force N_s and the moments M_y and M_z. However, we cannot evaluate the appropriate integrals in Eqs. (2.7) without knowing σ_s as a function of y and z. Obviously, with the equations of statics now exhausted, we must turn to considerations of deformation for additional information—a step in the derivation that brings us to the seldom-appreciated conclusion that a simple bar in bending is a statically indeterminate problem.

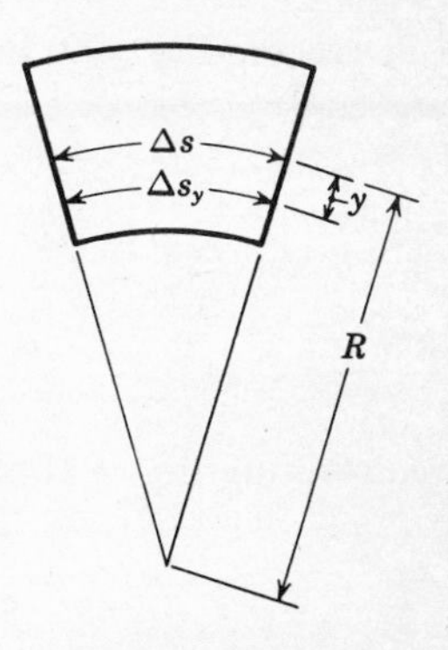

FIGURE 4.4 Geometry of beam element.

We avoid the unnecessary complexity of obtaining an "exact" description of the displacements by introducing Navier's assumption:[2] *plane sections normal to the bar's geometric axis before deformation remain plane and normal to this axis after deformation.* This is one of the most common hypotheses in structural analysis; its validity is well confirmed by both experiment and more rigorous theory[3] for a wide range of structural problems. If such a condition prevails, it follows that the components of displacement in a direction normal to a cross section must satisfy the equation of a plane. Thus, for a given value of s,

$$u = \tilde{a} + \tilde{b}y + \tilde{c}z \tag{4.9}$$

where $\tilde{a}$, $\tilde{b}$, and $\tilde{c}$ are functions of s only which, for a given cross section, may be regarded as constants.

Let us now examine the geometry of the bar element between cross sections at s and $s + \Delta s$ shown in Fig. 4.4. Now Δs is the length of an increment of arc measured along the bar's axis. Fibers located at a radial distance y from this axis are of a different length Δs_y. From the geometry element, we find

$$\frac{\Delta s}{R} = \frac{\Delta s_y}{R - y}$$

[2] This assumption was originally used by James Bernoulli (1654–1705), though Louis Navier (1785–1836) used it to develop the first complete theory of beam stresses. Theory for bars of small curvature based on this assumption was first introduced in 1858 by E. Winkler (1835–1888); it is sometimes called the "Winkler theory" of curved bars.

[3] See, for example, Ref. 61.

or, in the limit as Δs approaches zero,

$$\frac{ds}{ds_y} = \frac{1}{1 - y/R} \tag{4.10}$$

Referring to Eqs. (2.17), the longitudinal strain of any fiber is

$$\epsilon_s = \frac{\partial u}{\partial s_y}$$

Introducing Eq. (4.9), we have

$$\epsilon_s = a\frac{ds}{ds_y} + b\frac{ds}{ds_y}y + c\frac{ds}{ds_y}z = \frac{ds}{ds_y}(a + by + cz)$$

where $\qquad a = \dfrac{d\tilde{a}}{ds} \qquad b = \dfrac{d\tilde{b}}{ds} \qquad c = \dfrac{d\tilde{c}}{ds}$

Substituting Eq. (4.10) into this result gives

$$\epsilon_s = \frac{1}{1 - y/R}(a + by + cz) \tag{4.11}$$

This strain variation is considerably different from that of a straight bar. Although the displacement in Eq. (4.9) is linear in y and z, the resulting strain is definitely nonlinear; in fact, it is hyperbolic in the y direction. This is due to the initial curvature of the bar. The fibers in the bar element are of different lengths Δs_y before deformation; hence, although their changes in length are linear functions, they are strained nonlinearly.

Finally, since we have neglected the influence of σ_y and σ_z on σ_s, the normal stress is given by

$$\sigma_s = E\epsilon_s = \frac{E}{1 - y/R}(a + by + cz) \tag{4.12}$$

The problem now reduces to one of determining a, b, and c. To this end, we refer to Eqs. (2.7) and obtain the following equations:

$$\begin{aligned}
N_s &= \int_A \sigma_s\, dA = aE\int_A \frac{dA}{1 - y/R} + bE\int_A \frac{y\, dA}{1 - y/R} + cE\int_A \frac{z\, dA}{1 - y/R} \\
M_z &= \int_A \sigma_s y\, dA = aE\int_A \frac{y\, dA}{1 - y/R} + bE\int_A \frac{y^2\, dA}{1 - y/R} + cE\int_A \frac{zy\, dA}{1 - y/R} \\
M_y &= \int_A \sigma_s z\, dA = aE\int_A \frac{z\, dA}{1 - y/R} + bE\int_A \frac{yz\, dA}{1 - y/R} + cE\int_A \frac{z^2\, dA}{1 - y/R}
\end{aligned} \tag{4.13}$$

in which the integral coefficients are obviously functions of the geometry of the cross section. For simplicity, we introduce the notation

$$
\begin{aligned}
J_y &= \int_A \frac{z^2}{1 - y/R}\, dA \\
J_{yz} &= \int_A \frac{yz}{1 - y/R}\, dA \qquad (4.14) \\
J_z &= \int_A \frac{y^2}{1 - y/R}\, dA
\end{aligned}
$$

and note after some algebraic manipulations that

$$
\begin{aligned}
\int_A \frac{1}{1 - y/R}\, dA &= \int_A dA + \frac{1}{R}\int_A y\, dA + \frac{1}{R^2}\int_A \frac{y^2}{1 - y/R}\, dA \\
&= A + \frac{1}{R}\int_A y\, dA + \frac{1}{R^2} J_z \\
\int_A \frac{z}{1 - y/R}\, dA &= \int_A z\, dA + \frac{1}{R}\int_A \frac{yz}{1 - y/R}\, dA = \int_A z\, dA + \frac{1}{R} J_{yz} \\
\int_A \frac{y}{1 - y/R}\, dA &= \int_A y\, dA + \frac{1}{R}\int_A \frac{y^2}{1 - y/R}\, dA = \int_A y\, dA + \frac{1}{R} J_z
\end{aligned}
$$

Furthermore, since the origin of the coordinate system is at the centroid of the section,

$$
\int_A y\, dA = \int_A z\, dA = 0
$$

Thus, Eqs. (4.13) reduce to

$$
\begin{aligned}
\frac{N_s}{E} &= \left(A + \frac{J_z}{R^2}\right) a + \frac{J_z}{R} b + \frac{J_{yz}}{R} c \\
\frac{M_z}{E} &= \frac{J_{yz}}{R} a + J_{yz} b + J_y c \\
\frac{M_y}{E} &= \frac{J_z}{R} a + J_z b + J_{yz} c
\end{aligned}
$$

Solving these equations, we find

$$
\begin{aligned}
Ea &= \frac{N_s}{A} - \frac{M_z}{AR} \\
Eb &= \frac{M_z J_y - M_y J_{yz}}{J_y J_z - J_{yz}^2} - \frac{N_s}{RA} + \frac{M_z}{AR^2} \qquad (4.15) \\
Ec &= \frac{M_y J_z - M_z J_{yz}}{J_y J_z - J_{yz}^2}
\end{aligned}
$$

Finally, substituting Eqs. (4.15) into Eq. (4.12), we arrive at the equation

$$\sigma_s = \frac{N_s}{A} - \frac{M_z}{RA} + \frac{M_zJ_y - M_yJ_{yz}}{J_yJ_z - J_{yz}^2}\frac{y}{1 - y/R} + \frac{M_yJ_z - M_zJ_{yz}}{J_yJ_z - J_{yz}^2}\frac{z}{1 - y/R} \tag{4.16}$$

uniform normal stress   non uniform

The first two terms in the above equation represent a uniform normal stress over the section. We see that even in the case of pure bending ($N_s = 0$) the curvature of the bar causes a normal stress to be developed at the centroid of magnitude $-M_z/RA$. For curved bars it appears that M_z is statically equivalent to a normal force $-M_z/R$ acting at the centroid of the section. The remaining terms represent a nonuniform stress distribution which, owing to the initial curvature of the bar, is vastly different from that given by the elementary theory. We shall find that Eq. (4.16) reduces to the formula for normal stress in a straight bar when R becomes infinitely large.

The lack of symmetry of the section is identified through the constant J_{yz}. This constant is zero if the section is symmetrical with respect to the y axis. Because of this nonsymmetry, we see that even when M_y is zero, the moment M_z results in a stress variation in the z direction.

As an example of this nonuniform normal stress distribution, let us consider the unsymmetrical curved bar shown in Fig. 4.5. The bar is bent by equal moments $M_y = M_z = M$ and $N_s = 0$, so that Eq. (4.16) becomes

$$\sigma_s = M\left(-\frac{1}{RA} + \frac{J_y - J_{yz}}{J_yJ_z - J_{yz}^2}\frac{y}{1 - y/R} + \frac{J_z - J_{yz}}{J_yJ_z - J_{yz}^2}\frac{z}{1 - y/R}\right) \tag{a}$$

Note that for the structure shown twisting moments are also necessary on some sections to provide equilibrium; but again we neglect their influence on the normal stresses. Section constants for the section in Fig. 4.5*a* are obtained from Eqs. (4.14), which, after some lengthy integration,[4] give

$$\begin{aligned} J_z &= -R^2[A + R(b\alpha + h\beta)] \\ J_y &= -\frac{R}{3}[(c_1^3 - c_3^3)\alpha + (c_1^3 + c_2^3)\beta] \\ J_{yz} &= -\frac{R}{2}[(c_3^2 - c_1^2)\alpha + (c_2^2 - c_1^2)\beta] \end{aligned} \tag{b}$$

where $\alpha = \ln \dfrac{R - c_2}{R + c_3}$ and $\beta = \ln \dfrac{R + c_3}{R + c_1}$

For the section shown in Fig. 4.5*b* and an R of 10.0 in., these equations give $J_z = 6.422$ in.^{4}, $J_y = 5.430$ in.^{4}, and $J_{yz} = -3.581$ in.^{4}; the area of

[4] For complicated geometries, the constants J_y, J_z, and J_{yz} are more easily evaluated by numerical integration. See Probs. 4.10 to 4.13.

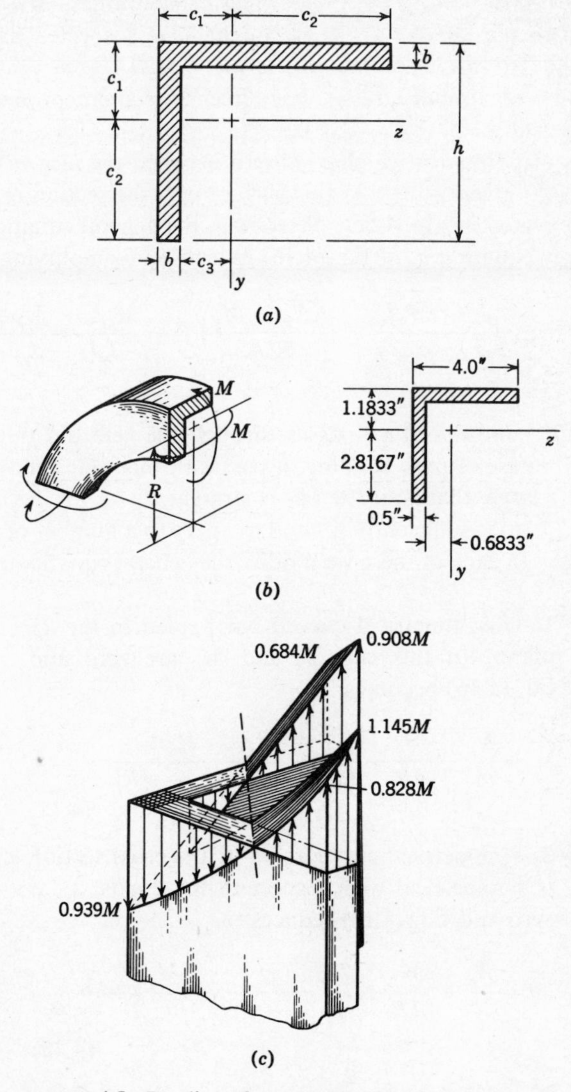

FIGURE 4.5 Bending of a curved bar with an unsymmetrical cross section.

the section is 3.75 sq in. Introducing these values into Eq. (*a*) and simplifying gives

$$\sigma_s = M\left(-0.027 + \frac{0.408y + 0.453z}{1 - 0.1y}\right) \qquad (c)$$

A plot of this equation is shown in Fig. 4.5*c*. The variation is linear in the z direction and hyperbolic in the y direction; the maximum stress occurs at the corner $y = c_2$, $z = -c_3$. The distribution is not symmetrical with respect to the line $y = z$, as we might have suspected had the bar been straight. In this example, the term $-M/RA$ contributes only 2 percent to the maximum stress. Its influence is more pronounced for smaller values of R.

The locus of points for which σ_s is zero is called the *neutral surface* of the bar. For a given value of s this surface cuts the section in a straight line, as is indicated in Fig. 4.5*c*. We obtain the general equation for this line by simply equating σ_s of Eq. (4.16) to zero and simplifying:

$$\frac{RN_s - M_z}{RA} + \left(\frac{M_z J_y - M_y J_{yz}}{J_y J_z - J_{yz}^2} - \frac{RN_s - M_z}{R^2 A}\right) y + \frac{M_y J_z - M_z J_{yz}}{J_y J_z - J_{yz}^2} z = 0 \tag{4.17}$$

Note that the neutral surface passes through the centroid only if $N_s = M_z/R$. In the case of pure bending, it passes through the centroid only if R is infinitely large, that is, if the bar is straight.

Equation (4.16) is sufficiently general to apply to a number of important special cases. In each of these we modify the general equation as follows:

1. Unsymmetrical curved bar loaded in the $x\bar{y}$ plane—In this case p_z and M_y are zero and Eq. (4.16) becomes

$$\sigma_s = \frac{N_s}{A} - \frac{M_z}{AR} + \frac{M_z(J_y y - J_{yz} z)}{(J_y J_z - J_{yz}^2)(1 - y/R)} \tag{4.18}$$

2. Symmetrical curved bar—If the cross section is symmetrical with respect to the y axis, J_{yz} is zero and Eq. (4.16) reduces to

$$\sigma_s = \frac{N_s}{A} - \frac{M_z}{AR} + \frac{M_z}{J_z}\frac{y}{1 - y/R} + \frac{M_y}{J_y}\frac{z}{1 - y/R} \tag{4.19}$$

3. Symmetrical curved bar loaded in the $x\bar{y}$ plane—Now both M_y and J_{yz} are zero, so that

$$\sigma_s = \frac{N_s}{A} - \frac{M_z}{AR} + \frac{M_z}{J_z}\frac{y}{1 - y/R} \tag{4.20}$$

4. Unsymmetrical straight bar or bar with small curvature—The nonlinearity of Eq. (4.16) is due to the quantity $1 - y/R$ in the denominators of the last two terms. Obviously, if y/R is small in comparison with unity, the stress distribution is essentially a linear function of y and z. For the straight bar, of course, R is infinite and y/R is zero. The section constants J_y, J_z, and J_{yz} also become modified; in fact, we see that when y/R is negligible Eqs. (4.14) reduce to

$$J_y = \int_A z^2\, dA = I_y$$

$$J_z = \int_A y^2\, dA = I_z$$

$$J_{yz} = \int_A yz\, dA = I_{yz}$$

where I_y and I_z are the moments of inertia of the cross-sectional area with respect to the y and z axes and I_{yz} is the product of inertia. Furthermore, the term $-M_z/RA$ in Eq. (4.16) represents a small fraction of the final stress distribution when R is large. An inspection of Eqs. (4.13) and (4.15) confirms that the absence of this term in the expression for σ_s is consistent with neglecting y/R in comparison with unity. It follows that in this case Eq. (4.16) becomes

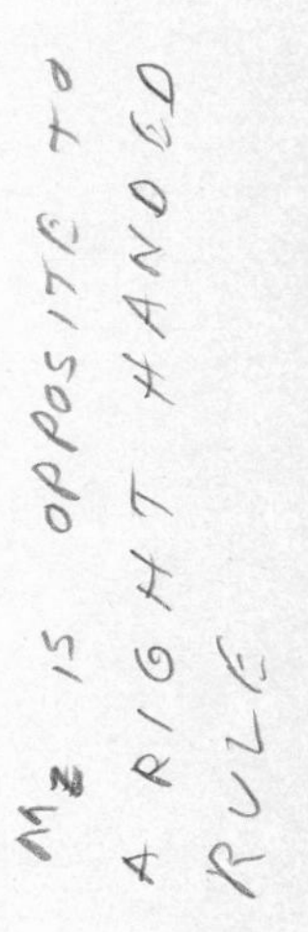

$$\sigma_s = \frac{N_s}{A} + \frac{M_z I_y - M_y I_{yz}}{I_y I_z - I_{yz}^2}\, y + \frac{M_y I_z - M_z I_{yz}}{I_y I_z - I_{yz}^2}\, z \tag{4.21}$$

Note that for the straight bar the coordinates s and y reduce to x and $\bar{y}$, respectively.

5. Unsymmetrical straight bar loaded in the xy plane—In this case σ_s and N_s become σ_x and N_x and M_y is zero:

$$\sigma_x = \frac{N_x}{A} + \frac{M_z(I_y y - I_{yz} z)}{I_y I_z - I_{yz}^2} \tag{4.22}$$

6. Symmetrical straight bar—If y and z are principal axes, I_{yz} is zero and Eq. (4.21) becomes

$$\sigma_x = \frac{N_x}{A} + \frac{M_z}{I_z}\, y \tag{4.23}$$

These cases are illustrated in Fig. 4.6. It is clear from the above developments that the elementary formula for normal stress is applicable only in very special cases of geometry and loading.

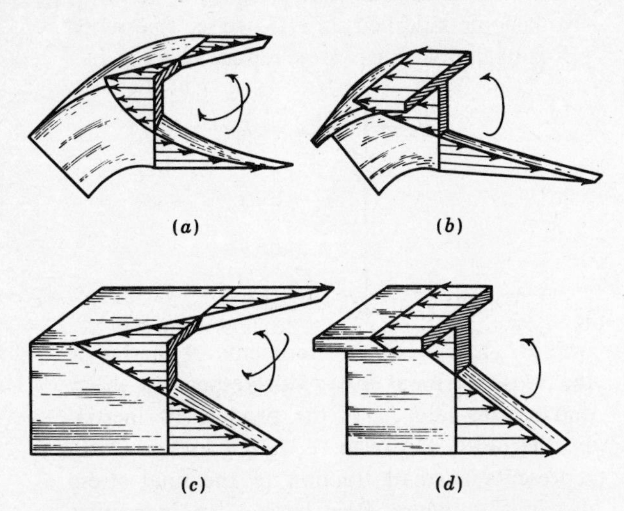

FIGURE 4.6 Examples of stress distributions due to various loadings and cross-section geometries. (*a*) Curved bar with unsymmetrical cross section. This type of distribution is developed even when the bar is loaded only in one plane, due to the lack of symmetry of the section. (*b*) Curved bar with symmetrical cross section loaded in the plane of symmetry. (*c*) Unsymmetrical straight bar or bar with small curvature loaded in the plane of symmetry. (*d*) Symmetrical straight bar or bar with small curvature loaded in the plane of symmetry.

4.4 Bending of symmetrical curved bars. To obtain some idea of the influence of curvature on the normal stress distribution, we now consider the behavior of a curved bar of symmetrical cross section subjected to pure bending in the $x\bar{y}$ plane. According to Eq. (4.20), in this case

$$\sigma_s = -\frac{M_z}{AR} + \frac{M_z}{J_z}\frac{y}{1 - y/R} \tag{4.24}$$

since N_s is zero.

To illustrate the influence of the initial curvature, let us consider the curved bar of rectangular cross section in Fig. 4.7. Here M_z is a constant moment of magnitude M. If b is the width of the section and h is the depth, we find

$$J_z = \int_{-h/2}^{h/2} \frac{y^2 b\, dy}{1 - y/R} = -R^2bh - R^3b \ln \frac{2R - h}{2R + h} \tag{4.25}$$

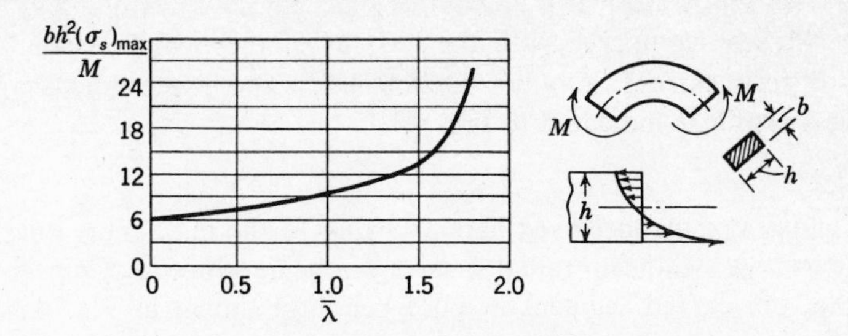

FIGURE 4.7 Influence of initial curvature on the maximum bending stress in a curved bar of rectangular cross section.

When h/R is less than approximately 0.3, the above equation for J_z becomes quite sensitive and six or seven significant figures must be used in its calculations.[5]

Since positive y in Eq. (4.24) is directed inward, toward the center of curvature, the maximum normal stress occurs at $y = +h/2$. Thus, introducing Eq. (4.25) into Eq. (4.24) and simplifying, we find for the maximum stress

$$(\sigma_s)_{max} = -\frac{M}{bh^2}\left\{\bar{\lambda} + \frac{\bar{\lambda}^3}{(2-\bar{\lambda})[\bar{\lambda} + \ln(2-\bar{\lambda})/(2+\bar{\lambda})]}\right\}$$

where $\bar{\lambda} = h/R$. A graph of this equation is given in Fig. 4.7. We see that the normal stress becomes infinitely large for $\bar{\lambda} = 2$, which, of course, is the limiting case of the complete solid circle. As $\bar{\lambda}$ decreases, $(\sigma_s)_{max}$ rapidly approaches the value for the straight bar $6M/bh^2$. Machine parts, hooks, chain links, and gears may have h/R ratios of near unity or larger, in which case Eq. (4.24) must be used in evaluating normal stresses. In a great many applications, however, curved bars with h/R ratios much smaller than this are used, and values of h/R of $\frac{1}{50}$ to $\frac{1}{100}$ are not uncommon. In such cases it appears that the formulas for bars of small curvature or straight bars [Eqs. (4.21) to (4.23), for example] can be used without introducing a significant error.

Taking, for illustration purposes, the case in which h/R is $\frac{1}{4}$, Eq. (4.25) gives

$$J_z = 0.08412bh^3$$

so that Eq. (4.24) becomes

$$\sigma_s = \frac{M}{bh^2}\left(-\frac{1}{4} + \frac{1}{0.08412h}\frac{y}{1 - y/4h}\right)$$

[5] For use in such calculations see, for example, "Table of Natural Logarithms," National Bureau of Standards, 1941.

In this case the stresses at the extreme fibers $y = \pm h/2$ are $-5.53M/bh^2$ and $6.54M/bh^2$, compared with the $6M/bh^2$ of the straight beam. The stress of the centroid is only $-0.25M/bh^2$. The hyperbolic shape of the stress profile is indicated in Fig. 4.7.

4.5 Radial stresses in curved bars. Owing to the initial curvature of a bar in bending, significant radial stresses can be developed. Consider, for example, the curved segment in pure bending[6] shown in Fig. 4.8*a*. If

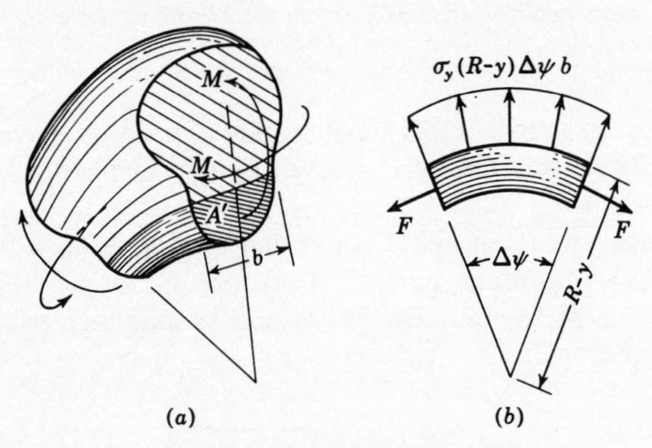

FIGURE 4.8 Radial stresses due to pure unsymmetrical bending of a curved bar.

we isolate a portion of the segment of area A', as shown in Fig. 4.8, tensile forces of magnitude F are developed on this area, where

$$F = \int_{A'} \sigma_s \, dA$$

Assuming that N_s is zero, we have from Eq. (4.16)

$$F = -\frac{M_z A'}{AR} + \frac{M_z J_y - M_y J_{yz}}{J_y J_z - J_{yz}^2} \bar{Q}_z + \frac{M_y J_z - M_z J_{yz}}{J_y J_z - J_{yz}^2} \bar{Q}_y \qquad (4.26)$$

where $\bar{Q}_z = \int_{A'} \frac{y}{1 - y/R} \, dA$ and $\bar{Q}_y = \int_{A'} \frac{z}{1 - y/R} \, dA \qquad (4.27)$

Now if σ_y is the average radial stress on the element and b is the dimension indicated in the figure, a force of magnitude $\sigma_y(R - y)\,\Delta\psi\, b$ must be developed to balance the vertical component of F (here $\Delta\psi = \Delta s/R$). Referring to Fig. 4.8*b* and summing forces in the vertical direction, we find

$$2F \sin \frac{\Delta\psi}{2} = \sigma_y b(R - y)\,\Delta\psi$$

[6] The influence of shearing stresses on the radial stress is discussed in Art. 5.2.

Again noting that sin $(\Delta\psi/2)$ approaches $\Delta\psi/2$ as $\Delta\psi$ becomes arbitrarily small, we find in the limit that

$$\sigma_y = \frac{F}{b(R-y)}$$

Introducing Eq. (4.26) into this result gives

$$\sigma_y = \frac{1}{b(R-y)}\left(-\frac{M_zA'}{RA} + \frac{M_zJ_y - M_yJ_{yz}}{J_yJ_z - J_{yz}^2}\,\bar{Q}_z + \frac{M_yJ_z - M_zJ_{yz}}{J_yJ_z - J_{yz}^2}\,\bar{Q}_y\right) \tag{4.28}$$

If the section is symmetrical with respect to the y axis and M_y is zero, Eq. (4.28) reduces to

$$\sigma_y = \frac{M_z}{b(R-y)}\left(-\frac{A'}{RA} + \frac{\bar{Q}_z}{J_z}\right) \tag{4.29}$$

We note that as R becomes larger, σ_y decreases and, thus, is often negligible compared with σ_s. Such is not the case in the analysis of elements such as hooks, chain links, and many machine parts, in which the ratio h/R is relatively large.

Consider, for example, a sharply curved bar with a rectangular section of dimensions $b \times h$ (Fig. 4.9). Noting that $\bar{Q}_z$ in Eq. (4.27) can also be written

$$\bar{Q}_z = \int_{A'} \frac{y\,dA}{1 - y/R} = -R\int_{A'} dA + R\int_{A'} \frac{dA}{1 - y/R} \tag{a}$$

we find that for the rectangular section

$$\bar{Q}_z = -Rb\left(\frac{h}{2} - y\right) + R^2b \ln \frac{R-y}{R-h/2} \tag{b}$$

Introducing (*b*) into Eq. (4.29) gives for the radial stress

$$\sigma_y = \frac{-M}{b(R-y)}\left[\left(\frac{1}{hR} + \frac{Rb}{J_z}\right)\left(\frac{h}{2} - y\right) - \frac{R^2b}{J_z} \ln \frac{R-y}{R-h/2}\right] \tag{c}$$

For the case in which $R = h$, (*c*) becomes

$$\sigma_y = \frac{-M}{bh^2(1 - y/h)}\left[11.140\left(\frac{1}{2} - \frac{y}{h}\right) - 10.141 \ln 2\left(1 - \frac{y}{h}\right)\right] \tag{d}$$

This distribution is shown in Fig. 4.9 compared with the corresponding normal stress. Note that positive moments produce a tensile transverse stress. We see that the maximum value of σ_y amounts to about 19 percent of the maximum normal stress. In sections of irregular shape, however, σ_y may be of the same order of magnitude as σ_s. As h/R decreases, σ_y becomes less significant, and for a rectangular section with $h/R = \frac{1}{4}$, the maximum radial stress is only 6 percent of the maximum normal stress.

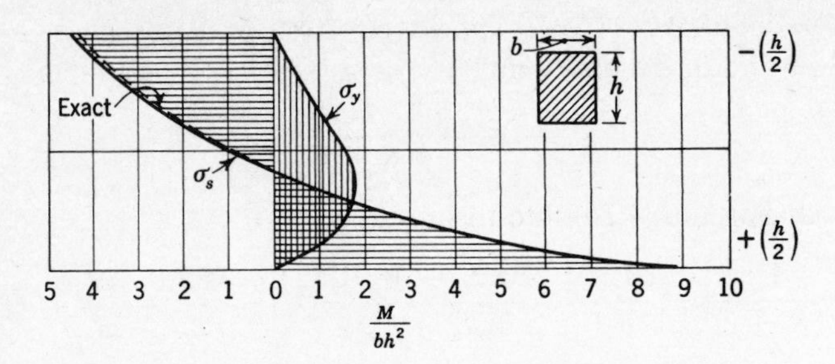

FIGURE 4.9 Comparison of radial and normal stresses in a rectangular curved bar.

It is also interesting to note that σ_y acquires its maximum value slightly below the centroid of the section and is zero at $y = \pm h/2$, while σ_s is maximum at the extreme fibers and is zero near the point of maximum radial stress. This explains why neglecting the term $-\nu\sigma_y/E$ in the expression for the strain component ϵ_s seldom leads to an appreciable error. As further proof that this is true, the "exact" stress profile of σ_s, calculated using the more rigorous theory[7] (which does not neglect $-\nu\sigma_y/E$ in ϵ_s and which does not employ the simplifying assumption that planes remain plane after deformation), is also indicated in Fig. 4.9. In this case the rigorous theory gives a value for the maximum normal stress which differs from that obtained from Eq. (4.24) by only 0.31 percent.

4.6 Bending of straight bars. Equation (4.21) gives the distribution of normal stress in bars of small curvature as well as in straight bars. In the case of the straight bar, it is more meaningful to replace σ_s and N_s with σ_x and N_x, since the s coordinate degenerates into the x axis as R becomes infinite. Thus,

$$\sigma_x = \frac{N_x}{A} + \frac{M_z I_y - M_y I_{yz}}{I_y I_z - I_{yz}^2} y + \frac{M_y I_z - M_z I_{yz}}{I_y I_z - I_{yz}^2} z \tag{4.30}$$

In case of pure bending ($N_x = 0$), the neutral surface passes through the centroid of the section and intersects the cross-sectional plane along a straight line. We obtain the equation of this line by simply equating σ_x to zero:

$$y = -\frac{M_y I_z - M_z I_{yz}}{M_z I_y - M_y I_{yz}} z \tag{4.31}$$

An interesting relationship is obtained if we denote the resultant of M_z and M_y by M and assume that it acts in a plane oriented at an angle β

[7] See, for example, Ref. 61, p. 64.

with respect to the y axis as shown in Fig. 4.10. Since M_z and M_y are equal to $M \cos \beta$ and $M \sin \beta$, respectively, Eq. (4.31) becomes

$$-\frac{y}{z} = \tan \alpha = \frac{I_z \tan \beta - I_{yz}}{I_y - I_{yz} \tan \beta} \tag{4.32}$$

where α is the angle between the z axis and the neutral axis. If y and z are principal axes, I_{yz} is zero and

$$\tan \alpha = \frac{I_z}{I_y} \tan \beta \tag{4.33}$$

From Eqs. (4.32) and (4.33) we conclude that the *neutral axis is not, in general, perpendicular to the plane of the resultant moment.* In fact, for a 2- by 10-in. rectangular section for which $I_z = 167$ in.^{4} and $I_y = 6.67$ in.^{4}, if β is only 1°

$$\alpha = \tan^{-1} \frac{167}{6.67} (0.01746) = 23.6°$$

Hence, a slight inclination of the loading plane may drastically affect the stress distribution. Furthermore, if β is zero, the unsymmetry of a section also influences the stress distribution. In this case, Eq. (4.32) becomes

$$\tan \alpha = -\frac{I_{yz}}{I_y} \tag{4.34}$$

For symmetrical sections, the neutral axis is perpendicular to the plane of the resultant moment only when I_y and I_z are equal or β is zero.

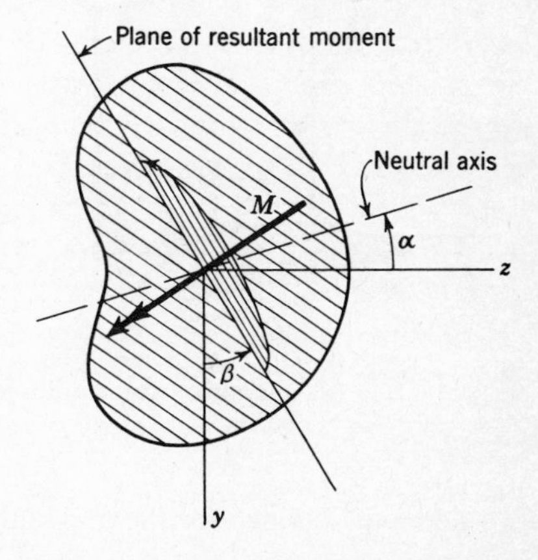

FIGURE 4.10 Resultant moment M acting on an unsymmetrical section.

It is clear that the elementary formula, Eq. (4.23), is never to be used unless y and z are principal axes and, of course, M_y is zero. To emphasize this point, let us consider the cantilever beam loaded by a concentrated end force P shown in Fig. 4.11*a*. Suppose that we are required to evaluate the normal stress distribution at the fixed end $x = 0$. The area properties of the section are

$$I_y = \tfrac{2}{3}ta^3 \qquad I_z = \tfrac{8}{3}ta^3 \qquad I_{yz} = -ta^3 \tag{a}$$

where t, the thickness of the section, is assumed to be small compared with a, the dimension indicated in the figure.

At $x = 0$, $M_z = -PL$, $M_y = 0$, and Eq. (4.30) gives

$$(\sigma_x)_{x=0} = -\frac{6PL}{7ta^3} y - \frac{9PL}{7ta^3} z \tag{b}$$

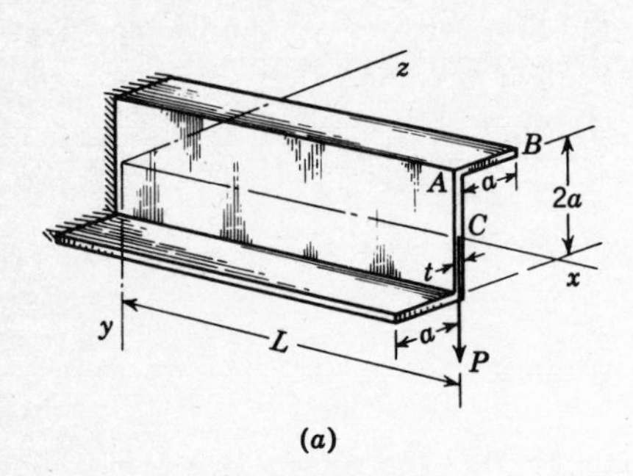

(a)

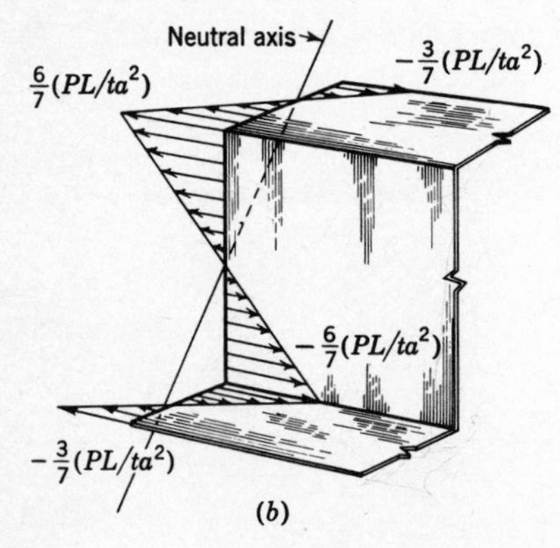

(b)

FIGURE 4.11 An unsymmetrical cantilever beam.

Again referring to Fig. 4.11*a*, we see that the stress at *A* is obtained by setting $y = -a$, $z = 0$ in the above expression. This gives

$$(\sigma_x)_A = \frac{6PL}{7ta^2} \qquad (c)$$

Similarly, the stress at *B*, $y = -a$, $z = a$, is

$$(\sigma_x)_B = \frac{6PL}{7ta^2} - \frac{9PL}{7ta^2} = -\frac{3PL}{7ta^2} \qquad (d)$$

The final stress distribution is shown in Fig. 4.11*b*. The angle α at which the neutral axis is inclined with respect to the z axis is obtained from Eq. (4.34):

$$\alpha = \tan^{-1}\frac{-I_{yz}}{I_y}$$

$$= \tan^{-1} 1.5 = 56.3°$$

Had we ignored the lack of symmetry of the section [i.e., used Eq. (4.23) rather than (4.30)] we would have obtained for the maximum stress $3PL/8ta^2$, a value 56 percent in error.

PROBLEMS

4.1. Evaluate the necessary cross-sectional constants and the normal stress distribution at section *AA* of the beam shown below. Plot the normal stress profile and compute the maximum value of σ_s.

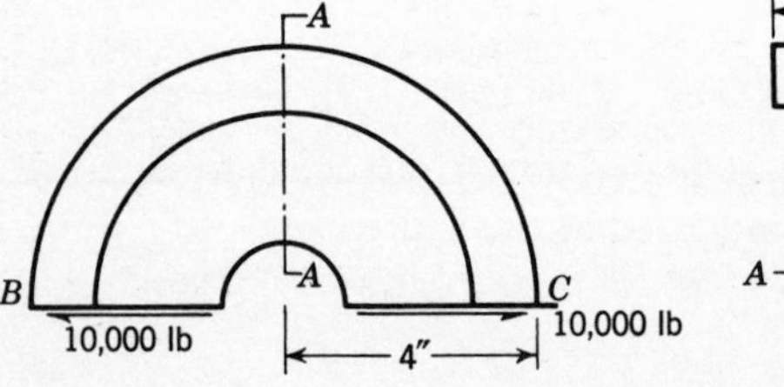

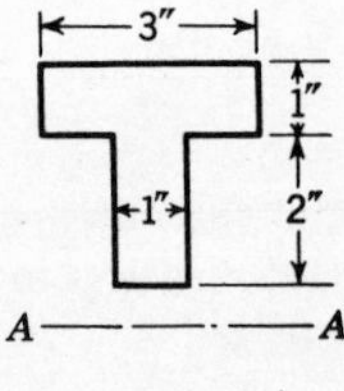

4.2. Evaluate the normal stress distribution at the fixed end of the beam shown. Plot the stress profile and indicate the maximum value of σ_s.

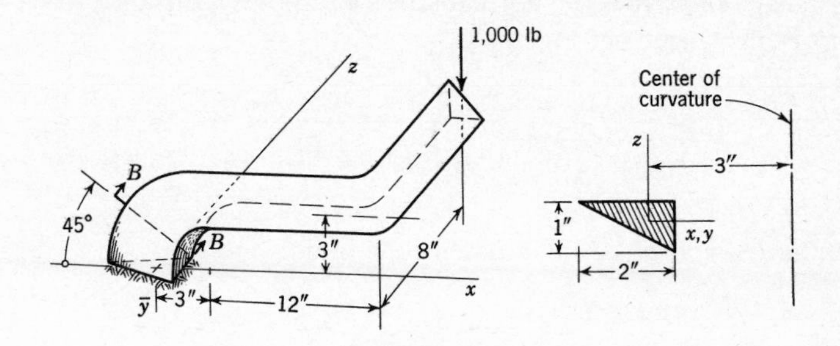

4.3. The curved bar shown has a circular cross section and is acted upon by equal couples $M_y = M_z = M$.

(*a*) Verify that

$$J_z = \pi\left\{2R^4\left[1 - \sqrt{1 - \left(\frac{r}{R}\right)^2}\right] - R^2r^2\right\}$$

(*b*) Evaluate the maximum normal stress in terms of M and $\lambda = r/R$ and plot $(\sigma_s)_{\max}$ versus λ.

(*c*) Locate the neutral surface of the bar.

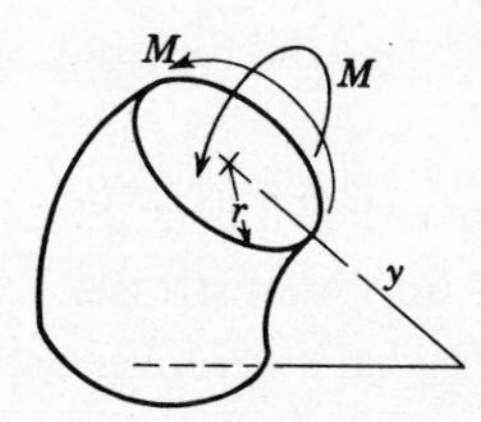

4.4. Show that the formula for normal stress in a symmetrical curved bar bending in the $x\bar{y}$ plane [Eq. (4.20)] can be written in the more compact form

$$\sigma_s = \frac{M_z y'}{(\bar{R} - y')Ad}$$

where d is the distance from the centroid to the neutral axis, $\bar{R} = R - d$, and $y' = y - d$.

4.5. Verify that the equation for the neutral surface in a symmetrical curved bar loaded in the $x\bar{y}$ plane is

$$y = \frac{R}{1 + R^2A/J_z}$$

provided no axial forces are present.

4.6. The cross section of a curved bar is composed of n rectangular elements as indicated in the figure below. The bar is in pure bending in its own plane. Show that the normal stress at a point a distance R_i from the center of curvature is given by

$$\sigma_s = -M_z\left[\frac{1}{RA} + \frac{R - R_i}{R_i\left(A + R\sum_{m=1}^{n} b_m \ln \frac{R_m}{R_{m+1}}\right)}\right]$$

where the R_m and b_m are the dimensions indicated.

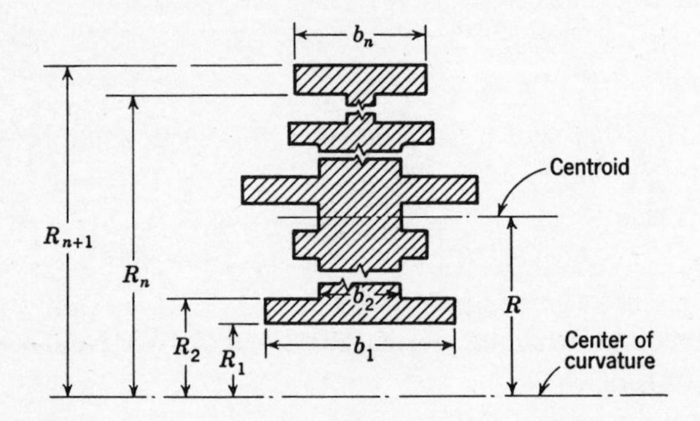

4.7. Evaluate the radial stress at the centroid of the curved bar in Prob. 4.3.

4.8. The curved coplanar bar shown is in pure bending.

(*a*) Derive the equations for radial stresses.

(*b*) Plot the variation of radial stress over the depth of the section.

(*c*) What is the maximum radial stress?

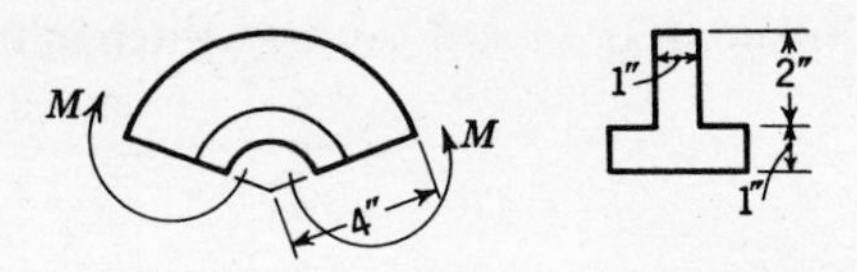

4.9. Repeat Prob. 4.8 for the bar in Prob. 4.2. Assume that the bar is in pure bending. Compare the results with those obtained in Prob. 4.8.

4.10–4.13. To evaluate the cross-sectional constants J_y, J_z, and J_{yz} for irregular shapes, it is customary to represent a given cross section by a finite number of elemental areas ΔA_i. The coordinates of the centroid of ΔA_i with respect to the centroidal axes of the cross section are denoted

y_i, z_i. Then the integrals in Eqs. (4.14) are approximated by the finite sums

$$J_y \approx \sum_{i=1}^{n} \frac{z_i^2}{1 - y_i/R} \Delta A_i$$

$$J_z \approx \sum_{i=1}^{n} \frac{y_i^2}{1 - y_i/R} \Delta A_i$$

$$J_{yz} \approx \sum_{i=1}^{n} \frac{y_i z_i}{1 - y_i/R} \Delta A_i$$

where n is the number of areas ΔA used in the approximation. Similarly, the approximate location of the centroid of the cross section is evaluated by means of the formulas

$$\bar{y} \approx \frac{1}{A} \sum_{i=1}^{n} y_i' \, \Delta A_i \qquad \bar{z} \approx \frac{1}{A} \sum_{i=1}^{n} z_i' \, \Delta A_i$$

where y_i' and z_i' are coordinates of the centroids of the ΔA_i with respect to an arbitrary set of axes parallel to y and z. Use the above relations to obtain approximate values of J_y, J_z, and J_{yz} of the cross sections shown.

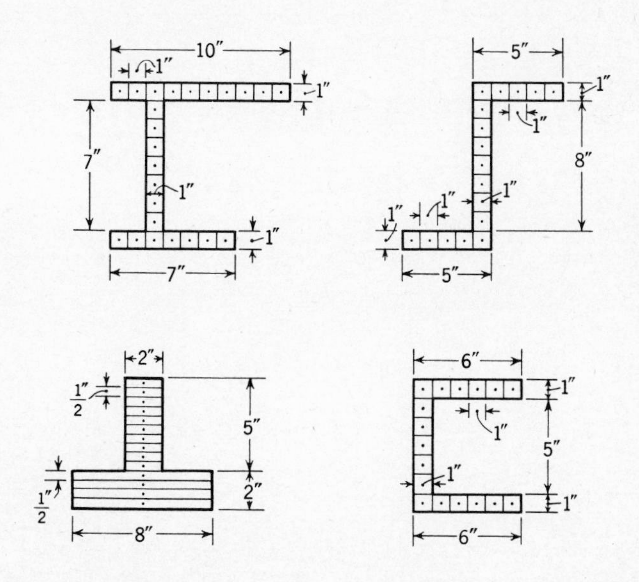

4.14. The area properties of the cross section shown are as follows: $I_y = 3.90$ in.^{4}, $I_z = 35.70$ in.^{4}, and $I_{yz} = -3.59$ in.^{4} This section is that of a 10-ft cantilever beam which is subjected to an inclined 10,000-lb force at its free end. Locate the neutral surface for the case in which the angle of inclination β is (*a*) zero; (*b*) 1°; (*c*) 30°. Neglect warping and twisting of the section.

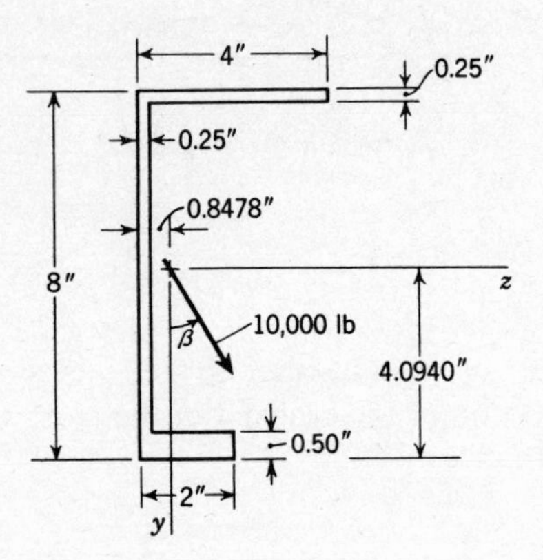

4.15. Compute the maximum normal bending stress developed in the beam of Prob. 4.14 for $\beta = 0$, 1°, and 30°.

five

SHEARING STRESSES AND SHEAR FLOW IN BEAMS

5.1 Introduction. Once the normal forces and bending moments developed on every section of a bar have been determined, we can, with the aid of the equations presented in the preceding chapter, evaluate the components of normal stress. We have not succeeded in completely defining the stress distribution, however, until we have also determined the shearing stresses at every point. To this end, in this chapter we establish relationships between the components of shearing stress and the remaining stress resultants V_y and V_z.

5.2 Shearing stresses in curved bars. To begin our investigation, let us consider a curved coplanar bar of constant cross section under general loading. Again, we neglect the effects of torsion and confine our attention to stress distributions which on each section result in the five stress resultants indicated in Fig. 4.1. A typical element of such a bar is shown in Fig. 5.1*a*. We assume that all of the stress resultants and loads on the element are known functions of the coordinates of points on the bar's axis which satisfy the equilibrium conditions [Eqs. (4.1) to (4.5)].

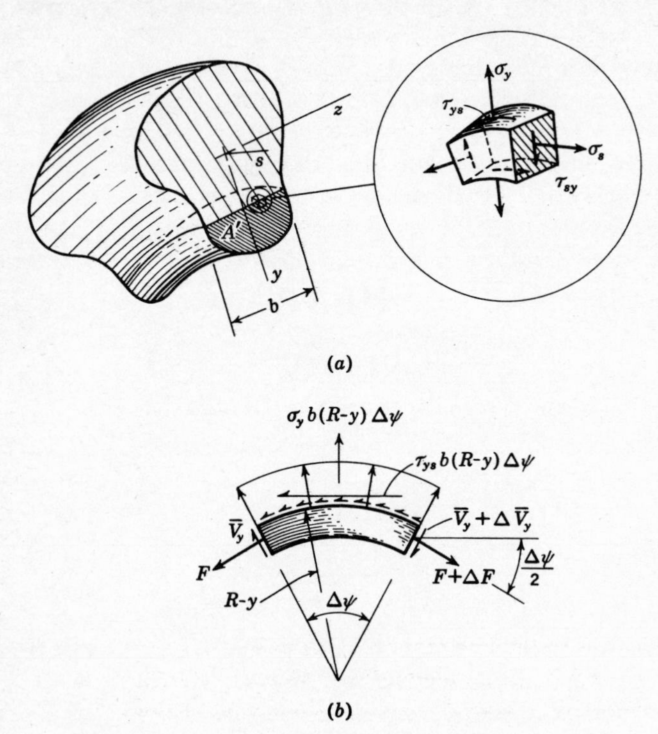

FIGURE 5.1 (*a*) Stresses on a curved bar element; (*b*) forces acting on a portion of the element of cross-sectional area A'.

Let us now examine the equilibrium of a slice of the element of area A', as shown in the figure. For simplicity, we assume that the dimension b of A' is parallel to the z axis. Upon application of loads, the bar deforms until the stresses are such that each element is in equilibrium. The normal stresses developed on A' result in a normal force

$$F = \int_{A'} \sigma_s \, dA$$

Introducing σ_s of Eq. (4.16) into this integral, we find

$$F = \left(\frac{N_s}{A} - \frac{M_z}{AR}\right)A' + \frac{M_z J_y - M_y J_{yz}}{J_y J_z - J_{yz}^{\,2}}\,\bar{Q}_z + \frac{M_y J_z - M_z J_{yz}}{J_y J_z - J_{yz}^{\,2}}\,\bar{Q}_y \quad (5.1)$$

where $\bar{Q}_z$ and $\bar{Q}_y$ are defined in Eqs. (4.27). Note that when N_s is zero, Eq. (5.1) becomes identical to Eq. (4.26). Similarly, the shearing stresses developed on A' result in a force $\bar{V}_y$ parallel to the cross section, where

$$\bar{V}_y = \int_{A'} \tau_{sy} \, dA \quad (5.2)$$

Clearly, if the integration is carried out over the total cross-sectional area, F and $\bar{V}_y$ become N_s and V_y, respectively.

We found in Art. 4.5 that radial stresses σ_y must be developed to balance the vertical components of the forces F (and $\bar{V}_y$). These stresses result in the force $\sigma_y b(R - y)\,\Delta\psi$ shown in the figure. Equation (4.28) for the radial stress, however, was derived on the assumption that the bar was in pure unsymmetrical bending, and in the present case it is no longer applicable because of the shearing forces $\bar{V}_y$. To determine the influence of $\bar{V}_y$ on σ_y, we refer to Fig. 5.1b and sum forces in the vertical direction:

$$(F + \Delta F + F)\sin\frac{\Delta\psi}{2} - \sigma_y b(R - y)\,\Delta\psi + (\bar{V}_y + \Delta\bar{V}_y - \bar{V}_y)\cos\frac{\Delta\psi}{2} = 0$$

Allowing Δs to approach zero, we find in the limit that

$$\frac{F}{R} + \frac{\partial\bar{V}_y}{\partial s} - \sigma_y b\left(1 - \frac{y}{R}\right) = 0 \tag{5.3}$$

Thus,

$$\sigma_y = \frac{1}{b(1 - y/R)}\left(\frac{F}{R} + \frac{\partial\bar{V}_y}{\partial s}\right) \tag{5.4}$$

Comparing this result with that obtained in Art. 4.5, we see that the shearing stresses alter σ_y by an amount $(\partial\bar{V}_y/\partial s)/b(1 - y/R)$.

Furthermore, if F and $\bar{V}_y$ acquire incremental changes ΔF and $\Delta\bar{V}_y$ in the interval Δs, an additional horizontal force must exist on the area $b(R - y)\,\Delta\psi$ to provide equilibrium. Since this force acts parallel to the area, it is the result of shearing stresses. Let τ_{ys} denote the *average shearing stress* on this area; then the total horizontal force developed is $\tau_{ys} b(R - y)\,\Delta\psi$ (see Fig. 5.1b). Thus, summing forces in the horizontal direction gives

$$\Delta F\cos\frac{\Delta\psi}{2} - (2\bar{V}_y + \Delta\bar{V}_y)\sin\frac{\Delta\psi}{2} - \tau_{ys} b\left(1 - \frac{y}{R}\right)R\,\Delta\psi = 0$$

or, in the limit,

$$\frac{\partial F}{\partial s} - \frac{1}{R}\bar{V}_y - \tau_{ys} b\left(1 - \frac{y}{R}\right) = 0 \tag{5.5}$$

Therefore

$$\tau_{ys} = \frac{1}{b(1 - y/R)}\left(\frac{\partial F}{\partial s} - \frac{1}{R}\bar{V}_y\right) \tag{5.6}$$

The quantity $\bar{V}_y$ in Eqs. (5.4) and (5.6) must be evaluated using Eq. (5.2). The integral in Eq. (5.2), however, involves the unknown function $\tau_{sy} = \tau_{ys}$. Thus, when we substitute Eq. (5.2) into Eq. (5.6) we obtain an integral equation in the dependent variable τ_{ys}.

In the following developments we avoid the complexity of solving such equations for various boundary conditions by introducing an assumption based on the relative magnitude of the terms in Eqs. (5.4) and (5.6). By virtue of Eq. (4.1), the term $\partial \bar{V}_y/\partial s$ in Eq. (5.4) can be, at most, equal in magnitude to $p_y + N_s/R$. When R is of the same order of magnitude as the depth of the beam, the dominant term in Eq. (5.4) is $F/Rb(1 - y/R)$. As R increases (and h/R decreases), this term becomes less significant and σ_y acquires a magnitude on the order of p_y/b, which is usually negligible in comparison with σ_s. Similarly, the term $\bar{V}_y/R$ in Eq. (5.6) is very small when R is large, and τ_{ys} is essentially due to a change in F. With these observations in mind, we introduce the approximation

$$\bar{V}_y \approx \frac{A'}{A} V_y \tag{5.7}$$

A being the total cross-sectional area, so that, from Eq. (4.1),

$$\frac{\partial \bar{V}_y}{\partial s} = -\frac{A'}{A}\left(p_y + \frac{N_s}{R}\right) \tag{5.8}$$

assuming that the cross-sectional dimensions do not vary with s. These approximations are equivalent to the assumption that the influence of the true shearing stresses on σ_y may be approximated by that of a uniform stress distribution over the cross section.

Finally, introducing Eqs. (5.1) and (5.8) into Eq. (5.4) gives

$$\sigma_y = \frac{1}{b(1 - y/R)} \times \left[\frac{1}{R}\left(-\frac{M_z A'}{AR} + \frac{M_z J_y - M_y J_{yz}}{J_y J_z - J_{yz}^2}\bar{Q}_z + \frac{M_y J_z - M_z J_{yz}}{J_y J_z - J_{yz}^2}\bar{Q}_y\right) - \frac{A'}{A}p_y\right] \tag{5.9}$$

Note that the terms containing N_s cancel and thus do not affect σ_y. According to this equation, σ_y is zero at the bottom of the section (the fibers nearest the center of curvature) and σ_y is equal to the applied force per unit surface area at the top fibers of the section. For example, in the case of a rectangular section of depth h and width b, $\sigma_y = 0$ at $y = h/2$ and $\sigma_y = -p_y/b(1 + h/2R)$ at $y = -h/2$. We recognize the quantity $b(1 + h/2R)$ as the area of the outside surface at $y = -h/2$ corresponding to a unit length of arc at the centroid. Thus, the applied load p_y is assumed to act on the outside surface of the bar ($y = -h/2$) and to be the result of a surface force $Y_s = -p_y/b(1 + h/2R)$ uniformly distributed over the width b.

Returning now to the evaluation of the shearing stress, we introduce Eqs. (5.1) and (5.7) into Eq. (5.6) and arrive at the equation

$$\tau_{ys} = \frac{1}{b(1 - y/R)}\left(\frac{A'}{A}\frac{dN_s}{ds} - \frac{A'}{AR}\frac{dM_z}{ds} + \frac{\bar{Q}_z J_y - \bar{Q}_y J_{yz}}{J_y J_z - J_{yz}^2}\frac{dM_z}{ds}\right.$$
$$\left. + \frac{\bar{Q}_y J_z - \bar{Q}_z J_{yz}}{J_y J_z - J_{yz}^2}\frac{dM_y}{ds} - \frac{A'}{AR}V_y\right) \qquad (5.10)$$

Substituting Eqs. (4.2), (4.3), and (4.5) into this equation gives

$$\tau_{ys} = \frac{1}{b(1 - y/R)}\left(\frac{\bar{Q}_z J_y - \bar{Q}_y J_{yz}}{J_y J_z - J_{yz}^2}V_y + \frac{\bar{Q}_y J_z - \bar{Q}_z J_{yz}}{J_y J_z - J_{yz}^2}V_z - \frac{A'}{AR}V_y\right) \qquad (5.11)$$

This is the final equation for shearing stresses in circular bars. In the case of bars of small curvature, the term $-A'V_y/AR$ inside the parentheses is usually small and can be neglected. Also, we may now obtain a better approximation to dV_y/ds in Eq. (5.3) by introducing Eq. (5.11) into the integral in Eq. (5.2) and then substituting the result into Eq. (5.4). This lengthy refinement is seldom worthwhile, however, since in cases in which σ_y is significant the shear term in Eq. (5.4) is usually small in comparison with the term due to σ_s.

We note that if the loading is such that M_y and V_z are zero, Eq. (5.11) reduces to

$$\tau_{ys} = \frac{V_y}{b(1 - y/R)}\left(\frac{\bar{Q}_z J_y - \bar{Q}_y J_{yz}}{J_y J_z - J_{yz}^2} - \frac{A'}{AR}\right) \qquad (5.12)$$

and if, in addition, the section is symmetrical with respect to the y axis, Eq. (5.12) becomes

$$\tau_{ys} = \frac{V_y}{b(1 - y/R)}\left(\frac{\bar{Q}_z}{J_z} - \frac{A'}{AR}\right) \qquad (5.13)$$

As a simple example, let us consider the problem of determining the distribution of shearing stress in the curved bar of rectangular cross section shown in Fig. 5.2*a*. Since the cross section of the bar is symmetrical with respect to the xy plane and since the bar is loaded in this plane, Eq. (5.13) is applicable. The function $\bar{Q}_z$ for this section is given in Eq. (*b*) of Art. 4.5, and J_z for a rectangular section is given by Eq. (4.25). Further, in this case,

$$\frac{A'}{A} = \frac{1}{2}\left(1 - \frac{2y}{h}\right) \qquad (a)$$

For demonstration purposes, let us take $R = h$. Then

$$\frac{\bar{Q}_z}{J_z} = \frac{10.141}{h}\left[\ln 2\left(1 - \frac{y}{h}\right) - \left(0.500 - \frac{y}{h}\right)\right] \qquad (b)$$

Substituting Eqs. (*a*) and (*b*) into Eq. (5.13) and simplifying, we get

$$\tau_{sy} = \frac{V_y}{2bh(1 - y/h)} \left\{ 10.141 \left[2 \ln 2 \left(1 - \frac{y}{h} \right) - \left(1 - \frac{2y}{h} \right) \right] - \left(1 - \frac{2y}{h} \right) \right\} \tag{c}$$

where, in this case, $V_y = P \cos \psi$, ψ being the coordinate angle indicated in the figure. The maximum shearing stress occurs on the section at which the force P is applied.

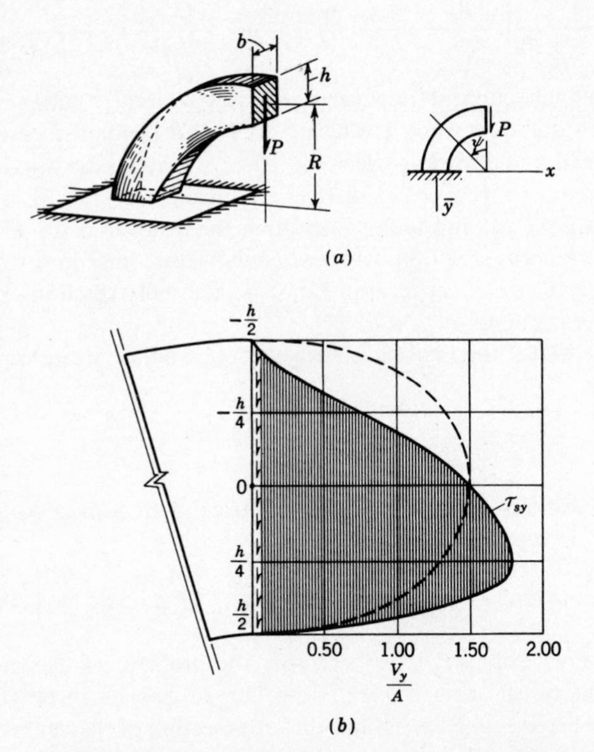

FIGURE 5.2 (*a*) Circular bar with rectangular cross section subjected to a force P at its free end; (*b*) shearing-stress distribution on a typical section for the case in which $R = h$. The dashed line indicates the parabolic distribution given by the elementary straight-bar formula.

A plot of Eq. (*c*) for a typical cross section is shown in Fig. 5.2*b* for comparison with the stress distribution given by the elementary straight-bar formula (indicated by the dashed line in the figure). We observe that the shearing stress at the centroid of the bar is only 2.8 percent less than the value of $1.5V_y/A$ obtained for straight bars. However, in the case of curved bars the maximum shearing stress does not occur at the centroid. Because of the sharp curvature of the bar, the point of maximum shearing stress is shifted toward the center of curvature, and, in this example, it is located a distance of approximately $0.26h$ below the centroid. Moreover, the magnitude of the maximum shearing stress exceeds that given by the straight-bar formula by 18 percent.

5.3 Shearing stresses in unsymmetrical straight bars. As the radius of curvature R becomes infinitely large, the curved s axis degenerates to the straight x axis, J_y, J_z, and J_{yz} become I_y, I_z, and I_{yz}, the curved area $b(R - y)\,\Delta\psi$ in Fig. 5.1*b* reduces to simply $b\,\Delta x$, and the quantities $\bar{Q}_y$ and $\bar{Q}_z$ of Eq. (4.27) become Q_y and Q_z, respectively, where Q_y and Q_z are the first moments of area of A' with respect to the y and z axes:

$$Q_y = \int_{A'} z\,dA \qquad \text{and} \qquad Q_z = \int_{A'} y\,dA \tag{5.14}$$

In addition, the equilibrium equations for the element [Eqs. (4.1) to (4.5)] reduce to

$$\frac{dV_y}{dx} = -p_y \qquad \frac{dV_z}{dx} = -p_z \tag{5.15}$$

$$\frac{dM_z}{dx} = V_y \qquad \frac{dM_y}{dx} = V_z \tag{5.16}$$

Eqs. (4.6) and (4.8) become

$$\frac{d^2M_z}{dx^2} + p_y = 0 \qquad \frac{d^2M_y}{dx^2} + p_z = 0 \tag{5.17}$$

and the equilibrium equations for the slice of area A' [Eqs. (5.3) and (5.5)] reduce to

$$\sigma_y b - \frac{\partial \bar{V}_y}{\partial x} = 0 \tag{5.18}$$

and

$$\tau_{yx} b - \frac{\partial F}{\partial x} = 0 \tag{5.19}$$

The force F now acts parallel to the x axis and the transverse normal stresses σ_y no longer influence the horizontal equilibrium of the slice. Finally, Eq. (5.11) becomes

$$\tau_{yx} = \frac{Q_z I_y - Q_y I_{yz}}{b(I_y I_z - I_{yz}^2)} V_y + \frac{Q_y I_z - Q_z I_{yz}}{b(I_y I_z - I_{yz}^2)} V_z \tag{5.20}$$

Equation (5.20) gives the stress components τ_{yx} and τ_{xy} only because we selected the dimension b to be parallel to the z axis. Actually, by the proper choice of b this equation can be used to evaluate the average shearing stress on the section in *any* direction, as indicated in Fig. 5.3. By taking the slice so that b is parallel to the y axis, for example, Eq. (5.20) gives τ_{xz} and τ_{zx}. By using Eq. (5.2) we may also easily verify that

$$\int \tau_{yx}\, dA = V_y$$

and

$$\int \tau_{zx}\, dA = V_z$$

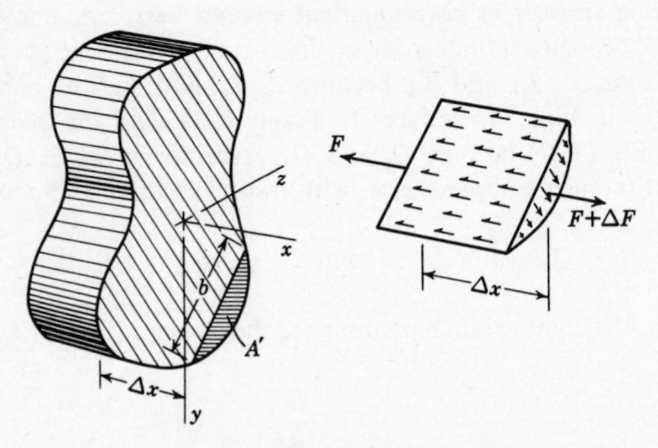

FIGURE 5.3 Forces on a portion of an element of a straight bar. ΔF is balanced by a force $\tau b\, \Delta x$, τ being the average shearing stress on area $b\, \Delta x$.

In the case in which the loading is in the xy plane ($M_y = V_z = 0$) and y and z are principal axes ($I_{yz} = 0$), Eq. (5.20) reduces to the familiar elementary formula

$$\tau_{xy} = \frac{V_y Q_z}{I_z b} \tag{5.21}$$

Again, it is important to note that the elementary formula is never applicable unless the coordinate axes are the principal axes of the cross section and the plane of the loading is parallel to the xy or xz planes.

When the cross section is composed of thin-walled elements, the shearing stress is essentially uniform over the thickness and is negligible in directions perpendicular to the center line of the walls. In such cases it is customary to treat the product τb in Eq. (5.19) as the unknown rather than τ. This product is referred to as q, *the shear flow due to transverse shear*, in analogy with the torsional shear flow studied in Chap. 3. For simplicity, we shall refer to q as the shear flow due to *bending*, since $V_y = dM_z/dx$ and $V_z = dM_y/dx$.

As a simple example of the application of Eq. (5.20), let us consider the shearing-stress distribution in the z section in Fig. 5.4a. The area properties of the section are given in Eq. (*a*) of Art. 4.6. The shearing force on any section is of magnitude P, and V_z, obviously, is zero. Thus, from Eq. (5.20) we have

$$\tau t = q = \frac{6Q_z + 9Q_y}{7ta^3} P \tag{a}$$

For the portion of the flange between point A and B shown in Fig. 4.11*a*, we find

$$Q_y = \frac{t}{2}(a^2 - z^2) \qquad \text{and} \qquad Q_z = -at(a - z) \tag{b}$$

Hence, the shearing stress is given by the parabolic law

$$\tau_{A-B} = -\frac{3P}{14ta^3}(a^2 - 4az + 3z^2) \tag{c}$$

Similarly, for the portion of the web AC we find

$$Q_y = a^2\frac{t}{2} \qquad Q_z = -a^2t - \frac{t}{2}(a^2 - y^2) \tag{d}$$

and

$$\tau_{A-C} = -\frac{3P}{14ta^3}(3a^2 - 2y^2) \tag{e}$$

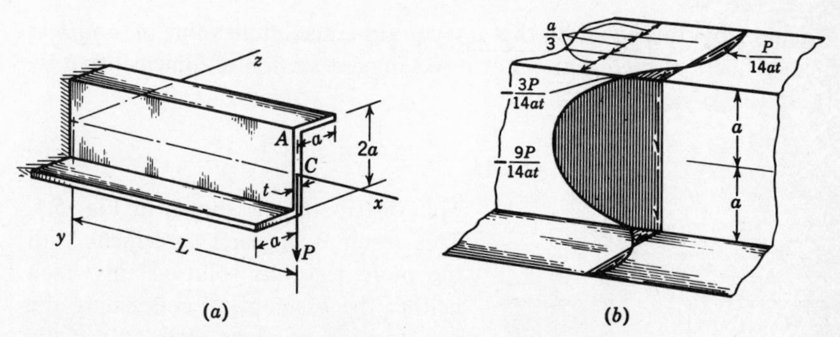

FIGURE 5.4 Shearing-stress distribution in z section.

The final distribution is shown in Fig. 5.4*b*. The maximum shearing stress of $9P/14at$ occurs at the centroid of the section, as expected. The stress in the web is negative since it results in a downward shearing force on a section a distance x from the origin; V_y in Fig. 5.4*a* is positive when upward on this section. Had the lack of symmetry of the section been ignored [that is, had we used Eq. (5.21) instead of Eq. (5.20)] the maximum shearing stress obtained would have been 12.5 percent in error.

Although σ_y is generally negligible in the case of straight beams, it can be easily calculated from Eq. (5.18). Consider, for example, the simple case in which V_z is zero and y and z are principal axes. Then $\bar{V}_y$ is evaluated by substituting Eq. (5.21) into Eq. (5.2):

$$\bar{V}_y = \int_{A'} \frac{V_y Q_z}{I_z b}\, dA = \frac{V_y}{I_z} \int_{A'} Q_z \frac{dA}{b}$$

Integrating by parts, we find

$$\int_{A'} Q_z \frac{dA}{b} = \int_y^c Q_z\, dy = yQ_z \Big|_y^c - \int_{A'} y\, dQ_z$$

where c is the distance from the centroid to the bottom fiber. Noting that $dQ_z = y\, dA$, we find

$$\int_{A'} Q_z \frac{dA}{b} = I'_z - yQ_z$$

where I'_z is the second moment of area A' with respect to the z axis. Thus,

$$\bar{V}_y = \frac{V_y}{I_z}(I'_z - yQ_z) \tag{5.22}$$

Substituting Eq. (5.22) into Eq. (5.18) and then introducing the first of Eqs. (5.15) into the result, we arrive at the equation[1]

$$\sigma_y = \frac{p_y}{I_z b}(yQ_z - I'_z) \tag{5.23}$$

We find from this equation that σ_y acquires a maximum value of $-p_y/b$ at the top fibers of the beam. For a rectangular section of dimensions b by h, Eq. (5.23) yields

$$\sigma_y = -\frac{p_y}{24 I_z}(h^3 - 3h^2 y + 4y^3) \tag{5.24}$$

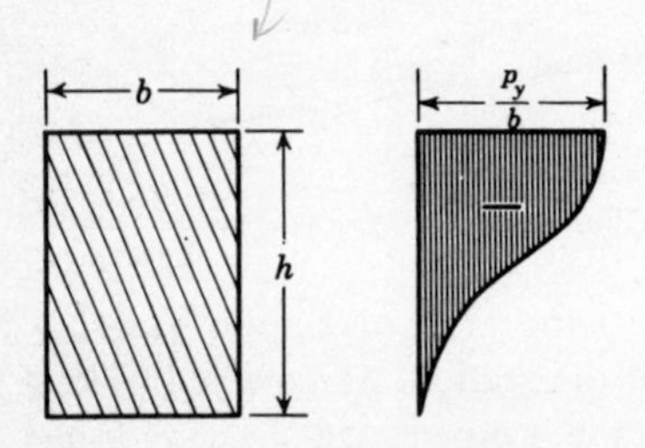

FIGURE 5.5 Transverse normal stress distribution in a straight rectangular beam.

This distribution is shown in Fig. 5.5. This result is in exact agreement with the more rigorous solution[2] in which neither the assumption concerning the conservation of plane sections nor the assumption that $-\nu\sigma_y/E$ and $-\nu\sigma_x/E$ are negligible in ϵ_x and ϵ_y is made. In fact, for the beam in Fig. 5.5 with $b = 1$, Eqs. (4.23) and (5.21) give

$$\sigma_x = \frac{M_z y}{I_z} \quad \text{and} \quad \tau_{xy} = \frac{V_y}{2I_z}\left[\left(\frac{h}{2}\right)^2 - y^2\right]$$

[1] A similar derivation of this equation s given in Ref. 52.

[2] See, for example, Ref. 61, p. 46.

If these equations, along with Eq. (5.24), are substituted into the general equilibrium conditions [Eqs. (2.4)], we find

$$\frac{dM_z}{dx}\frac{y}{I_z} - \frac{V_y}{I_z}y = 0$$

and

$$\frac{dV_y}{dx}\frac{Q_z}{I_z} + p_y\frac{Q_z}{I_z} = 0$$

which, by virtue of Eqs. (5.15) and (5.16), are identically satisfied. The rigorous solution for σ_x, however, contains an additional term due to a slight warping[3] of each cross section which is not accounted for in the elementary theory; as a result, the above stresses do not satisfy the general compatibility conditions. Fortunately, when the depth h is small compared with the length of the beam, these warping terms become very small and can be neglected.

5.4 Stresses in beams of variable depth. The formulas that we derived in the preceding articles are applicable only to beams of uniform cross section. To investigate the influence of a variation in the depth of the section on the stress distribution, we consider a beam of the form shown in Fig. 5.6*a*. We assume that the section is of constant width b and that it is symmetrical with respect to the vertical axis through its centroid. The depth h_x is assumed to change at a rate $\alpha_x = \tan^{-1}(dh_x/dx)$, as indicated in the figure. Again, we take $\sigma_z = \tau_{xz} = \tau_{yz} = 0$, so that the beam is in a two-dimensional state of stress and, as an additional simplification, we assume that no external longitudinal forces act on the beam except, perhaps, at its end points. Thus, the axial force N_x is assumed to be a constant throughout the length of the beam. Although the transverse normal stress σ_y is generally small, we shall find that we cannot neglect it as in the analysis of uniform bars. Its presence is essential for equilibrium.

According to Eq. (4.23), if plane sections remain plane after deformation, the normal stress due to bending is $M_z y/I_z$. It can be shown that this formula gives quite accurate results provided the variation in cross section is not too rapid.[4] The distribution of shearing stress, however, is considerably altered.

Since the location of the centroid changes from point to point along the beam, it is convenient to choose, instead of y, a cross-sectional coordinate whose origin does not vary with x. With this in mind, we note that all of the top fibers of the beam are parallel to the x axis. Thus, we introduce the new coordinate ζ indicated in Fig. 5.6*b*. From the geometry of

[3] The effects of warping on σ_x are discussed in Chap. 7.

[4] For a tapered cantilever, Timoshenko has shown that the bending stress is given by $\beta M_z y/I_z$. For taper angles even as high as 15°, $\beta = 0.95$ and the elementary formula is in error only 5 percent. See Ref. 62, p. 62.

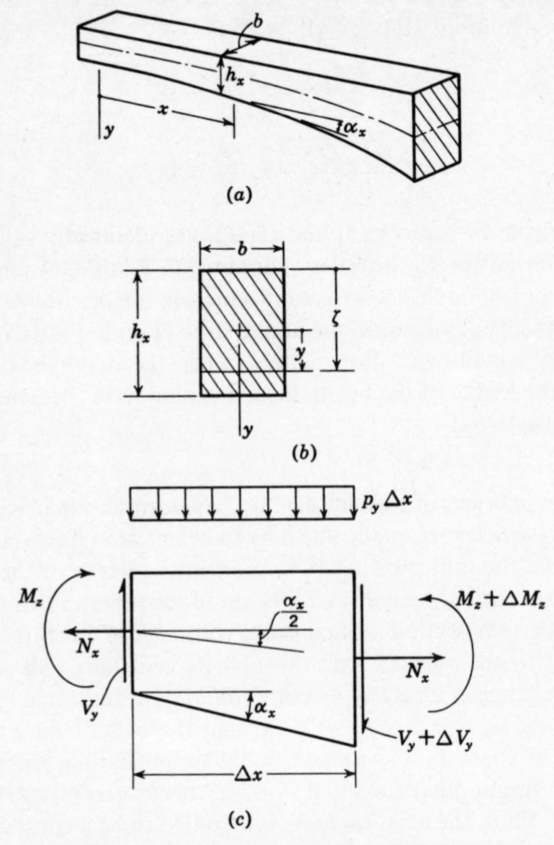

FIGURE 5.6 (*a*) Beam of variable depth; (*b*) cross-sectional geometry; (*c*) stress resultants on a typical beam segment.

this figure, it is clear that

$$\zeta = \frac{h_x}{2} + y$$

Thus, instead of Eq. (4.23), we write for the normal stress

$$\sigma_x = \frac{N_x}{A_x} + \frac{M_z}{I_z}\left(\zeta - \frac{h_x}{2}\right) \tag{5.25}$$

Assuming that no body forces are present, we obtain from Eqs. (2.4) the equilibrium conditions

$$\begin{aligned} \frac{\partial \sigma_x}{\partial x} + \frac{\partial \tau_{xy}}{\partial \zeta} &= 0 \\ \frac{\partial \tau_{xy}}{\partial x} + \frac{\partial \sigma_y}{\partial \zeta} &= 0 \end{aligned} \tag{5.26}$$

Thus
$$\tau_{xy} = -\int \frac{\partial \sigma_x}{\partial x}\, d\zeta + g(x) \tag{5.27}$$

and
$$\sigma_y = -\int \frac{\partial \tau_{xy}}{\partial x}\, d\zeta + f(x) \tag{5.28}$$

where $g(x)$ and $f(x)$ are functions only of x which are to be determined from boundary conditions.

Substituting Eq. (5.25) into Eq. (5.27) and remembering that N_x is assumed to be constant, we carry out the indicated integration and get

$$\tau_{xy} = \zeta \frac{N_x}{A_x^2}\frac{dA_x}{dx} + \frac{\zeta}{2}(\zeta - h_x)\left(\frac{M_z}{I_z^2}\frac{dI_z}{dx} - \frac{1}{I_z}\frac{dM_z}{dx}\right) + \zeta \frac{M_z}{2I_z}\frac{dh_x}{dx} + g(x) \tag{5.29}$$

Since τ_{xy} must vanish at $\zeta = 0$, it follows that $g(x) = 0$. Further, by considering the equilibrium of moments on a typical beam segment (Fig. 5.6*c*), we obtain the relation

$$\frac{dM_z}{dx} = V_y - N_x \tan\frac{\alpha_x}{2} = V_y - \tfrac{1}{2} N_x \frac{dh_x}{dx} \tag{5.30}$$

which replaces Eqs. (5.16) for uniform beams. Therefore

$$\tau_{xy} = \zeta N_x\left(\frac{1}{A_x^2}\frac{dA_x}{dx} + \frac{\zeta - h_x}{4I_z}\frac{dh_x}{dx}\right) + \frac{\zeta}{2}(\zeta - h_x)\left(\frac{M_z}{I_z^2}\frac{dI_z}{dx} - \frac{V_y}{I_z}\right) + \zeta\frac{M_z}{2I_z}\frac{dh_x}{dx} \tag{5.31}$$

We can simplify this equation by noting that $A_x = bh_x$ and $I_z = bh_x^3/12$ for the rectangular section. Thus

$$\tau_{xy} = -\frac{\zeta N_x}{bh_x^2}\frac{dh_x}{dx}\left(2 - 3\frac{\zeta}{h_x}\right) + \frac{6\zeta}{bh_x^2}\left[V_y\left(1 - \frac{\zeta}{h_x}\right) - \frac{M_z}{h_x}\frac{dh_x}{dx}\left(2 - 3\frac{\zeta}{h_x}\right)\right] \tag{5.32}$$

Returning now to the calculation of σ_y, we introduce Eq. (5.32) into Eq. (5.28) and integrate. This results in the relation

$$\begin{aligned}\sigma_y = {} & \frac{\zeta^2 N_x}{bh_x^2}\left[\frac{d^2h_x}{dx^2}\left(1 - \frac{\zeta}{h_x}\right) - \frac{1}{h_x}\left(\frac{dh_x}{dx}\right)^2\left(5 - 6\frac{\zeta}{h_x}\right)\right] \\ & + \frac{6\zeta^2}{bh_x^3}\left[\tfrac{1}{2}h_x p_y - \frac{\zeta p_y}{3h_x} + 2V_y\frac{dh_x}{dx}\left(1 - \frac{\zeta}{h_x}\right) + M_z\frac{d^2h_x}{dx^2}\left(1 - \frac{\zeta}{h_x}\right)\right. \\ & \left. - \frac{M_z}{h_x}\left(\frac{dh_x}{dx}\right)^2\left(3 - 4\frac{\zeta}{h_x}\right)\right] - \frac{p_y}{b}\end{aligned} \tag{5.33}$$

where p_y is the applied load per unit length. The last term on the right side of this equation is the function $f(x)$ in Eq. (5.28). We obtain $f(x) = -p_y/b$ from the condition that σ_y must be equal to the intensity $-p_y/b$ of the applied forces at $\zeta = 0$, the top fibers of the beam. We can easily verify that the stress σ_y due to p_y vanishes at the extreme bottom fibers of the beam by setting $\zeta = h_x$ in the above equation.

In marked contrast to the case of a beam of constant depth, we note that neither τ_{xy} nor σ_y vanishes, in general, at the extreme bottom fibers of a beam of variable depth. In fact, if we set $\zeta = h_x$ in Eqs. (5.25), (5.32), and (5.33), we find

$$
\begin{aligned}
(\sigma_x)_b &= \frac{N_x}{bh_x} + \frac{6M_z}{bh_x^2} \\
(\tau_{xy})_b &= \frac{N_x}{bh_x}\frac{dh_x}{dx} + \frac{6M_z}{bh_x^2}\frac{dh_x}{dx} = (\sigma_x)_b\frac{dh_x}{dx} \\
(\sigma_y)_b &= \frac{N_x}{bh_x}\left(\frac{dh_x}{dx}\right)^2 + \frac{6M_z}{bh_x^2}\left(\frac{dh_x}{dx}\right)^2 = (\sigma_x)_b\left(\frac{dh_x}{dx}\right)^2 = (\tau_{xy})_b\frac{dh_x}{dx}
\end{aligned}
\tag{5.34}
$$

where $(\sigma_x)_b$, $(\tau_{xy})_b$, and $(\sigma_y)_b$ are the stresses at the bottom fibers of the beam. The existence of stresses on these lower fibers immediately suggests an important question: Do these stresses satisfy the static boundary conditions on the lower boundary surface of a beam of variable depth? To answer this, let us consider the beam shown in Fig. 5.7*a* and, in particular, a small element at the lower boundary of the beam. The stresses acting on this element are indicated in Fig. 5.7*b*. According to the general static boundary conditions given by Eqs. (2.6), the boundary element is in equilibrium if

$$
\begin{aligned}
l(\sigma_x)_b + m(\tau_{xy})_b &= 0 \\
l(\tau_{xy})_b + m(\sigma_y)_b &= 0
\end{aligned}
\tag{5.35}
$$

where l and m are the direction cosines of a normal to the boundary. Since we have defined the rate of change of depth so that $\tan \alpha_x = dh_x/dx$, it is clear from Fig. 5.7*b* that $l = -\sin \alpha_x$ and $m = \cos \alpha_x$. Thus, Eqs. (5.35) can be written

$$
\begin{aligned}
-(\sigma_x)_b\frac{dh_x}{dx} + (\tau_{xy})_b &= 0 \\
-(\tau_{xy})_b\frac{dh_x}{dx} + (\sigma_y)_b &= 0
\end{aligned}
\tag{5.36}
$$

which, in view of Eqs. (5.34), are identically satisfied. Hence, the stress components defined in Eqs. (5.25), (5.32), and (5.33) satisfy the equilibrium

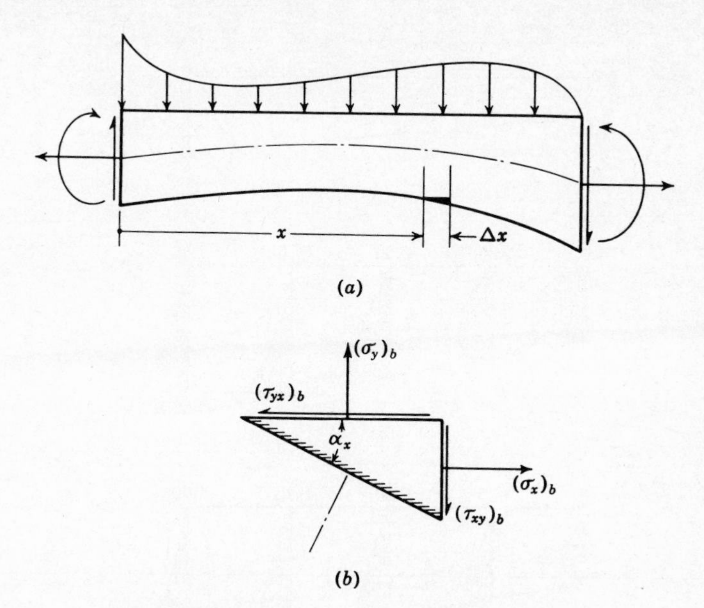

FIGURE 5.7 (*a*) Beam of variable depth; (*b*) stresses on a boundary element at bottom of beam.

conditions not only at all points within the beam but also at all points on the boundaries.

To illustrate the application of these equations, let us consider the cantilever beam of linearly varying depth shown in Fig. 5.8*a*. Here N_x and p_y are zero so that[5]

$$\sigma_x = \frac{12M_z}{bh_x^2}\left(\frac{\zeta}{h_x} - \frac{1}{2}\right) \tag{5.37a}$$

$$\tau_{xy} = \frac{6\zeta}{bh_x^2}\left[V_y\left(1 - \frac{\zeta}{h_x}\right) - \frac{M_z}{h_x}\frac{dh_x}{dx}\left(2 - 3\frac{\zeta}{h_x}\right)\right] \tag{5.37b}$$

$$\sigma_y = \frac{6\zeta^2}{bh_x^3}\left[2V_y\frac{dh_x}{dx}\left(1 - \frac{\zeta}{h_x}\right) + M_z\frac{d^2h_x}{dx^2}\left(1 - \frac{\zeta}{h_x}\right) - \frac{M_z}{h_x}\left(\frac{dh_x}{dx}\right)^2\left(3 - 4\frac{\zeta}{h_x}\right)\right] \tag{5.37c}$$

For the beam shown, $M_z = Px$, $V_y = P$, and

$$h_x = h_0(1 + \beta x) \tag{a}$$

where

$$\beta = \frac{\mu - 1}{L} \tag{b}$$

[5] A similar set of equations for stresses in beams of variable depth is found in Ref. 37, pp. 72–177.

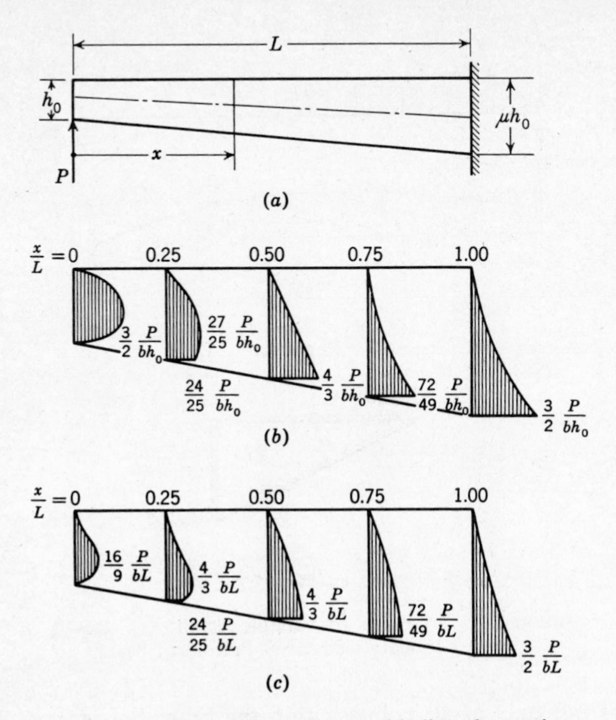

FIGURE 5.8 (*a*) Cantilever beam with linearly varying depth; (*b*) shearing-stress distributions; (*c*) transverse-normal-stress distributions for the case in which $\mu = 2$.

and μ is the parameter defined in the figure. Introducing these relations into Eqs. (5.37), we get

$$\sigma_x = \frac{6Px}{bh_0^3(1+\beta x)^3}\left[2\zeta - h_0(1+\beta x)\right] \tag{c}$$

$$\tau_{xy} = \frac{6P\zeta}{bh_0^3(1+\beta x)^4}\left[h_0(1-\beta^2x^2) - \zeta(1-2\beta x)\right] \tag{d}$$

$$\sigma_y = \frac{6P\beta\zeta^2}{bh_0^3(1+\beta x)^5}\left[h_0(2+\beta x-\beta^2x^2) - 2\zeta(1-\beta x)\right] \tag{e}$$

Because of our original hypothesis, the longitudinal stress σ_x varies linearly over the depth of each section, as in the case of uniform beams. Moreover, at the free end of the beam τ_{xy} has the same parabolic distribution as is found in a prismatic bar of depth h_0. At interior sections, however, the shearing-stress distribution is considerably different. Taking for illustration purpose $\mu = 2$, we find from Eq. (*d*) that

$$\tau_{xy} = \frac{6P\zeta}{bh_0^3(1+x/L)^4}\left[h_0\left(1-\frac{x^2}{L^2}\right) - \zeta\left(1-2\frac{x}{L}\right)\right] \tag{f}$$

This distribution is shown in Fig. 5.8*b* for various values of x. We see that the point of maximum shearing stress shifts below the centroid as we move inward from the free end of the beam. At $x = L/2$, the shearing stress varies linearly over the cross section, and for $x > L/2$ the maximum stress on each section occurs at the extreme bottom fibers of the beam. We see that the maximum value of τ_{xy}, $3P/2bh_0$ occurs at the centroid of the section at $x = 0$ and at the bottom of the section at $x = L$.

For this particular variation in depth, Eq. (*e*) becomes

$$\sigma_y = \frac{6P\zeta^2}{Lbh_0^3(1 + x/L)^5}\left[h_0\left(2 + \frac{x}{L} - \frac{x^2}{L^2}\right) - 2\zeta\left(1 - \frac{x}{L}\right)\right] \qquad (g)$$

This distribution is shown for various values of x in Fig. 5.8*c*. We see that the maximum transverse normal stress occurs at the free end of the beam. Note that while σ_y is necessary for equilibrium, it is generally small compared with τ_{xy} and σ_x. In the present case, for example, $\sigma_y = \tau_{xy}h_0/L$ at the bottom of each section.

Another example is provided by the bar shown in Fig. 5.9*a*. This bar has the same dimensions as the one shown in Fig. 5.8, but it is subjected

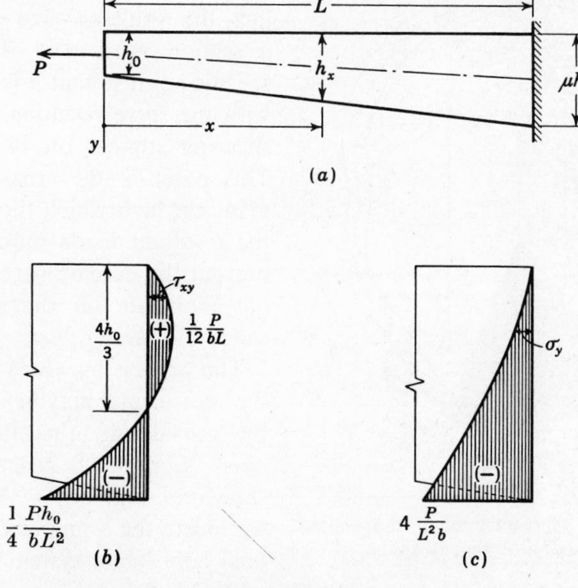

FIGURE 5.9 (*a*) Beam with linearly varying depth subjected to a longitudinal force P; (*b*) shearing-stress distribution; (*c*) transverse-normal-stress distribution at the fixed end for the case in which $\mu = 2$.

to a longitudinal force P at the centroid of the section at the free end. In this case, $N_x = P$, $V_y = 0$, and

$$M_z = -\frac{P}{2}(h_x - h_0) = -\frac{P}{2}h_0\beta x \tag{h}$$

Thus,

$$\begin{aligned}
\sigma_x &= \frac{P}{bh_0(1+\beta x)} - \frac{6P\beta x}{bh_0^2(1+\beta x)^2}[2\zeta - h_0(1+\beta x)] \\
\tau_{xy} &= \frac{\zeta P\beta(2\beta x - 1)}{bh_0^2(1+\beta x)^4}[2h_0(1+\beta x) - 3\zeta] \\
\sigma_y &= \frac{\zeta^2 P\beta^2}{bh_0^3(1+\beta x)^5}[6\zeta h_0(1-\beta x) + h_0^2(4\beta^2x^2 - \beta x - 5)]
\end{aligned} \tag{i}$$

Plots of the distributions of τ_{xy} and σ_y at the fixed end of the beam are given in Fig. 5.9*b* and *c*, respectively, for the case in which $\mu = 2$.

5.5 The shear center. In the previous developments we assumed that the stress resultants on each section were such that no twisting moments were developed. For such a situation to exist, we cannot apply the external loads at random. In fact, the resultant force parallel to a section must pass through a specific point so that it is collinear with the force resulting from the shearing stresses on the section. This point in the cross-sectional plane through which the plane of the resultant loads must pass to prevent the development of twisting moments on the section is called the *shear center*.

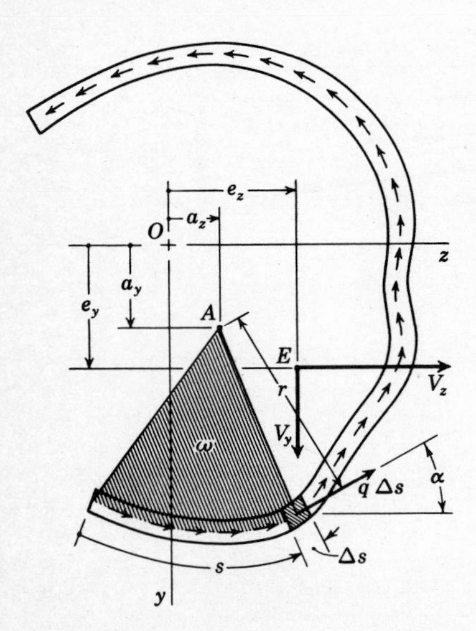

FIGURE 5.10 Geometry of a thin-walled open section of general shape.

The process by which we locate the shear center may be illustrated by considering the thin-walled open section of general shape shown in Fig. 5.10. As before, we locate the origin of an orthogonal coordinate system at O, the centroid of the section; V_y and V_z act at E, the shear center, which has coordinates e_y and e_z, as shown. The force developed by the shear flow on any element Δs of the wall is

$q\,\Delta s$ and A is an arbitrary point in the cross-sectional plane. From the above definition of the shear center, the total moment about A must vanish. Thus,

$$-V_y(e_z - a_z) + V_z(e_y - a_y) + \int_0^S qr\,ds = 0 \tag{5.38}$$

in which a_y and a_z are the coordinates of A, S is the total length of the tube wall, and r is the perpendicular distance from A to the line of action of $q\,ds$ (which is tangent to the tube wall).

The shear flow q is the product of t, the wall thickness, and shearing stress given by Eq. (5.20). Referring to Eq. (3.39), we recall that $r\,ds$ is equal to $2\,d\omega$, where ω is the sectorial area (shaded in Fig. 5.10). It follows that

$$\int_0^S qr\,ds = 2\frac{I_yI_{\omega z} - I_{yz}I_{\omega y}}{I_yI_z - I_{yz}^2}V_y + 2\frac{I_zI_{\omega y} - I_{yz}I_{\omega z}}{I_yI_z - I_{yz}^2}V_z \tag{5.39}$$

in which $I_{\omega y}$ and $I_{\omega z}$ are the *sectorial products of inertia* of the section about A defined by

$$\begin{aligned} I_{\omega y} &= \int_0^S Q_y\,d\omega \\ I_{\omega z} &= \int_0^S Q_z\,d\omega \end{aligned} \tag{5.40}$$

Introducing Eq. (5.39) into Eq. (5.38) and equating the coefficients of V_y and V_z to zero, we find

$$\begin{aligned} e_y &= a_y - 2\frac{I_zI_{\omega y} - I_{yz}I_{\omega z}}{I_yI_z - I_{yz}^2} \\ e_z &= a_z + 2\frac{I_yI_{\omega z} - I_{yz}I_{\omega y}}{I_yI_z - I_{yz}^2} \end{aligned} \tag{5.41}$$

Let us now move the pole A to O, the centroid. In this case a_y and a_z are zero and

$$r\,ds = y\cos\alpha\,ds + z\sin\alpha\,ds$$

where α is the angle between the z axis and a tangent to the tube wall at s (Fig. 5.10). Noting that $\cos\alpha = dz/ds$ and $\sin\alpha = -dy/ds$, we have

$$r\,ds = y\,dz - z\,dy = 2\,d\omega \tag{5.42}$$

Therefore, Eqs. (5.41) become

$$\begin{aligned} e_y &= -2\frac{I_zI_{\omega y} - I_{yz}I_{\omega z}}{I_yI_z - I_{yz}^2} \\ e_z &= +2\frac{I_yI_{\omega z} - I_{yz}I_{\omega y}}{I_yI_z - I_{yz}^2} \end{aligned} \tag{5.43}$$

where, in this case,

$$\begin{aligned} I_{\omega y} &= \tfrac{1}{2} \int Q_y(y\,dz - z\,dy) \\ I_{\omega z} &= \tfrac{1}{2} \int Q_z(y\,dz - z\,dy) \end{aligned} \tag{5.44}$$

Though quite general, the above equations are used to locate the shear center only in special cases involving complicated geometry. In such instances $I_{\omega y}$ and $I_{\omega z}$ are often evaluated by some numerical integration technique. More often, we can locate the shear center much more easily by using simple statics.

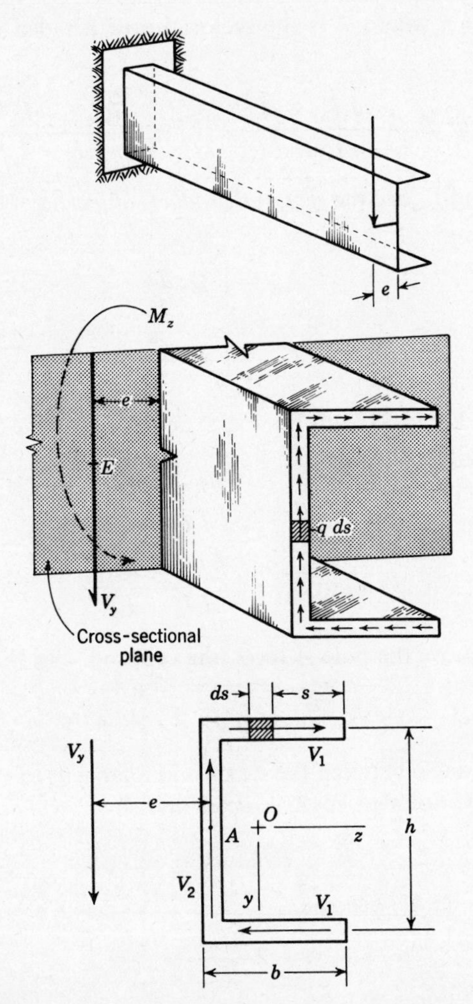

FIGURE 5.11 A channel loaded so that no twisting occurs. The resultant shear V passes through the shear center of the section.

The shear center of the channel shown in Fig. 5.11, for example, is easily found by taking moments about A:

$$eV_y = V_1 h = h \int q \, ds = \frac{hV_y}{I_z} \int Q_z \, ds$$

Since $Q_z = tsh/2$, the above equation reduces to

$$e = \frac{th^2b^2}{4I_z} \tag{5.45}$$

Similar calculations for a number of common sections lead us to three useful observations:

> **1.** The shear center for sections with two intersecting rectangular flanges is the point of intersection of the axes of the flanges.
> **2.** The shear center of sections with one axis of symmetry lies on that axis.
> **3.** The shear center of sections with two axes of symmetry is at the centroid of the section.

Examples of these cases are shown in Fig. 5.12.

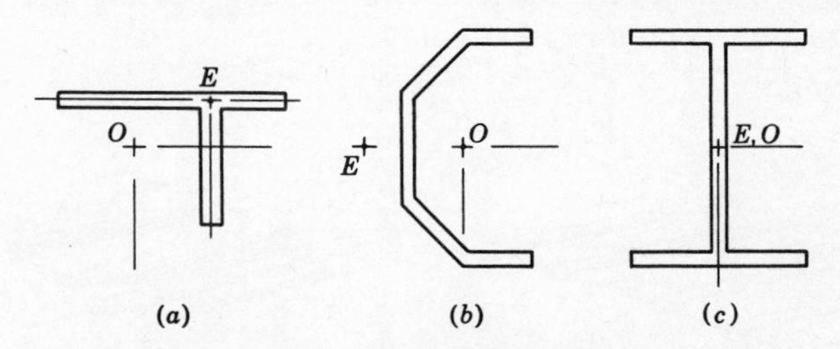

FIGURE 5.12 Location of the shear center for some typical sections. Points E and O indicate the shear center and the centroid of the section, respectively.

5.6 Thin-walled sections with longitudinal stringers. The location of the shear center of thin-walled open sections with "concentrated" flanges deserves special consideration. Such sections are often used in aircraft and missile structures and are characterized by thin metal "skins" connected to longitudinal flange elements called *stringers*. The stringers are assumed to develop the normal forces on the section needed for equilibrium and the thin skin carries the shear flow from one flange to another. Since the thin skin develops no significant normal stresses, the shear flow is constant between the stringers.

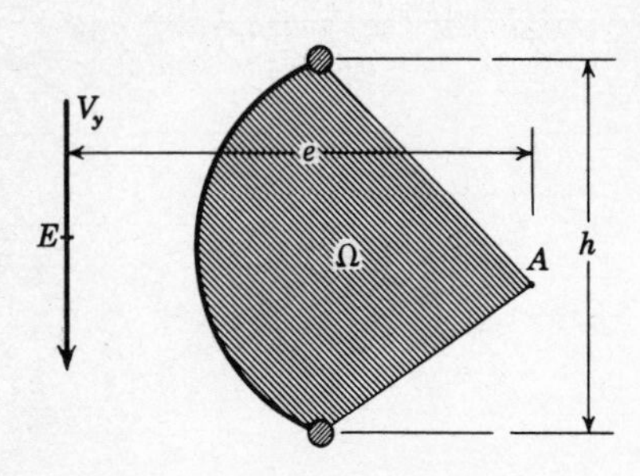

FIGURE 5.13 Two-flange open section.

Consider, for example, the two-flange beam in Fig. 5.13. If V_y acts through the shear center, the total moment about any point A must vanish. Hence,

$$V_y e = \int qr\, ds = q2\Omega$$

where, again, Ω is the sectorial area shown shaded in the figure. Simple statics shows that $q = V_y/h$, so that

$$e = \frac{2\Omega}{h} \tag{5.46}$$

The two-flange beam in Fig. 5.13 is capable of resisting only moments which act in the plane of the flanges. For any other type of loading it is unstable and secondary bending is developed in the skin. The structure then behaves as a shell rather than a beam and may undergo large displacements or fail. For this reason open sections are often provided a number of flanges (stringers), as is indicated in Fig. 5.14, to increase the stiffness and to stabilize the structure.

In analyzing such structures we again take advantage of Navier's assumption (planes remain plane) and, as before, assume that the thin skin panels between flanges are ineffective in resisting bending and that the stringer areas are concentrated at points.

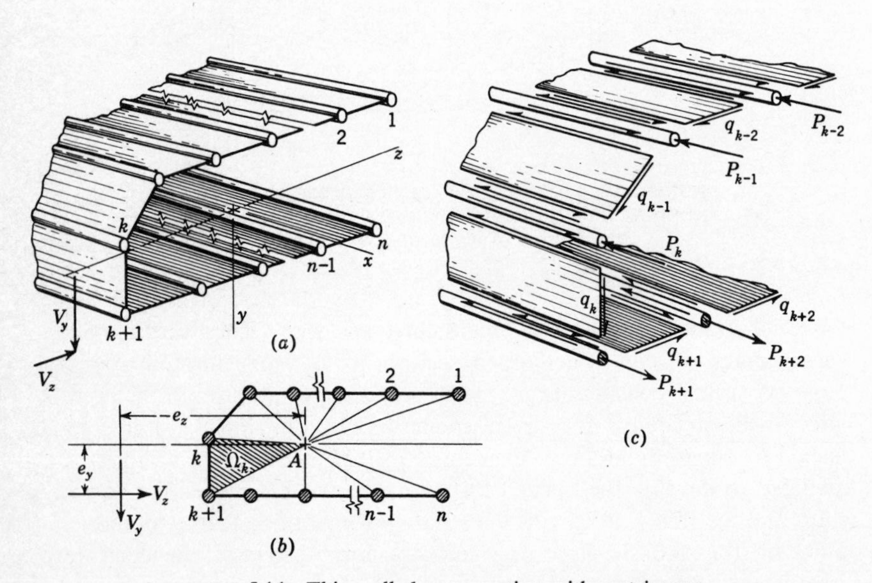

FIGURE 5.14 Thin-walled open section with n stringers.

Let us consider, for example, the bending of a thin-walled open section having n stringers. In view of the assumptions we have made, the section properties are given by

$$I_y = \sum_{i=1}^{n} z_i^2 a_i \qquad I_z = \sum_{i=1}^{n} y_i^2 a_i \qquad I_{yz} = \sum_{i=1}^{n} y_i z_i a_i$$

in which a_i denotes the area of the ith stringer which is located at the point (y_i, z_i). According to Eq. (5.20), the shear flow on the kth panel is

$$q_k = \frac{I_y \sum_{i=1}^{k} a_i y_i - I_{yz} \sum_{i=1}^{k} a_i z_i}{I_y I_z - I_{yz}^2} V_y + \frac{I_z \sum_{i=1}^{k} a_i z_i - I_{yz} \sum_{i=1}^{k} a_i y_i}{I_y I_z - I_{yz}^2} V_z \qquad (5.47)$$

The summations in the above equation are obviously Q_z and Q_y at panel k. Similarly, the axial force developed in the ith stringer is

$$P_i = a_i \left[\frac{N_x}{\sum_{i=1}^{n} a_i} + \frac{M_z I_y - M_y I_{yz}}{I_y I_z - I_{yz}^2} y_i + \frac{M_y I_z - M_z I_{yz}}{I_y I_z - I_{yz}^2} z_i \right] \qquad (5.48)$$

Furthermore, by examining the equilibrium of a longitudinal element of the ith stringer (Fig. 5.15), it is easily verified that

$$\frac{dP_i}{dx} = q_{i-1} - q_i \qquad (5.49)$$

where q_i and q_{i-1} are the shear flows in panels i and $i - 1$.

Since q_k is constant throughout panel k, we know from Eq. (3.40) that it creates a moment about an arbitrarily selected point A of magnitude $2q_k\Omega_k$, where Ω_k is the sectorial area through which a radius vector from A to the center line of the panel sweeps in moving from point k to point $k + 1$ (see Fig. 5.14b). The total moment due to all of the shear flows, then, is

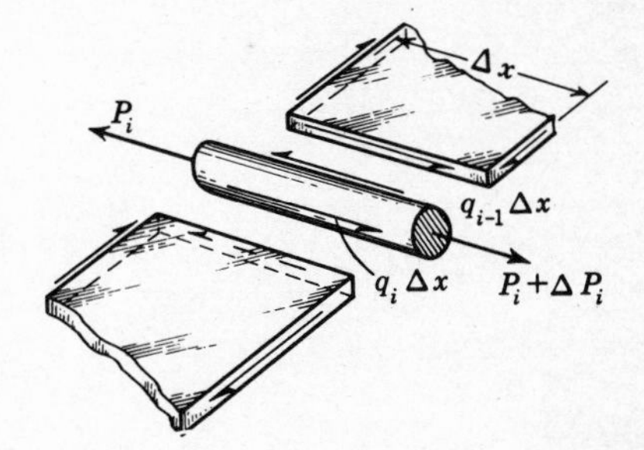

FIGURE 5.15 Change in stringer force P_i in length Δx.

$$2 \sum_{k=1}^{n-1} q_k \Omega_k$$

where q_k is given by Eq. (5.47). The sum ranges to $n - 1$ since the number of panels is one less than the number of stringers. The moment of V_y and V_z about A is $(-e_z V_y + e_y V_z)$. Thus,

$$-e_z V_y + e_y V_z + 2 \sum_{k=1}^{n-1} q_k \Omega_k = 0$$

from which we find

$$e_y = -\frac{2\sum_{k=1}^{n-1}\left[\Omega_k\left(I_z\sum_{i=1}^{k} a_i z_i - I_{yz}\sum_{i=1}^{k} a_i y_i\right)\right]}{I_y I_z - I_{yz}^{\,2}}$$

$$e_z = +\frac{2\sum_{k=1}^{n-1}\left[\Omega_k\left(I_y\sum_{i=1}^{k} a_i y_i - I_{yz}\sum_{i=1}^{k} a_i z_i\right)\right]}{I_y I_z - I_{yz}^{\,2}} \tag{5.50}$$

Comparing this result with Eqs. (5.43) it is clear that

$$I_{\omega y} = \sum_{k=1}^{n-1}\left(\Omega_k \sum_{i=1}^{k} a_i z_i\right)$$

$$I_{\omega z} = \sum_{k=1}^{n-1}\left(\Omega_k \sum_{i=1}^{k} a_i y_i\right) \tag{5.51}$$

A simple example is provided by the cantilevered open section shown in Fig. 5.16. The thin skins are of equal thickness, and a resultant vertical force V acts through the shear center at the free end. The stringers are of equal area A, so that I_{yz} is zero and the pole A is taken at the centroid. Other dimensions are given in the figure. The location of the shear center follows directly on applying Eqs. (5.50):

$$e_z = +\frac{2\sum_{k=1}^{5}\left(\Omega_k\sum_{i=1}^{k} a_i y_i\right)}{I_z}$$

$$= -\frac{2}{6Ah^2/4}\left[\frac{bh}{4}\left(\frac{Ah}{2}\right) + \frac{bh}{4}\left(\frac{Ah}{2}\cdot 2\right) + \frac{bh}{4}\left(\frac{Ah}{2}\cdot 3\right) + \frac{bh}{4}\left(\frac{3Ah}{2} - \frac{Ah}{2}\right) + \frac{bh}{4}\left(\frac{Ah}{2}\right)\right]$$

or

$$e_z = 0$$

$$e_y = -\tfrac{3}{2}b \tag{a}$$

The shear flows are calculated using Eq. (5.47):

$$q_1 = q_5 = -\frac{V}{I_z}A\frac{h}{2} = -\frac{V}{3h}$$

$$q_2 = q_4 = -\frac{V}{I_z}\left(A\frac{h}{2} + A\frac{h}{2}\right) = -\frac{2V}{3h} \tag{b}$$

$$q_3 = -\frac{V}{I_z}\left(A\frac{h}{2} + A\frac{h}{2} + A\frac{h}{2}\right) = -\frac{V}{h}$$

This distribution is shown in Fig. 5.16*c*.

From Eq. (5.48), the magnitude of the stringer forces at the fixed end is $\pm 2VL/3h$.

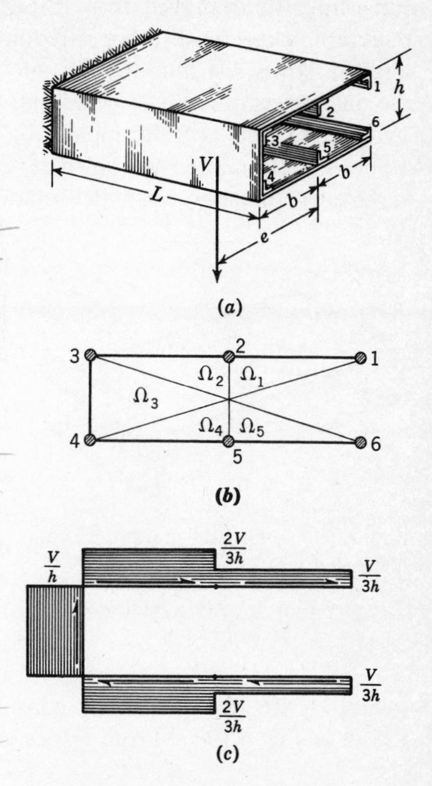

FIGURE 5.16 (*a*) A cantilevered thin-walled open section; (*b*) idealized cross section; (*c*) shear-flow distribution.

5.7 Shear lag. The formulas developed in the previous sections were based on the elementary theory of flexure; the influence of shear deformations on the final stress distribution was considered to be negligible. For a great many structures this assumption is justifiable and leads to no significant error. However, in the case of thin-walled structures, particularly those of the type shown in Fig. 5.14, the shear deformation of thin webs and panels between stringers may considerably alter the stress distribution given by the elementary theory. Experiments on "box" beams such as the one shown in Fig. 5.17, for example, have shown that the true normal-stress distribution "lags behind" that given by the elementary theory. This deviation from the elementary theory is known as *shear lag;* the phenomenon is also referred to as *diffusion.*

A classic example[6] of the influence of shear lag is given by the thin-walled panel shown in Fig. 5.18*a*. The panel consists of three "concentrated" flanges (stringers) with areas a_1, $2a_2$, and a_1 and thin shear panels of thickness t. The elementary theory predicts that the shear panels will not undergo shear deformation upon the application of the longitudinal loads. If this were true, the structure would elongate uniformly, as indicated in Fig. 5.18*b*, and each section would be subjected to a uniform stress obtained by simply dividing the total load $2P$ by the total cross-sectional area. Not only does this violate the assumption that the panels carry only shearing stresses, but it is obviously unrealistic. We know from Saint-Venant's principle[7] that the stress is uniform only on sections

[6] See, for example, Refs. 34, 45, and 76. This example, along with a detailed discussion of approximate methods of shear lag analysis, is given in Ref. 34.

[7] See Art. 2.4.

sufficiently far removed from the points of application of the loads; therefore, a certain distance is required for the loads to "diffuse" into a uniform stress distribution. It follows that the intensity of the stress in the outside stringers must be substantially higher than that in the central stringer at points near the application of the loads. Thus, a more realistic mode of deformation is shown in Fig. 5.18*c*. Here the thin webs are assumed to undergo shear deformations and to develop a shear flow q. This shear flow is solely responsible for the transmission of normal stress to the central stringer, and, since it varies with x, it leads to variable forces in the stringers.

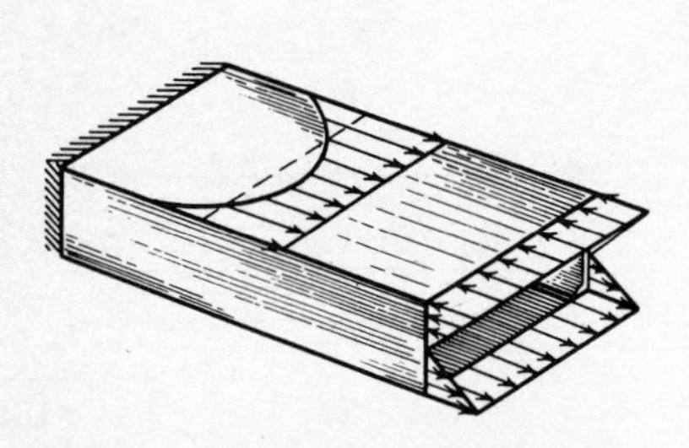

FIGURE 5.17 Shear lag in a box beam. The actual normal-stress distribution in the top flange panel is compared with that predicted by My/I, indicated by the dashed line.

Since both the structure and the loading are symmetrical with respect to the x axis, we need consider only half of the panel, as shown in Fig. 5.19*a*. Let N_1 denote the axial force developed in outside (lower) stringer and N_2 denote the force developed in half of the central stringer. The total force in the central stringer is then $2N_2$, and the normal stress in this stringer is $2N_2/2a_2$. We now consider the equilibrium of the segment shown in Fig. 5.19*b*. From statics we find that

$$\frac{dN_1}{dx} = q = -\frac{dN_2}{dx} \tag{5.52}$$

The shear panels are assumed to be infinitely stiff in the transverse (y) direction. Thus, transverse forces due to changes in q with x oppose one another on each side of the central stringer and do not enter the analysis.

Now the elongations of unit lengths of each stringer at a distance x from the free end due to N_1 and N_2 are simply N_1/Ea_1 and N_2/Ea_2, respectively. The longitudinal displacement of points in stringers 1 and 2 are denoted u_1 and u_2, as is indicated in Fig. 5.19*c*. Since these displacements are not equal, a shearing strain γ of magnitude $(u_1 - u_2)/h$ is developed in the web. From the basic strain-displacement relations, it follows that

$$\frac{d\gamma}{dx} = \frac{1}{h}\left(\frac{\partial u_1}{\partial x} - \frac{\partial u_2}{\partial x}\right) = \frac{1}{h}(\epsilon_1 - \epsilon_2)$$

or

$$\frac{d\gamma}{dx} = \frac{\sigma_1 - \sigma_2}{Eh} \tag{5.53}$$

where σ_1 and σ_2 are the normal stresses developed in stringers 1 and 2: $\sigma_1 = N_1/a_1$ and $\sigma_2 = N_2/a_2$.

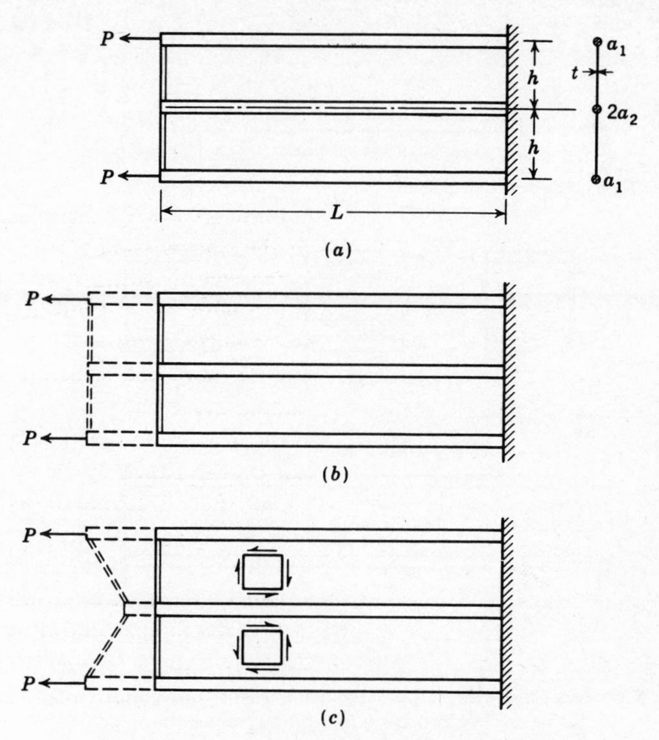

FIGURE 5.18 (*a*) A symmetrical thin-walled panel subjected to concentrated end loads. (*b*) Uniform elongation of the panel, as predicted by the elementary theory. No shearing stress is developed in the thin webs. (*c*) A more realistic mode of deformation. Here the shear panels are assumed to undergo shear deformations and normal stresses are assumed to be transmitted to the interior stringer by means of a shear flow q.

Since $q = \gamma t G$, Eq. (5.53) can be written

$$\frac{dq}{dx} - \frac{Gt}{Eh}\left(\frac{N_1}{a_1} - \frac{N_2}{a_2}\right) = 0 \tag{5.54}$$

Differentiating Eq. (5.54) with respect to x and substituting Eqs. (5.52) into the result gives the differential equation

$$\frac{d^2q}{dx^2} - \mu^2 q = 0 \tag{5.55}$$

where

$$\mu = \sqrt{\frac{Gt}{Eh}\left(\frac{1}{a_1} + \frac{1}{a_2}\right)} \tag{5.56}$$

The constant μ is called the *coefficient of diffusion*.

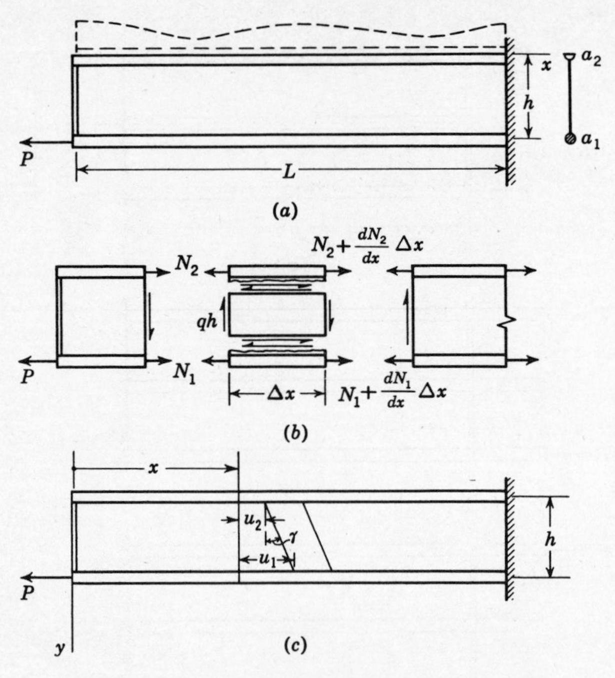

FIGURE 5.19 (*a*) Lower half of symmetrical three-stringer panel; (*b*) forces on a typical segment; (*c*) geometry of deformation of a typical segment.

Equation (5.55) is a homogeneous linear differential equation with constant coefficients. Since it is of second order, a linear combination of any two independent solutions is the complete solution to the equation. By direct substitution, we find that Ce^{mx}, C and m being constants, is a solution to Eq. (5.55) provided

$$m^2 - \mu^2 = 0$$

This equation is called the *characteristic polynomial* of Eq. (5.55), and its roots $m = \pm\mu$ are called the eigenvalues of the equation. Hence, both $C_1e^{\mu x}$ and $C_2e^{-\mu x}$ are solutions to the equation. Since they are also independent, it follows that the complete solution to Eq. (5.55) is

$$q = C_1e^{\mu x} + C_2e^{-\mu x} \tag{5.57}$$

where C_1 and C_2 are constants. The shear-flow distribution must be such that the following static boundary conditions are satisfied:

$$N_1 = P \quad \text{and} \quad N_2 = 0 \quad \text{at } x = 0 \tag{5.58}$$

and

$$q = 0 \quad \text{at } x = L \tag{5.59}$$

Introducing Eqs. (5.58) into Eq. (5.54) gives the additional condition

$$\frac{dq}{dx} = \frac{GtP}{Eha_1} \qquad \text{at } x = 0 \tag{5.60}$$

Substituting Eq. (5.57) into Eqs. (5.59) and (5.60) yields two equations in C_1 and C_2. Solving these, we find

$$C_1 = -C_2 e^{-2\mu L} = \frac{GtP}{\mu Eha_1(1 + e^{2\mu L})}$$

Incorporating this result into Eq. (5.57) and simplifying, we get

$$q = -\frac{P\mu a_2}{A \cosh \mu L} \sinh \mu x' \tag{5.61}$$

where $x' = L - x$ and $A = a_1 + a_2$. Finally, substituting Eq. (5.61) into Eqs. (5.52) and using Eqs. (5.58) to evaluate the constants of integration gives

$$\begin{aligned} N_1 &= \frac{Pa_1}{A}\left(1 + \frac{a_2 \cosh \mu x'}{a_1 \cosh \mu L}\right) \\ N_2 &= \frac{Pa_2}{A}\left(1 - \frac{\cosh \mu x'}{\cosh \mu L}\right) \end{aligned} \tag{5.62}$$

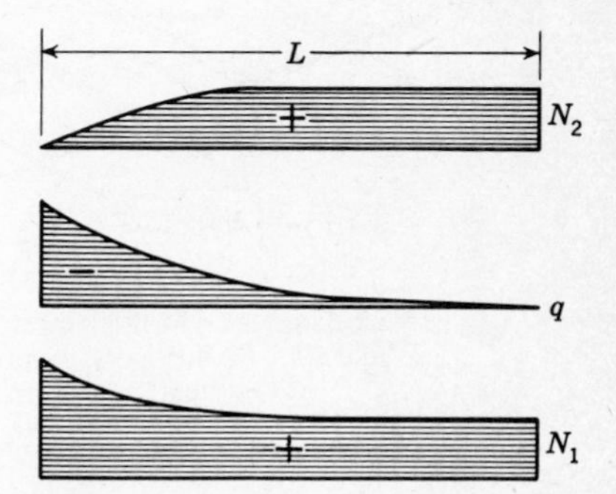

FIGURE 5.20 Normal forces and shear flow in the half panel shown in Fig. 5.19.

The variations in the flange forces and in the shear flow with x are illustrated in Fig. 5.20. Note that the shear-lag effects are considerably more pronounced at the free end where q is a maximum; the axial forces N_1 and N_2 quickly approach the elementary values of Pa_1/A and Pa_2/A, respectively, as x increases. In fact, for long panels for which $\mu L \geq 3.0$, approximately, it is permissible to neglect $e^{-\mu L}$ in comparison with $e^{\mu L}$ so that Eqs. (5.61) and (5.62) become simply

$$\begin{aligned} q &= -\frac{P\mu a_2}{A} e^{-\mu x} \\ N_1 &= \frac{Pa_1}{A}\left(1 + \frac{a_2}{a_1} e^{-\mu x}\right) \\ N_2 &= \frac{Pa_2}{A}(1 - e^{-\mu x}) \end{aligned} \tag{5.63}$$

We now investigate the extension of the above procedure to symmetric multistringer panels. Consider a symmetrical multistringer panel with

$2n$ panels and $2n + 1$ stringers which is loaded symmetrically with respect to the axis of symmetry, the center line of stringer $n + 1$. As before, we need consider only half of the panel, as shown in Fig. 5.21. The panel skins are assumed to be infinitely stiff in the transverse (y) direction, and the n shear flows are taken as unknowns. The thickness of each panel and the stringer areas are assumed to be constant. The area of stringer $n + 1$ is denoted $2a_{n+1}$, and its force is $2N_{n+1}$. Pertinent dimensions are shown in the figure.

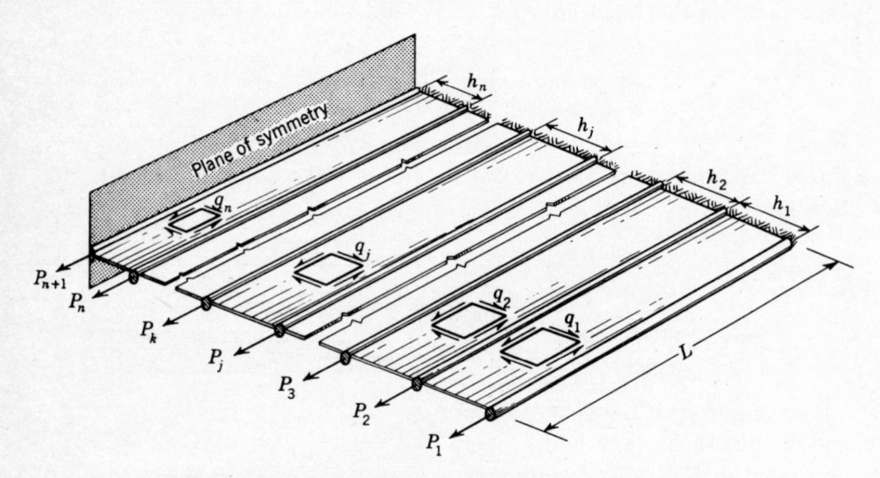

FIGURE 5.21 Multistringer panel with $2n$ shear panels and $2n + 1$ stringers, symmetric about stringer $n + 1$. Only half of the structure is shown here.

Proceeding as before, we isolate a number of segments of length Δx and, using statics, we arrive at the system of equations:

$$\begin{aligned}
\frac{dN_1}{dx} &= q_1 \\
\frac{dN_2}{dx} &= q_2 - q_1 \\
&\cdots\cdots \\
\frac{dN_j}{dx} &= q_j - q_i \\
&\cdots\cdots \\
\frac{dN_n}{dx} &= q_n - q_{n-1} \\
\frac{dN_{n+1}}{dx} &= q_n
\end{aligned} \tag{5.64}$$

Here N_j is the axial force in the jth stringer. Further, if γ_j is the shearing strain in panel j (between stringers j and k) it follows that [see Eq. (5.53)]

$$\frac{1}{Gt_j}\frac{dq_j}{dx} = \frac{d\gamma_j}{dx} = \frac{N_j/a_j - N_k/a_k}{Eh_j} \tag{5.65}$$

where q_j is the shear flow in panel j.

Finally, differentiating each of Eqs. (5.65) and substituting Eqs. (5.64) into the result leads to the system of n second-order differential equations for the shear flows:

$$\begin{aligned}
&\frac{d^2q_1}{dx^2} - \mu_{11}^2 q_1 + \mu_{12}^2 q_2 = 0 \\
&\frac{d^2q_2}{dx^2} - \mu_{22}^2 q_2 + \mu_{21}^2 q_1 + \mu_{23}^2 q_3 = 0 \\
&\cdots\cdots\cdots\cdots\cdots \\
&\frac{d^2q_j}{dx^2} - \mu_{jj}^2 q_j + \mu_{ji}^2 q_i + \mu_{jk}^2 q_k = 0 \\
&\cdots\cdots\cdots\cdots\cdots \\
&\frac{d^2q_n}{dx^2} - \mu_{nn}^2 q + \mu_{n,n-1}^2 q_{n-1} = 0
\end{aligned} \tag{5.66}$$

where

$$\begin{aligned}
\mu_{jj} &= \sqrt{\frac{Gt_j}{Eh_j}\left(\frac{1}{a_j} + \frac{1}{a_k}\right)} \\
\mu_{ji} &= \sqrt{\frac{Gt_j}{Eh_j a_j}} \\
\mu_{jk} &= \sqrt{\frac{Gt_j}{Eh_j a_k}}
\end{aligned} \tag{5.67}$$

We now assume solutions to Eqs. (5.66) of the form $q_j = C_j e^{mx}$, where C_j and m are constants. If the assumed solutions are introduced into Eqs. (5.66), there results a set of n homogeneous equations in the constants C_j. For a nontrivial solution to exist, the determinant of the coefficients of these equations must vanish. Thus,

$$\begin{vmatrix}
m^2 - \mu_{11}^2 & \mu_{12}^2 & 0 & \cdots & 0 & 0 \\
\mu_{21}^2 & m^2 - \mu_{22}^2 & \mu_{23}^2 & \cdots & 0 & 0 \\
\cdot & \cdot & \cdot & \cdot & \cdot & \cdot \\
0 & 0 & 0 & \cdots & \mu_{n,n-1}^2 & m^2 - \mu_{nn}^2
\end{vmatrix} = 0$$

Expanding the above determinant gives an nth-degree polynomial in m^2 having n real roots. If $\bar{\mu}_j^2$ is the jth root of this polynomial, the shear flow in panel j may be written

$$q_j = C_{1j} e^{\bar{\mu}_j x} + C_{2j} e^{-\bar{\mu}_j x} \tag{5.68}$$

The n shear flows must satisfy the static boundary conditions

$$N_j = P_j \qquad \text{at } x = 0 \tag{5.69}$$

$$q_j = 0 \qquad \text{at } x = L \tag{5.70}$$

and the n additional conditions obtained by substituting Eqs. (5.69) into (5.65),

$$\frac{dq_j}{dx} = \frac{Gt_j}{Eh_j}\left(\frac{P_j}{a_j} - \frac{P_k}{a_k}\right) \qquad \text{at } x = 0 \tag{5.71}$$

The $2n$ constants C_{1j} and C_{2j} are determined from Eqs. (5.70) and (5.71), and the $n + 1$ stringer forces are then evaluated using Eqs. (5.64) and (5.69).

5.8 Bending of single-cell thin-walled tubes. The evaluation of normal stresses developed in single-cell thin-walled closed sections presents no special problems. Assuming that plane sections remain plane, the normal stress distribution is given, as before, by Eq. (4.30). The evaluation of shearing stresses, however, is more involved. Equation (5.20) is not directly applicable in this case because the quantities Q_y and Q_z have meaning only for open sections.

To remedy this situation, let us consider the closed thin-walled section of general shape in Fig. 5.22. The resultant shearing forces V_y and V_z act at E, the shear center, so that no twisting occurs. Point O is the centroid of the section and m is an arbitrarily selected point on the periphery of the tube which serves as the origin for the coordinate s. If we isolate a slice of the section, as shown in the figure, we see that the resultant longitudinal

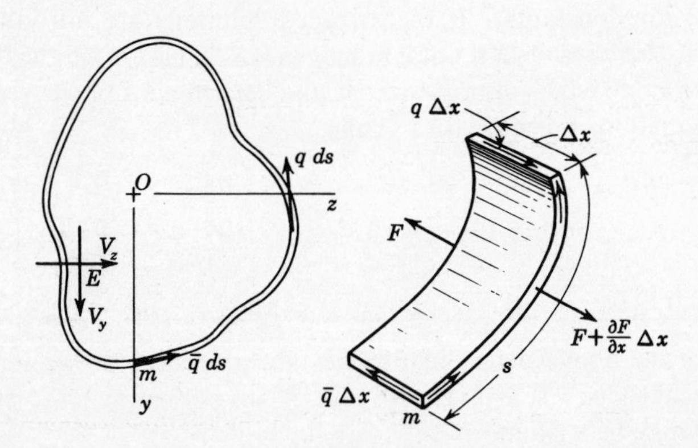

FIGURE 5.22 Thin-walled closed section of general shape.

force $(\partial F/\partial x)\,\Delta x$ is balanced by shearing forces at m and some point s. Hence,

$$q = \bar{q} + q_0 \tag{5.72}$$

where $\bar{q}$ is the shear flow at m and

$$q_0 = -\frac{\partial F}{\partial x} = -\int_m^s \frac{\partial \sigma_x}{\partial x}\, t\, ds$$

q_0 is the shear flow that would be developed if the section were open at m, that is, if $\bar{q}$ were zero. Thus, q_0 is obtained from Eq. (5.20). [The minus sign is present here because we have taken the total shear flow q to be positive if it acts counterclockwise, in the direction of s, while in Eq. (5.20), q is positive when in the direction of positive V_y and V_z.]

We can obtain the magnitude of q in either of two ways: (1) by equating to zero the sum of moments about a longitudinal axis, since no torque is developed, or (2) by equating to zero the total twist of the section. Both methods, of course, yield identical results. The first method is simpler in concept; but the second method involves ideas that will prove helpful when we study the multicell tube in bending. Therefore, we adopt the second method. Referring to Eq. (3.45), we have for the rate of twist

$$\theta = \frac{1}{2\Omega G}\oint q\frac{ds}{t} = 0$$

Substituting Eq. (5.72) into this equation and eliminating the common term $\frac{1}{2}\Omega$ gives

$$\delta_0 + \bar{q}\delta_1 = 0 \tag{5.73}$$

where

$$\delta_0 = \frac{1}{G}\oint q_0\frac{ds}{t} \qquad \delta_1 = \frac{1}{G}\oint\frac{ds}{t} \tag{5.74}$$

The quantities δ_0 and δ_1 have important physical interpretations. Comparing Eqs. (5.74) with Eq. (3.53) or (3.55) with $\theta = 0$, we see that each coefficient represents a relative warping displacement $\bar{u}$. Equation (5.73), in fact, can be written

$$\bar{u} = \frac{1}{G}\oint q\frac{ds}{t} = \frac{1}{G}\oint q_0\frac{ds}{t} + \bar{q}\frac{1}{G}\oint\frac{ds}{t} = 0$$

Thus, if we imagine that the tube is temporarily opened by introducing a longitudinal cut or slit at m, the term δ_0 of Eq. (5.74) is obviously the *relative warping of opposite faces of the slit at m of the open tube due to the shear flow* q_0 (Fig. 5.23*a*). Similarly, δ_1 is the *relative warping of opposite faces of the slit at m of the open tube due to a constant shear flow of unit magnitude* $\bar{q} = 1$ (Fig. 5.23*b*); $\bar{q}\delta_1$, then, is the relative warping due to $\bar{q}$.

The quantities δ_0 and δ_1 are called *influence coefficients*, or more specifically, *flexibilities*, and Eq. (5.73) is an *equation of consistent deformation* physically stating that the total relative displacement of the faces at the hypothetical slit is zero for the closed tube. Thus, according to Eq. (5.73), if these displacements are to be compatible,

$$\bar{q} = -\frac{\delta_0}{\delta_1} = -\frac{\oint q_0(ds/t)}{\oint (ds/t)} \tag{5.75}$$

It is worth noting that we have used a condition of compatibility to solve a statically determinate problem. While perfectly legitimate, this was certainly unnecessary. As mentioned earlier, the same result is obtained from equilibrium considerations.

The total shear flow [Eq. (5.72)] may now be expressed in terms of V_y and V_z by introducing Eq. (5.20). The result is

$$q = -\frac{Q_zI_y - Q_yI_{yz} - K_z}{I_yI_z - I_{yz}^2}V_y - \frac{Q_yI_z - Q_zI_{yz} - K_y}{I_yI_z - I_{yz}^2}V_z \tag{5.76}$$

where

$$K_z = \frac{\oint (Q_zI_y - Q_yI_{yz})(ds/t)}{\oint (ds/t)}$$

$$K_y = \frac{\oint (Q_yI_z - Q_zI_{yz})(ds/t)}{\oint (ds/t)} \tag{5.77}$$

Here, q is positive when it acts in the s (counterclockwise) direction.

Single-celled closed tubes are often referred to as *monocoque* structures, from the French word meaning "single shell." In many practical cases it is not feasible to use a "pure" monocoque tube to resist bending loads; the thin walls offer little resistance to local buckling, and longitudinal stiffeners (stringers) must be provided. Such stiffened structures are called *semimonocoque* tubes. We shall use this term to denote tubes constructed of a thin skin reinforced by longitudinal stringers, as in Fig. 5.14. Axial forces and bending moments are developed by concentrated forces in each stringer and the shearing forces are developed by shear flows in the intermediate panels.

The evaluation of shearing stresses in such structures is accomplished in the same manner as the pure monocoque tube except that, since q is constant between stringers, the integrals in Eqs. (5.77) are replaced by finite sums. The shear flow q_0, for example, is now given by Eq. (5.47). Thus, if the thin-walled beam in Fig. 5.16 were closed by adding a panel

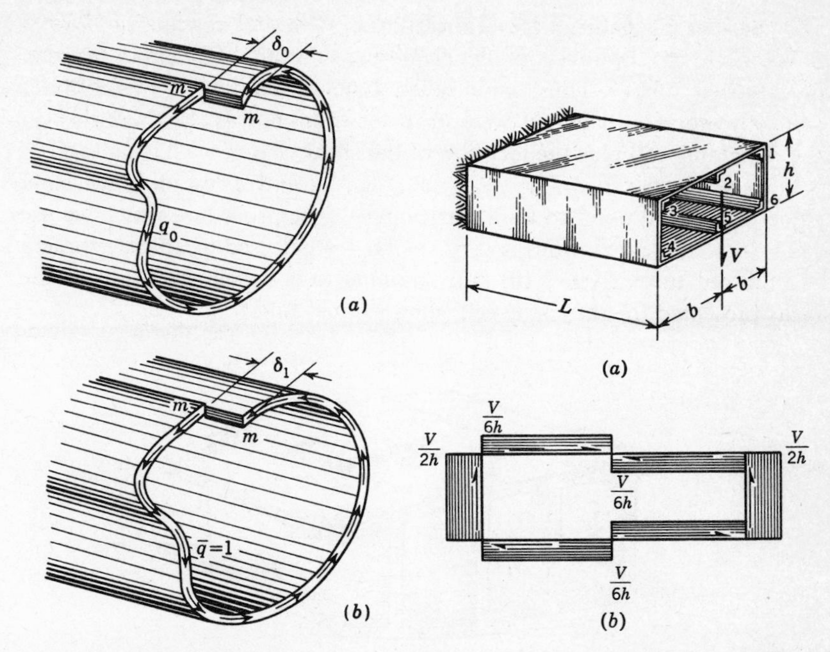

FIGURE 5.23 Physical interpretation of coefficients in Eq. (5.73).

FIGURE 5.24 (*a*) Cantilevered thin-walled closed section obtained by introducing a panel between flanges 1 and 6 of the section shown in Fig. 5.16*a*; (*b*) shear-flow distribution.

between flanges 1 and 6 (see Fig. 5.24*a*), we use Eqs. (*b*) of Art. 5.6 to evaluate δ_0. Assuming the slit m to be below flange 6, we find

$$\delta_0 = \frac{1}{G}\oint q_0 \frac{ds}{t} = -\frac{1}{Gt}\left(\frac{V}{3h}2b + \frac{2V}{3h}2b + \frac{V}{h}h\right)$$

or
$$\delta_0 = -\frac{V}{Gth}(2b + h) \tag{a}$$

Similarly,
$$\delta_1 = \frac{1}{G}\oint \frac{ds}{t} = \frac{2}{Gt}(2b + h) \tag{b}$$

Thus, from Eq. (5.75), the shear flow in the new panel is

$$\bar{q} = -\frac{\delta_0}{\delta_1} = +\frac{V}{2h} \tag{c}$$

which is also obvious from symmetry (the shear center is now at the centroid). The final shear flow is obtained by superimposing the constant flow given in Eq. (*c*) on that shown in Fig. 5.16*c*. The resulting shear flow is shown in Fig. 5.24*b*.

5.9 Bending of multicell thin-walled tubes. One unknown shear flow is introduced with the addition of each closed cell in the problem of bending of multicell tubes. Thus, an n-celled tube in bending is $n - 1$ times statically indeterminate. If we add to these unknowns the shear flow in the remaining cell plus the location of the shear center, we have a total of $n + 1$ unknowns: $\bar{q}_1, \bar{q}_2, \ldots, \bar{q}_i, q_j, \bar{q}_k, \ldots, \bar{q}_n$, and d, the distance from some preselected point to the shear center. The procedure which we use to evaluate these quantities is very similar to that used to analyze torsion of multicell tubes (Art. 3.10) and amounts to a simple extension of the ideas presented for the single-cell (monocoque) tube.

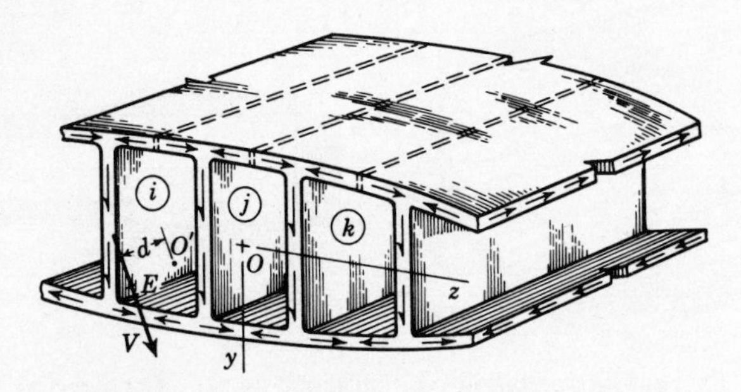

FIGURE 5.25 Typical cells i, j, k of an n-celled tube in bending.

To demonstrate this procedure, let us consider the three typical cells i, j, and k of a n-celled tube shown in Fig. 5.25. The shearing forces V_y and V_z are shown resolved into a resultant shear $V = (V_y^2 + V_z^2)^{\frac{1}{2}}$ acting through the shear center a distance d from some arbitrary reference point O'. Proceeding as before, we introduce a longitudinal slit in each cell, indicated by the dashed lines in the figure, and thereby reduce the structure to a thin-walled open section. Let m_i, m_j, and m_k denote the points at which these slits are introduced in cells i, j, and k, respectively. The location of these points is arbitrary. If continuity is restored at m_j (that is, cell j becomes closed, the remaining cells being open), the shear flow developed in cell j is

$$q_j = q_{j0} + \bar{q}_j \tag{5.78}$$

where q_{j0} is the shear flow at any point in cell j, assuming slits at all points $m_1, m_2, \ldots, m_i, m_j, m_k, \ldots, m_n$, and $\bar{q}_j$ is the constant shear flow at m_j which is required to close the slit at m_j. q_{j0} is given by Eq. (5.20) [or by Eq. (5.47) for semimonocoque tubes]. Thus, the final shear flow can be interpreted as the algebraic sum of the shear flows q_{j0} in the opened tube and n corrective, constant shear flows $\bar{q}_j$ applied independently in each cell. Now the total relative warping displacement at m_j is obviously zero if all cells are closed. Thus, from Eq. (3.53) (with θ zero),

$$\bar{u}_j = 0 = \frac{1}{G}\oint_{s_j} q_j \frac{ds}{t} = \frac{1}{G}\left(-\int_{s_{ji}} \bar{q}_i \frac{ds}{t} + \oint_{s_j} \bar{q}_j \frac{ds}{t} - \int_{s_{jk}} \bar{q}_k \frac{ds}{t} + \oint_{s_j} q_{j0} \frac{ds}{t}\right) \tag{5.79}$$

The integrals containing $\bar{q}_i$ and $\bar{q}_k$ account for the fact that when the positive corrective shear flows (counterclockwise) are introduced to close m_i and m_k in cells i and k, they oppose $\bar{q}_j$ in the webs s_{ji} and s_{jk} common to cells i and j and j and k.

We now denote

$$\delta_{j0} = \frac{1}{G}\oint_{s_j} q_{j0} \frac{ds}{t} \tag{5.80}$$

and

$$\delta_{jj} = \frac{1}{G}\oint_{s_j} \frac{ds}{t}$$

$$\delta_{ji} = -\frac{1}{G}\int_{s_{ij}} \frac{ds}{t} \tag{5.81}$$

$$\delta_{jk} = -\frac{1}{G}\int_{s_{jk}} \frac{ds}{t}$$

so that Eq. (5.79) becomes

$$\delta_{ji}\bar{q}_i + \delta_{jj}\bar{q}_j + \delta_{jk}\bar{q}_k + \delta_{j0} = 0 \tag{5.82}$$

Comparing Eqs. (5.81) with (3.63) and (3.64), we find that the coefficients of the corrective shear flows are identical to the warping flexibilities for the problem of torsion of multicell tubes. This is because both q_j due to torsion and $\bar{q}_j$ are constant shear flows, and we have seen that any constant shear flow q_j produces a relative warping at m_j of $q_j\delta_{jj}$, etc., regardless of the type of loading. It follows that physically δ_{j0}, δ_{ji}, δ_{jj}, and δ_{jk}, respectively, are interpreted as the relative warping displacements of opposite faces of the slit at m_j due to q_{j0}, and unit values of $\bar{q}_i$, $\bar{q}_j$, and $\bar{q}_k$. Finally, for continuity to exist at all n slits,

$$\begin{aligned}
\delta_{11}\bar{q}_1 + \delta_{12}\bar{q}_2 + \delta_{10} &= 0 \\
\delta_{12}\bar{q}_1 + \delta_{22}\bar{q}_2 + \delta_{23}\bar{q}_3 + \delta_{20} &= 0 \\
\cdots\cdots\cdots\cdots\cdots & \\
\delta_{ji}\bar{q}_i + \delta_{jj}\bar{q}_j + \delta_{jk}\bar{q}_k + \delta_{j0} &= 0 \\
\cdots\cdots\cdots\cdots\cdots & \\
\delta_{n,n-1}\bar{q}_{n-1} + \delta_{nn}\bar{q}_n + \delta_{n0} &= 0
\end{aligned} \tag{5.83}$$

Equations (5.83) are n linearly independent *equations of consistent deformation*, which we may solve for the n corrective flows at the hypothetical slits, $m_1, \ldots, m_n$. The final shear flow in cell j is given by Eq.

(5.78). The same general procedure is followed in the case of "layered" tubes (tubes in which a given cell is adjacent to more than two cells), the only difference being that more terms per equation are necessary.

We can now locate the shear center. Let M_0 denote the moment about a longitudinal axis through some arbitrary point due to the determinate shear flows q_{j0}. Referring to Eq. (3.40), we recall the torque developed by a constant shear flow q_j about some point O is $2\Omega_j q_j$, where Ω_j is the sectorial area of cell j. Thus, the total moment developed by the flows $\bar{q}_j$ is $2\sum_{j=1}^{n} \bar{q}_j\Omega_j$. It follows that the requirement of equilibrium of moments

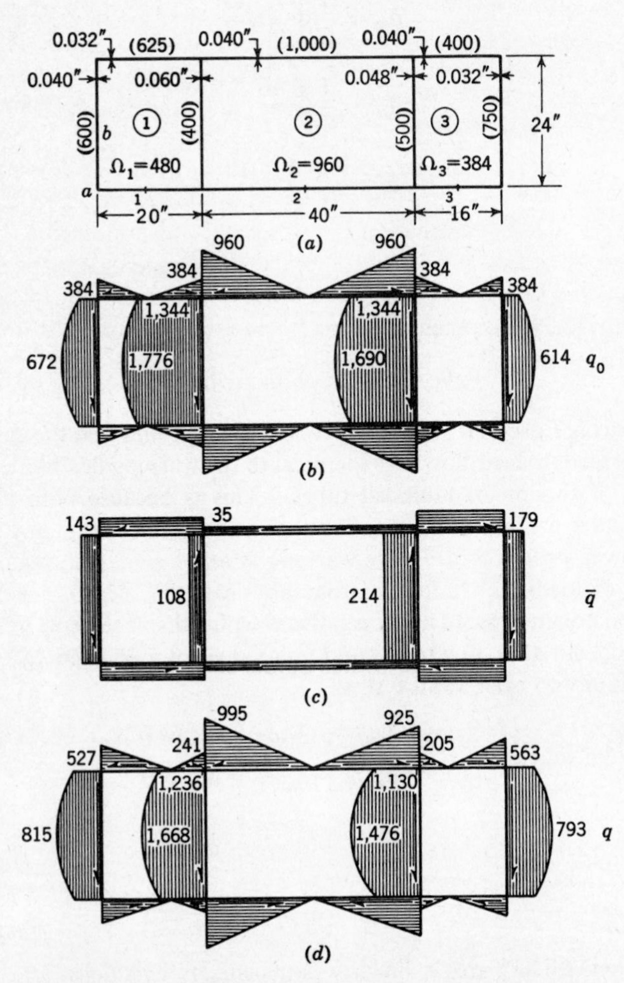

FIGURE 5.26 Numerical example of shear flow in a multicell tube in bending. The shear-flow values shown have units of pounds per inch.

about a longitudinal axis is satisfied provided

$$Vd + M_0 + 2\sum_{j=1}^{n} \bar{q}_j \Omega_j = 0 \tag{5.84}$$

from which the distance d is determined.

To illustrate the procedure of analysis, consider the three-celled tube shown in Fig. 5.26*a*. For simplicity, let the structure be loaded only in the xy plane ($V_z = M_y = 0$) and assume that V_y is such that $V_y/I_z = 100$. Note that the section is symmetrical with respect to the z axis so that $I_{yz} = 0$. The numbers in parentheses along each segment indicate the value of s/t for that segment.

We begin by cutting cells 1, 2, and 3 in order to create an open section. Though the positions of these cuts are arbitrary from a theoretical point of view, they are quite important from a numerical point of view. In the case of a large number of cells a poor choice of the redundant shear flows $\bar{q}$ could result in an "ill-conditioned" system of equations. A large number of significant figures must be used when handling such a system in order to obtain reliable results. For the structure under consideration, this situation is avoided by locating the cuts near the center of the bottom skin. Here both the determinate flows q_{j0} and the redundants $\bar{q}_j$ are small and the final shear flows differ only slightly from that of the determinate structure. The resulting equations of consistent deformation are, therefore, well conditioned.

With this in mind, we cut the structure at the midpoints of the bottom skin (points 1, 2, and 3 in Fig. 5.26*a*) and refer to Eq. (5.20) to calculate q_0:

$$q_0 = \frac{V_y}{I_z} Q_z = 100 Q_z \tag{a}$$

The values of q_0 at points a and b of cell 1, for example, are

$$(q_0)_a = 100 \times 10 \times 0.032 \times 12 = 384 \text{ lb/in.}$$

and

$$(q_0)_b = 384 + 100 \times 12 \times 0.040 \times 6 = 672 \text{ lb/in.}$$

directed as shown. Values at other points calculated in this manner are shown in Fig. 5.26*b*. The directions of q_0 are determined using the sign convention of Eq. (5.20) (positive q_0 acts in the direction of a positive V_y); however, when evaluating δ_{j0}, counterclockwise flows around any cell are taken as positive.

Coefficients δ_{j0} are now calculated by means of Eq. (5.80):

$$\delta_{10} = \frac{1}{G}[(384 + \tfrac{2}{3}288)(600) - (1{,}344 + \tfrac{2}{3}432)(400)] = -\frac{307{,}200}{G}$$

$$\delta_{20} = \frac{1}{G}[(1{,}344 + \tfrac{2}{3}432)(400) - (1{,}344 + \tfrac{2}{3}346)(500)] = -\frac{134{,}400}{G}$$

$$\delta_{30} = \frac{1}{G}[(1{,}344 + \tfrac{2}{3}346)(500) - (384 + \tfrac{2}{3}230)(750)] = \frac{384{,}000}{G}$$

From Eqs. (5.81),

$$\delta_{11} = \frac{1}{G}[600 + 2(625) + 400] = \frac{2{,}250}{G}$$

$$\delta_{22} = \frac{1}{G}[400 + 2(1{,}000) + 500] = \frac{2{,}900}{G}$$

$$\delta_{33} = \frac{1}{G}[500 + 2(400) + 750] = \frac{2{,}050}{G}$$

$$\delta_{12} = \delta_{21} = -\frac{400}{G}$$

and

$$\delta_{23} = \delta_{32} = -\frac{500}{G}$$

Thus, the equations of consistent deformation [Eqs. (5.83)] become (after eliminating the common factor $1/G$)

$$\begin{aligned} 2{,}250\bar{q}_1 - 400\bar{q}_2 - 307{,}200 &= 0 \\ -400\bar{q}_1 + 2{,}900\bar{q}_2 - 500\bar{q}_3 - 134{,}400 &= 0 \\ -500\bar{q}_2 + 2{,}050\bar{q}_3 + 384{,}000 &= 0 \end{aligned}$$

Solving these equations, we find

$$\bar{q}_1 = 143 \text{ lb/in.} \qquad \bar{q}_2 = 35 \text{ lb/in.} \qquad \bar{q}_3 = -179 \text{ lb/in.}$$

The shear-flow distribution due to these is shown in Fig. 5.26*c*. The final shear-flow distribution, shown in Fig. 5.26*d*, is obtained by superimposing the distributions in Fig. 5.26*b* and *c*.

The sectorial areas of each cell, in square inches, are indicated in Fig. 5.26*a*, and I_z for the section may be shown to be 1,036.8 in.⁴, so that $V = V_y = 103{,}680$ lb. The moment M_0 produced by the shear flows q_{j0} about the centroid of cell 2 is

$$\begin{aligned} M_0 &= +(384 + \tfrac{2}{3}288)(24)(40) + (1{,}344 + \tfrac{2}{3}432)(24)(20) \\ &\qquad -(1{,}344 + \tfrac{2}{3}346)(24)(20) - (384 + \tfrac{2}{3}230)(24)(36) \\ &= 116{,}224 \text{ in.-lb} \end{aligned}$$

The shear flows $\bar{q}_j$ produce a moment of

$$2\sum_{j=1}^{3}\bar{q}_j\Omega_j = 2(143 \times 480 + 35 \times 960 - 179 \times 384) = 67{,}008 \text{ in.-lb}$$

Thus, using Eq. (5.84), we find

$$103{,}680d + 116{,}224 + 67{,}008 = 0$$

or

$$d = -1.77 \text{ in.}$$

The minus sign indicates that the shear center is 1.77 in. to the right of the center of cell 2.

5.10 Combined bending and torsion of multicell tubes. It is a simple matter to extend the ideas covered previously to apply to the problem of combined bending and torsion of multicell thin-walled tubes, unrestrained against warping. To do this, we regard the total shear flow developed at any point as the sum of two components: q_b, the shear flow due to bending, and q_t, the shear flow due to torsion. According to Eq. (5.78), the bending shear flow in cell j may likewise be considered to be the sum of a constant shear flow $\bar{q}_j$ and the flow in the open tube q_{j0}. Thus, if q_j is the total shear flow in cell j,

$$q_j = [(q_t)_j + \bar{q}_j] + q_{j0} \tag{5.85}$$

Both components inside the brackets of this equation are constants; hence, by introducing the notation

$$\tilde{q}_j = (q_t)_j + \bar{q}_j$$

Eq. (5.85) can be written

$$q_j = \tilde{q}_j + q_{j0} \tag{5.86}$$

This means that the final shear flow in any cell is the sum of the bending shear flow in the opened tube and a constant flow $\tilde{q}_j$. Therefore, once $\tilde{q}_j$ has been determined the final flows are evaluated using Eq. (5.86) and it is never necessary to distinguish between $(q_t)_j$ and $\bar{q}_j$ since they are both constants.

Proceeding as before, we introduce slits in each cell at points $m_1, m_2, \ldots, m_n$ and use Eq. (3.53) to evaluate the relative warping displacements of opposite sides of the slit. In this case, however, the rate of twist θ is not zero. Thus

$$\bar{u}_j = \frac{1}{G}\oint_{s_j} q\frac{ds}{t} - 2\Omega_j\theta \tag{5.87}$$

Clearly, $-2\Omega_j\theta$ is the relative warping due to a pure twist of the tube. We note that $\bar{u}_j$ is zero for the closed cell and that

$$\frac{1}{G}\oint_{s_j} q\frac{ds}{t} = -\tilde{q}_i\frac{1}{G}\int_{s_{ji}}\frac{ds}{t} + \tilde{q}_j\frac{1}{G}\oint_{s_j}\frac{ds}{t} - \tilde{q}_k\frac{1}{G}\int_{s_{jk}}\frac{ds}{t} + \frac{1}{G}\oint_{s_j} q_{j0}\frac{ds}{t}$$

where $\tilde{q}_j$ and $\tilde{q}_k$ are the corrective shear flows in adjacent cells i and k, respectively. Or, using the notation of Eqs. (5.80) and (5.81),

$$\frac{1}{G}\oint_{s_j} q\frac{ds}{t} = \tilde{q}_i\delta_{ji} + \tilde{q}_j\delta_{jj} + \tilde{q}_k\delta_{jk} + \delta_{j0}$$

Therefore, Eq. (5.87) becomes

$$\tilde{q}_i\delta_{ji} + \tilde{q}_j\delta_{jj} + \tilde{q}_k\delta_{jk} + \delta_{j0} - 2\Omega_j\theta = 0 \tag{5.88}$$

Equation (5.88) is the equation of consistent deformation for cell j. As before, one such equation is written for each cell and the resulting set is solved for the corrective shear flows $\tilde{q}_j$ in terms of θ. To evaluate θ we sum moments about an arbitrary longitudinal axis normal to the cross section at some point O'. If T is the applied twisting moment on the section and d is the perpendicular distance from O' to the line of action of V, the resultant shear, then the total external moment in the plane of the cross section about O' is

$$M_t = Vd + T$$

Thus, equilibrium of moments about O' is satisfied provided

$$M_t + M_0 + 2 \sum_{j=1}^{n} \tilde{q}_j \Omega_j = 0 \tag{5.89}$$

where M_0 is the moment due to the q_{j0} shear flows. We may now evaluate θ by introducing the values of $\tilde{q}$ from Eqs. (5.88) into Eq. (5.89). Again, final shear flows are obtained by the superposition indicated in Eq. (5.86).

We find that in this case it is not necessary to determine the location of the shear center. Its location can be evaluated, however, by noting that solutions to Eqs. (5.88) are of the form

$$\tilde{q}_j = \bar{q}_j + K_j\theta = \bar{q}_j + (q_t)_j \tag{5.90}$$

where K_j is some constant. The distance d can be obtained by introducing the $\bar{q}_j$ components into Eq. (5.84).

PROBLEMS

5.1. Compute the variation in τ_{sy} and σ_y over section AA of the beam shown. Compare the magnitudes of $(\sigma_y)_{\max}$ and $(\sigma_s)_{\max}$ for this section.

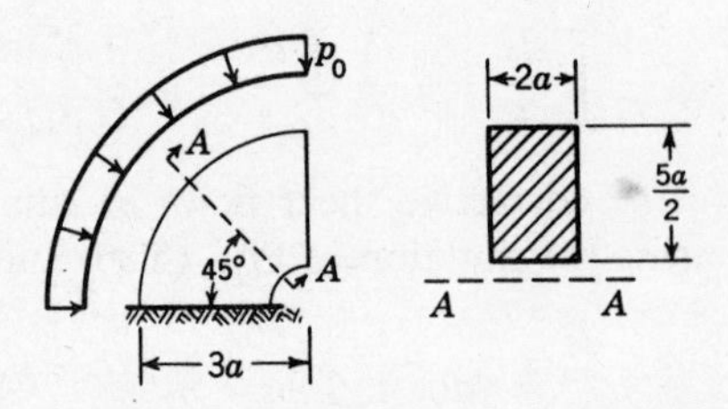

5.2. Compute the shearing stress and plot its variation at section B of the beam in Prob. 4.2.

5.3. Evaluate the maximum transverse normal stress σ_y developed in the beam shown in Fig. 4.11. Compare this result with $(\sigma_x)_{\max}$ given in Fig. 4.11b.

5.4. The unsymmetrical straight beam shown is subjected to a uniformly varying transverse load $p_y = p_0(1 - x/L)$, which passes through its shear center. The wall thickness t is assumed to be small in comparison with the other dimensions.

(*a*) Evaluate the shear flow due to bending at the fixed end of the beam in terms of I_y, I_z, and I_{yz} and plot its variation along the contour of the section.

(*b*) Evaluate the maximum transverse normal stress σ_y at $x = 0$ and at $x = L/2$.

(*c*) Evaluate $(\sigma_x)_{\max}$ and compare this value with $(\sigma_y)_{\max}$ of part (*b*) taking $t = 0.25$ in., $a = 4.0$ in., $L = 100$ in., and $p_0 = 100$ lb/in.

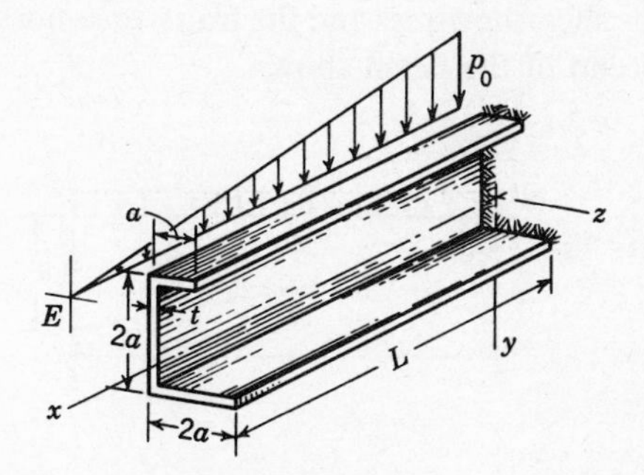

5.5. The depth of the beam shown varies as a second-degree parabola. Compute the shearing-stress and the transverse-normal-stress distributions at the fixed end assuming $h_0 = 10.0$ in., $L = 120.0$ in., $b = 3.0$ in., and $P = 20{,}000$ lb.

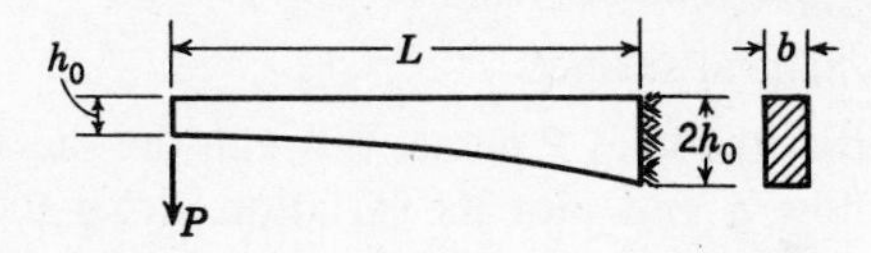

5.6. The tapered beam shown is composed of a thin shear-carrying skin and two flanges which develop the normal stresses. Owing to the inclination of the flanges, components of the flange forces aid in resisting the shear V_y. Neglect shear lag.

(*a*) Show that the flange-force contribution to the transverse shear is

$$V_f = P\,\frac{h_x - h_0}{h_x}$$

where h_0 is the depth at $x = 0$ and h_x is the depth at $x = x$.

(*b*) Show that the shear flow in the web is

$$q = \frac{Ph_0}{h_x^{\,2}}$$

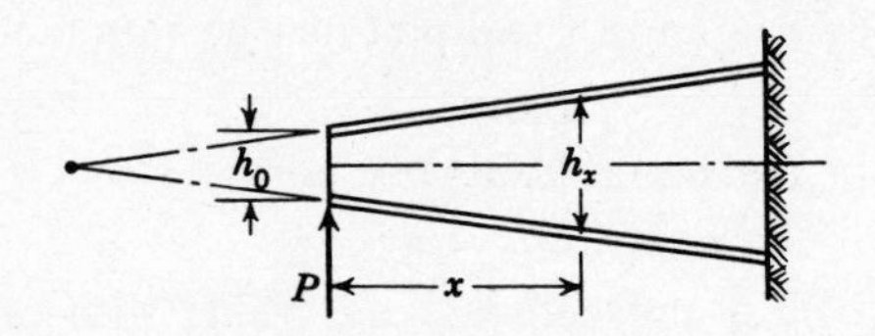

5.7. Compute the shearing-stress and the transverse-normal-stress distribution at the fixed end of the beam shown.

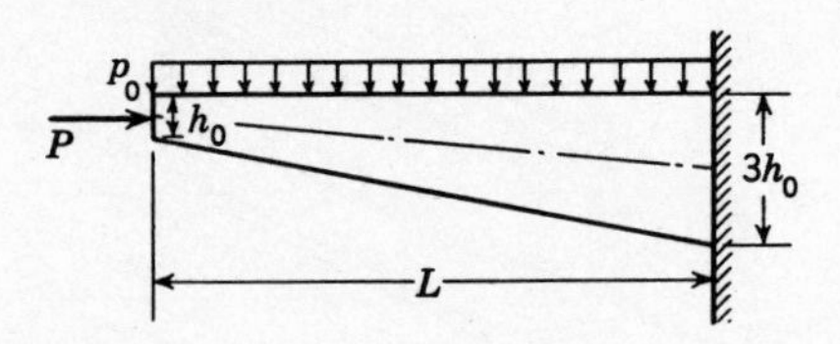

5.8. The cantilevered beam shown has the following cross-sectional properties:

$$I_y = 7.67 \text{ in.}^4 \qquad I_z = 143.0 \text{ in.}^4 \qquad I_{yz} = -13.30 \text{ in.}^4$$

(*a*) Locate the shear center of the cross section.

(*b*) Locate the neutral surface.

(*c*) Assuming that the load P passes through the shear center, compute the shear flow q and plot its variation along the contour of the section.

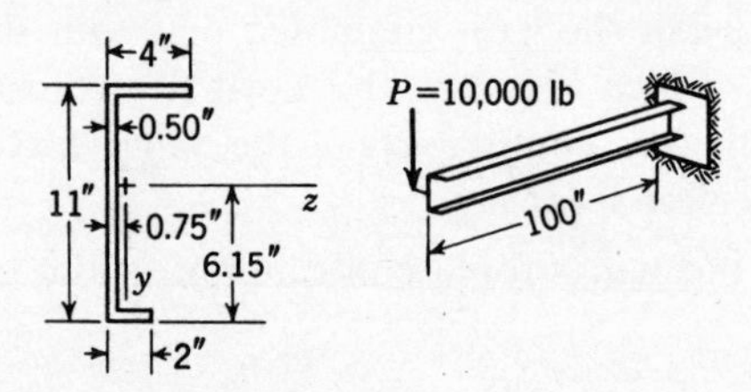

5.9. Locate the shear center of the section shown.

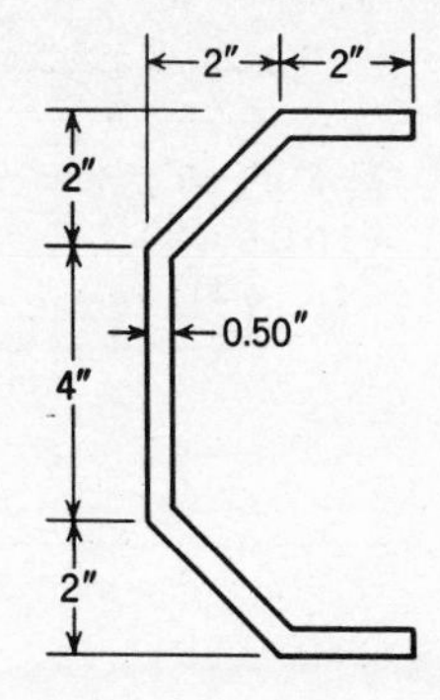

5.10. Show that the shear center of the thin-walled circular section below is located a distance

$$e = 4r\,\frac{\sin(\theta/2) - (\theta/2)\cos(\theta/2)}{\theta - \sin\theta}$$

from the center of curvature.

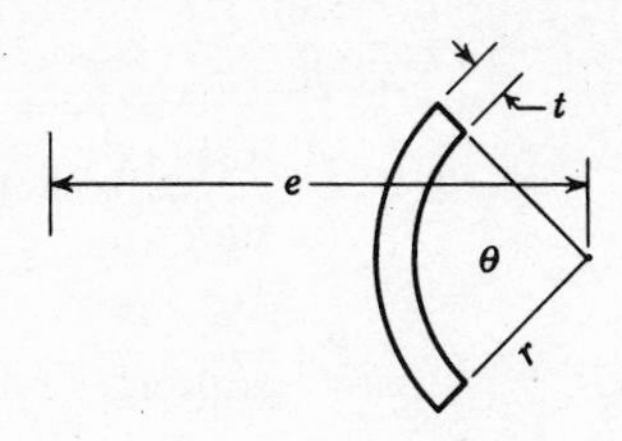

5.11–5.14. Locate the shear centers of the thin-walled sections shown and plot the shear flows due to bending. The concentrated flanges develop only normal stresses, and the skins (webs) carry the shear flow. All skins are of constant thickness and the areas of all flanges are 0.5 sq in.

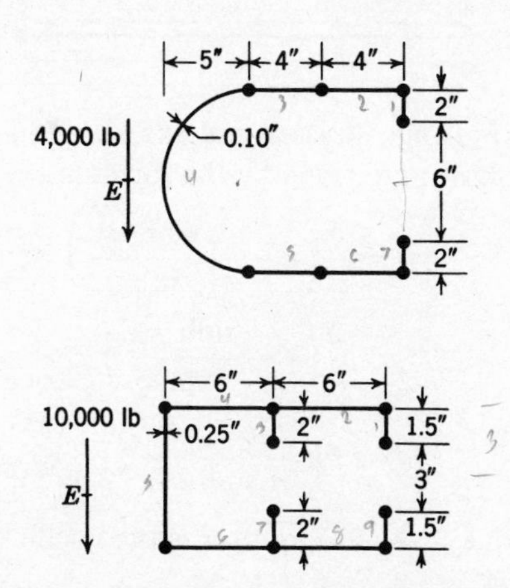

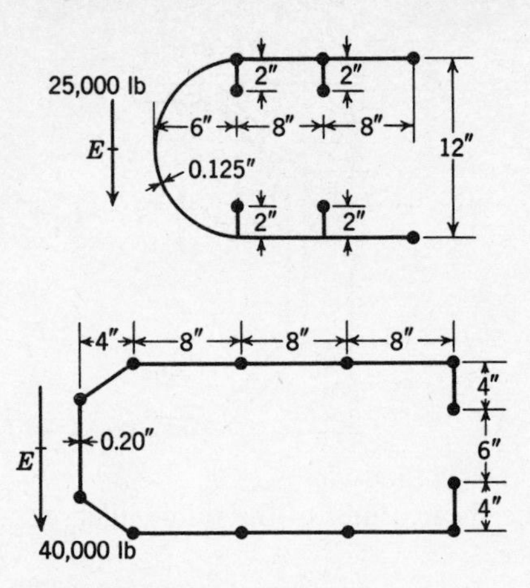

5.15. Show that the flange stresses and the shear flow developed in the thin-walled panel shown are given by the formulas

$$\sigma_1 = \frac{P}{a_1} \frac{\sinh \mu x'}{\sinh \mu L}$$

$$\sigma_2 = -\frac{P}{a_2} \frac{\cosh \mu x'}{\sinh \mu L} + 1 - \coth \mu L$$

$$q = -P\mu \frac{\cosh \mu x'}{\sinh \mu L}$$

where $x' = L - x$ and μ is the coefficient of diffusion defined in Eq. (5.56). Use the condition $N_1 = 0$ at $x = L$.

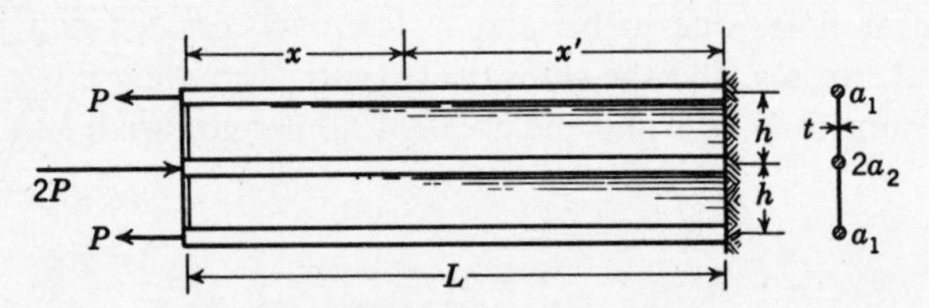

5.16. Show that the flange stresses and the shear flow developed in the thin-walled panel shown are given by the formulas

$$\sigma_1 = \frac{P}{A}\left(1 + \frac{a_2}{a_1} \frac{\sinh \mu x'}{\sinh \mu L}\right)$$

$$\sigma_2 = \frac{P}{A}\left(1 - \frac{\sinh \mu x'}{\sinh \mu L}\right)$$

$$q = P\mu \frac{a_2}{A} \frac{\cosh \mu x'}{\sinh \mu L}$$

where $x' = L - x$, $A = a_1 + a_2$, and μ is the coefficient of diffusion. Use the condition $N_1 = N_2$ at $x = L$.

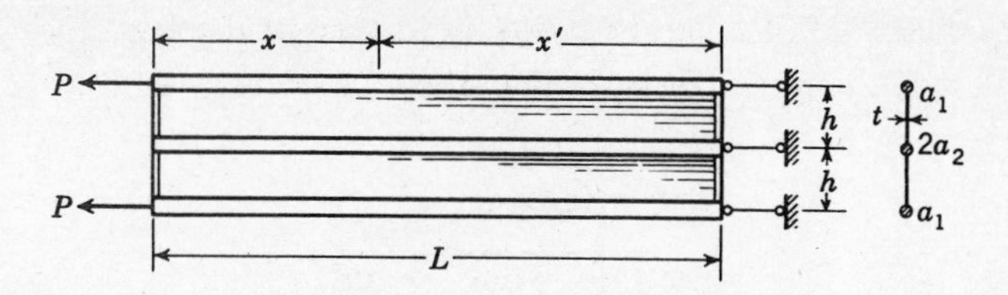

5.17. Compute the shear flow and the flange stresses in the four-stringer panel shown at the following sections: $x = 0$, 10.0, 20.0, 30.0, and 40.0 in. All stringer areas are 1.0 sq in., and the thickness of each panel is 0.05 in.

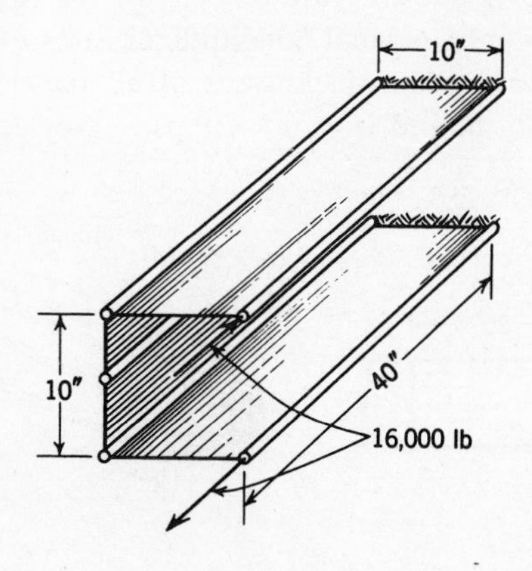

5.18. The transverse load acting on the tube shown produces no twisting. Evaluate the shear flow due to bending at the fixed end and plot its variation along the contour of the section. Then locate the shear center of the section.

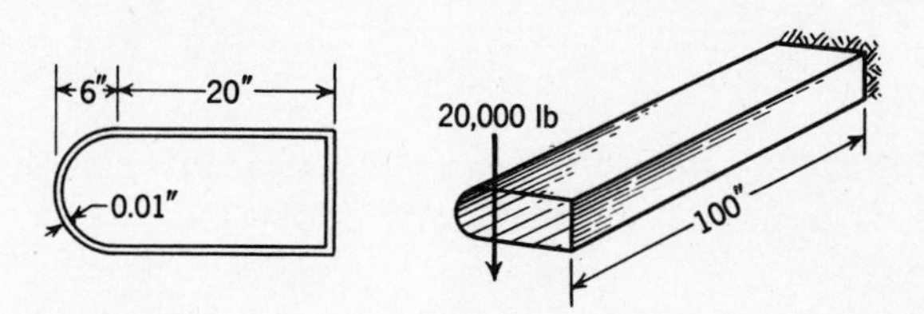

5.19. The cross section of a two-celled cantilevered tube is shown. The free end of the tube is subjected to a transverse force of 10,000 lb applied at the corner indicated. The length of the tube is 80.0 in. and $G = 5.4 \times 10^6$ psi. Warping effects are to be neglected.

(*a*) Compute and plot the normal stress profile at the fixed end.

(*b*) Compute and plot the shear flow developed at the fixed end.

(*c*) Evaluate the rate of twist of the tube.

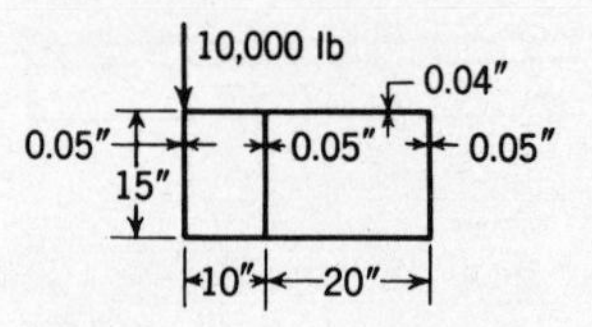

5.20. Compute the flange stresses and the shear flows at the fixed end of the tube shown. Also evaluate the rate of twist of the tube. Assume that the flanges develop only normal longitudinal stresses and the panels develop the shear flow. The thicknesses of all panels are 0.05 in.; all flange areas are 1.0 sq in. and $G = 6 \times 10^6$ psi. Neglect warping effects.

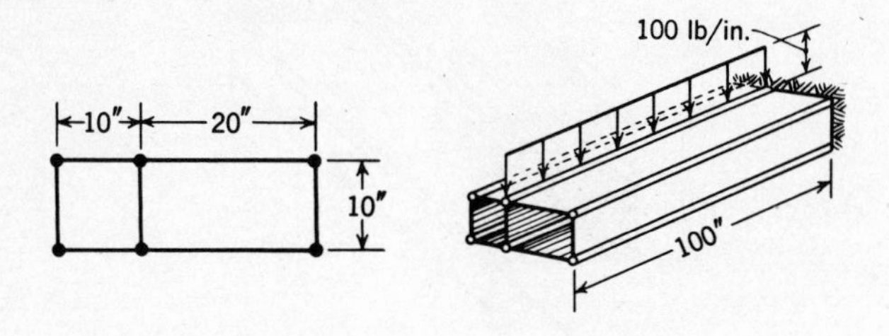

six

THE ELASTIC CURVE

6.1 Introduction. In developing the formulas for stresses given in the preceding chapters, we usually found it necessary to consider only the deformation of representative elements of the bar such as the displacement of a typical plane section. We did not consider the shape of the elastic curve; in fact, the boundary conditions of the bar seldom entered the problem. At the beginning of each derivation the stress resultants on any section were assumed to be known quantities obtained, perhaps, through the use of the equations of statics.

The manner in which a structure deforms under loads, however, is an essential characteristic of its behavior; hence, our task of describing structural behavior is far from finished. Moreover, the equations of statics are not always sufficient to determine the stress resultants on every section, and in any indeterminate system it is absolutely necessary to consider deformations of the structure. For these reasons we study in this chapter the deflections of coplanar bars. Force and moment-deflection relationships are established so that we can evaluate both stresses and deformations once the stress resultants are known.

6.2 Deflections of a curved bar. Let us begin our investigation by considering the small deformations of an elastic coplanar curved bar. Any point on a typical cross section of such a bar is displaced from its unstrained position through small components of displacement u, v, and w to its deformed configuration. We assume, as in Chap 2, that u, v, and w are continuous single-valued functions of the coordinates and, in the case of a curved bar, we let u and v represent tangential and radial displacements of points on the cross sections and w represent the displacement in a direction normal to the plane of the bar (the z direction). We continue to adopt the hypothesis that plane cross sections remain plane after deformation and that the bar material is linearly elastic, homogeneous, isotropic, and continuous. This allows us to take advantage of the principle of superposition and to study each of these displacement components separately. The absolute displacement of a point is, of course, the vector sum of the three independent components.

As before, the extensional strain is given by

$$\epsilon_s = \frac{\partial u}{\partial s_y}$$

but owing to our assumption that plane sections remain plane after deformation, the displacement u can be obtained as a function of derivatives of v and w through geometric considerations. In fact, the complete deformation of a bar element can be considered to be due to three types of deformation: u_0, a displacement of points on the cross section due to a stretching of the centroidal axis; u_1, a shortening of the fibers due to a decrease in the radius of curvature R by an amount v; and u_2, a normal displacement due to rotations of the cross-sectional plane about the y and z axes. These displacements are illustrated in Fig. 6.1.

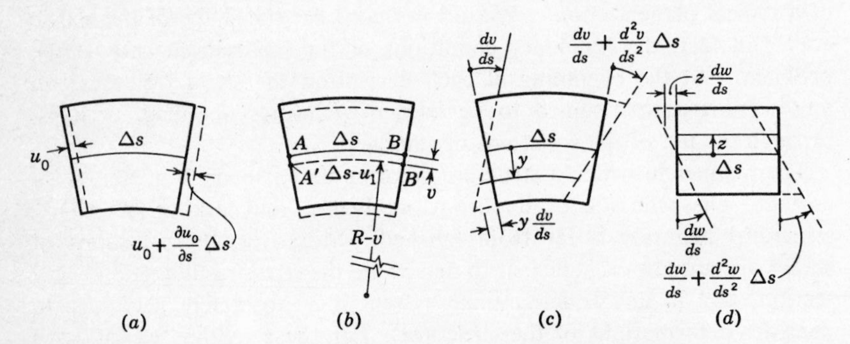

FIGURE 6.1 Deformations of a curved bar element. Uniform normal (*a*) and uniform radial (*b*) displacements of cross sections plus rotations of the cross-sectional plane about the y axis (*c*) and about the z axis (*d*).

When the centroid undergoes the tangential displacement u_0 fibers below the centroid are displaced $u_0(1 - y/R)$ owing to the initial curvature of the bar. Therefore, the extensional strain due to this displacement is

$$(\epsilon_s)_0 = \left(1 - \frac{y}{R}\right)\frac{du_0}{ds}\frac{ds}{ds_y}$$

Recalling that $ds/ds_y = 1/(1 - y/R)$, we find that

$$(\epsilon_s)_0 = \frac{du_0}{ds} \tag{6.1}$$

To visualize the strain due to u_1, consider the deformation of the element in Fig. 6.1*b*. Points A and B on the two faces of the element move through displacements of equal magnitude to positions A' and B'. This develops a strain in a fiber a distance y below the centroid of

$$(\epsilon_s)_1 = \lim_{\Delta s \to 0} \frac{\Delta s[1 - y/(R - v)] - \Delta s(1 - y/R)}{\Delta s(1 - y/R)}$$

$$= \frac{-y}{1 - y/R}\left(\frac{1}{R - v} - \frac{1}{R}\right) - \frac{v}{R}$$

The terms inside the final parentheses clearly represent a change in curvature due to v. Since v is very small, this expression may be simplified as follows:

$$\frac{1}{R - v} - \frac{1}{R} = \frac{v}{R^2(1 - v/R)} \cong \frac{v}{R^2}$$

Thus

$$(\epsilon_s)_1 = \frac{-y}{1 - y/R}\frac{v}{R^2} - \frac{v}{R} \tag{6.2}$$

The final portion of the strain is also due to a change in curvature. Sections rotate owing to a change in v and w with respect to s. From the geometry of Fig. 6.1*c* and *d*, we see that the rotations of a cross-sectional plane about its centroidal axes produce displacements,

$$u_2 = -y\frac{dv}{ds} - z\frac{dw}{ds}$$

The resulting strain, therefore, is

$$(\epsilon_s)_2 = \frac{\partial u_2}{\partial s}\frac{ds}{ds_y} = \frac{1}{1 - y/R}\left(-y\frac{d^2v}{ds^2} - z\frac{d^2w}{ds^2}\right) \tag{6.3}$$

We obtain the total extensional strain by simple addition:

$$\epsilon_s = (\epsilon_s)_0 + (\epsilon_s)_1 + (\epsilon_s)_2$$

Substituting Eqs. (6.1) to (6.3) into the above expression and comparing the result with that obtained directly from Eq. (4.16), we arrive at the equality

$$\frac{du_0}{ds} - \frac{V}{R} - \left(\frac{d^2v}{ds^2} + \frac{v}{R^2}\right)\frac{y}{1 - y/R} - \frac{d^2w}{ds^2}\frac{z}{1 - y/R} = \frac{\sigma_x}{E}$$

$$= \frac{N_s}{AE} - \frac{M_z}{RAE} + \frac{M_zJ_y - M_yJ_{yz}}{E(J_yJ_z - J_{yz}^2)}\frac{y}{1 - y/R} + \frac{M_yJ_z - M_zJ_{yz}}{E(J_yJ_z - J_{yz}^2)}\frac{z}{1 - y/R}$$

The quantities J_y, J_z, and J_{yz} are the section constants defined in Eqs. (4.14). Finally, equating like coefficients of unit $y/(1 - y/R)$ and $z/(1 - y/R)$ on each side of this equation, we obtain the differential equations of the elastic curve:

$$\begin{aligned} \frac{du_0}{ds} - \frac{V}{R} &= \frac{N_s}{AE} - \frac{M_z}{RAE} \\ \frac{d^2v}{ds^2} + \frac{v}{R^2} &= -\frac{M_zJ_y - M_yJ_{yz}}{E(J_yJ_z - J_{yz}^2)} \\ \frac{d^2w}{ds^2} &= -\frac{M_yJ_z - M_zJ_{yz}}{E(J_yJ_z - J_{yz}^2)} \end{aligned} \tag{6.4}$$

Equations (6.4) give the displacements in terms of the stress resultants. In effect, they represent a set of compatibility conditions based on our initial assumptions concerning the bar's behavior. To solve these equations, it is necessary that we first obtain N_s, M_y, and M_z as functions of the coordinate s. The solutions to Eqs. (6.4) then give the locus of points of the centroidal axis of the bar after deformation.

6.3 Alternative forms of the equations of the elastic curve. The last two of Eqs. (6.4) can be replaced by equations in terms of the loads rather than moments. Solving these two equations for the bending moments, we find

$$\begin{aligned} M_y &= -E\left[J_{yz}\left(\frac{d^2v}{ds^2} + \frac{v}{R^2}\right) + J_y\frac{d^2w}{ds^2}\right] \\ M_z &= -E\left[J_z\left(\frac{d^2v}{ds^2} + \frac{v}{R^2}\right) + J_{yz}\frac{d^2w}{ds^2}\right] \end{aligned} \tag{6.5}$$

Note that owing to the lack of symmetry of the cross section a change in curvature in the xy plane produces bending moments in the xz plane and vice versa.

We can also write the first of Eqs. (6.4) in the form

$$N_s^* = AE\left(\frac{du_0}{ds} - \frac{V}{R}\right) \tag{6.6}$$

where N_s^* is the equivalent axial force $(N_s - M_z/R)$.

The forms of Eqs. (6.5) and (6.6) are significant. They relate each stress resultant to a characteristic deformation—M_y and M_z to changes in curvature and N_s to an elongation of the bar's axis. The quantities EJ_y, EJ_z, and EJ_{yz} are called *elemental flexural stiffnesses*, or more often, *flexural rigidities* of the bar, and EA is called the *elemental axial stiffness* (or the *axial rigidity*) of the bar. The reciprocals of the elemental stiffnesses are called *elemental flexibilities.*

Finally, introducing Eqs. (6.5) into Eqs. (4.6) and (4.8), we obtain the alternative form of the differential equations of the elastic curve:

$$
\begin{gathered}
-E\frac{d^2}{ds^2}\left[J_{yz}\left(\frac{d^2v}{ds^2}+\frac{v}{R^2}\right)+J_y\frac{d^2w}{ds^2}\right]+p_z=0 \\
-E\left(\frac{d^3}{ds^3}+\frac{R'}{R}\frac{d^2}{ds^2}+\frac{1}{R^2}\frac{d}{ds}\right)\left[J_z\left(\frac{d^2v}{ds^2}+\frac{v}{R^2}\right)+J_{yz}\frac{d^2w}{ds^2}\right]+\frac{dp_y}{ds}+\frac{R'}{R}p_y=0
\end{gathered}
\tag{6.7}
$$

where $R' = dR/ds$.

If R is constant (that is, if the bar is circular), the terms containing R' obviously vanish in the above equations. Further, if neither R nor the cross-sectional dimensions vary with s, Eqs. (6.7) reduce to

$$
\begin{gathered}
-EJ_{yz}\left(\frac{d^4v}{ds^4}+\frac{1}{R^2}\frac{d^2v}{ds^2}\right)-EJ_y\frac{d^4w}{ds^4}+p_z=0 \\
-EJ_z\left(\frac{d^5v}{ds^5}+\frac{2}{R^2}\frac{d^3v}{ds^3}+\frac{1}{R^4}\frac{dv}{ds}\right)-EJ_{yz}\left(\frac{d^5w}{ds^5}+\frac{1}{R^2}\frac{d^3w}{ds^3}\right)+\frac{dp_y}{ds}=0
\end{gathered}
\tag{6.8}
$$

If the cross section is symmetrical with respect to the y axis, J_{yz} is zero and Eqs. (6.7) become

$$
\begin{gathered}
-E\frac{d^2}{ds^2}J_y\frac{d^2w}{ds^2}+p_z=0 \\
-E\left(\frac{d^3}{ds^3}+\frac{R'}{R}\frac{d^2}{ds^2}+\frac{1}{R^2}\frac{d}{ds}\right)\left[J_z\left(\frac{d^2v}{ds^2}+\frac{v}{R^2}\right)\right]+\frac{dp_y}{ds}+\frac{R'}{R}p_y=0
\end{gathered}
\tag{6.9}
$$

Clearly, we can introduce a number of additional modifications for various special cases.

Solutions to any of the above differential equations must satisfy a number of independent static and/or kinematic boundary conditions equal to the sum of the orders of the equations. Five boundary conditions, for example, are needed to specify unique solutions to Eqs. (6.4)—one involving u_0 at a boundary for the first equation and two involving v and dv/ds and w and dw/ds, respectively, for each of the remaining equations. In this case, the conditions are necessarily kinematic conditions since the functions on the right side of Eqs. (6.4) are determined from statics.

If Eqs. (6.6) and (6.7) are used, 10 conditions are required—one for Eq. (6.6), four for the first of Eqs. (6.7), and five for the remaining equation.

Consider, for example, a cantilevered curved bar which is free at $s = 0$ and fixed at $s = L$ (Fig. 6.2). According to Eqs. (4.3), (4.5), and (6.5), at the free end we have the static boundary conditions:

$$M_y = -E\left[J_{yz}\left(\frac{d^2v}{ds^2} + \frac{v}{R^2}\right) + J_y\frac{d^2w}{ds^2}\right]_{s=0} = 0$$

$$M_z = -E\left[J_z\left(\frac{d^2v}{ds^2} + \frac{v}{R^2}\right) + J_{yz}\frac{d^2w}{ds^2}\right]_{s=0} = 0$$

$$V_z = \frac{dM_y}{ds} = -E\frac{d}{ds}\left[J_{yz}\left(\frac{d^2v}{ds^2} + \frac{v}{R^2}\right) + J_y\frac{d^2w}{ds^2}\right]_{s=0} = 0$$

$$V_y = \frac{dM_z}{ds} = -E\frac{d}{ds}\left[J_z\left(\frac{d^2v}{ds^2} + \frac{v}{R^2}\right) + J_{yz}\frac{d^2w}{ds^2}\right]_{s=0} = 0 \tag{6.10}$$

$$\left(p_y + \int \frac{R'}{R} p_y \, ds\right)_{s=0} = E\left(\frac{d^2}{ds^2} + \frac{R'}{R}\frac{d}{ds} + \frac{1}{R^2}\right) \times \left[J_z\left(\frac{d^2v}{ds^2} + \frac{v}{R^2}\right) + J_{yz}\frac{d^2w}{ds^2}\right]_{s=0} = 0$$

The last condition is obtained by integrating the second of Eqs. (6.7) once. Similarly, at the fixed end we have the five kinematic boundary conditions:

$$(u_0)_{s=L} = (v)_{s=L} = (w)_{s=L} = \left(\frac{dv}{ds}\right)_{s=L} = \left(\frac{dw}{ds}\right)_{=L} = 0 \tag{6.11}$$

If Eqs. (6.4) are used instead of Eqs. (6.6) and (6.7), only the conditions in Eqs. (6.11) need be considered; Eqs. (6.10) are automatically satisfied.

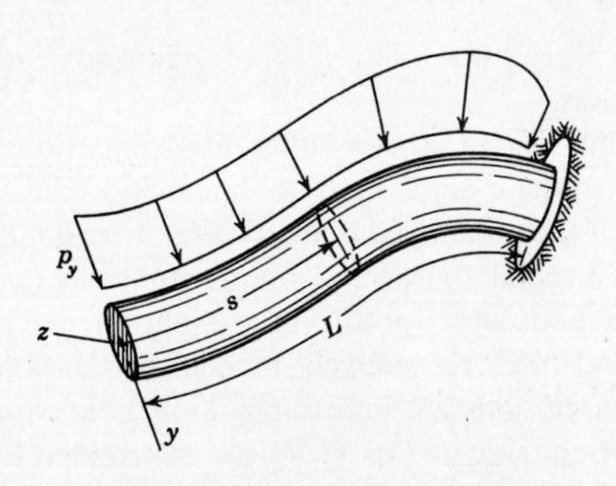

FIGURE 6.2 Cantilevered curved bar.

6.4 Deflections of a symmetrical curved bar. To illustrate the application of Eqs. (6.4), let us consider the simple example of a cantilevered curved bar of constant radius R with a circular cross section (Fig. 6.3a). The axis of the bar describes a quarter circle, as is indicated in the figure, and end A is completely constrained against displacement. The problem is to compute the elastic curve for the loading shown.

Since the section is symmetrical, $J_{yz} = 0$. Also, $M_y = w = 0$ for this case because the load P is applied in the $x\bar{y}$ plane. Thus, Eqs. (6.4) reduce to

$$\frac{du_0}{ds} - \frac{V}{R} = \frac{N_s}{AE} - \frac{M_z}{RAE} \qquad (a)$$

$$\frac{d^2v}{ds^2} + \frac{v}{R^2} = -\frac{M_z}{EJ_z}$$

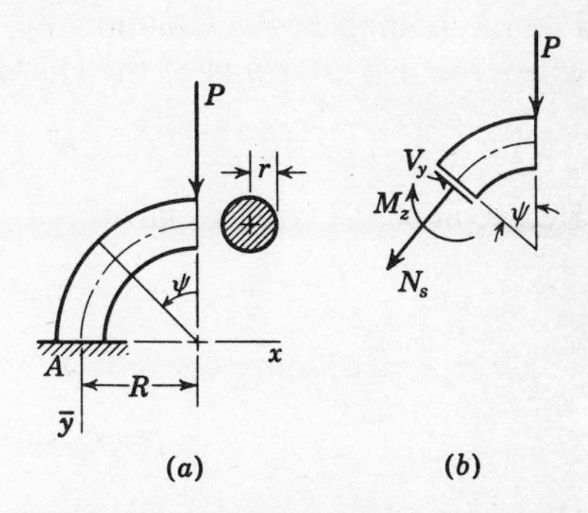

FIGURE 6.3 (a) Curved bar of constant radius R with a circular cross section; (b) segment of the bar.

Referring to the free body of a segment of the bar shown in Fig. 6.3b, we find from statics that

$$\begin{aligned} N_s &= -P \sin \psi \\ M_z &= -PR \sin \psi \end{aligned} \qquad (b)$$

where ψ is the angular coordinate indicated ($s = R\psi$). Hence,

$$\frac{1}{R}\frac{du_0}{d\psi} = \frac{V}{R} \qquad (c)$$

and

$$\frac{1}{R^2}\frac{d^2v}{d\psi^2} + \frac{v}{R^2} = \frac{PR}{EJ_z} \sin \psi \qquad (d)$$

The displacement u_0 is obtained from Eq. (c) by simple integration once V is determined from Eq. (d). The solution of Eq. (d) is more involved. This is a second-order ordinary nonhomogeneous differential equation with constant coefficients. We recall that the solution to such an equation is the sum of two functions v_c, the complementary solution to the homogeneous equation,

$$\frac{d^2v_c}{d\psi^2} + v_c = 0 \qquad (e)$$

and any particular solution v_p to

$$\frac{d^2v_p}{d\psi^2} + v_p = \frac{PR^3}{EJ_z} \sin \psi \qquad (f)$$

We may easily verify by direct substitution that a particular solution to Eq. (*d*) is

$$v_p = -\frac{PR^3}{2EJ_z}\psi\cos\psi \tag{g}$$

For the homogeneous equation we assume a solution of the form $Ae^{m\psi}$, where the m's are roots of the characteristic equation

$$m^2 + 1 = 0$$

Hence, $m = \pm i$, and we have

$$v_c = A_1e^{i\psi} + A_2e^{-i\psi}$$

which is written more conveniently as

$$v_c = C_0 \sin\psi + C_1\cos\psi \tag{h}$$

Here C_0 and C_1 are arbitrary constants.

Adding Eqs. (*g*) and (*h*), we obtain the complete solution to the differential equation

$$v = C_0\sin\psi + C_1\cos\psi - \frac{PR^3}{2EJ_z}\psi\cos\psi \tag{i}$$

We determine C_0 and C_1 from the two kinematic boundary conditions that the slope dv/ds and the displacement v are zero at A:

$$v\left(\frac{\pi}{2}\right) = 0 \qquad \text{thus} \qquad C_0 = 0$$

and $$\frac{1}{R}\frac{dv(\pi/2)}{d\psi} = 0 \qquad \text{so that} \qquad -C_1 + \frac{\pi PR^3}{4EJ_z} = 0$$

Hence, Eq. (*i*) becomes

$$v = -\frac{PR^3\cos\psi}{2EJ_z}\left(\psi - \frac{\pi}{2}\right) \tag{j}$$

It is important to point out the physical significance of the particular and complementary solutions. The complementary solution alone does not satisfy the equilibrium equations [Eqs. (*c*) and (*d*)]; in fact, it represents an elastic curve completely free of bending moments. It is the particular solution that represents the influence of applied loads and provides equilibrium for each element of the bar. Used by itself, however, the particular solution completely violates the boundary conditions. Thus, it is the job of the complementary solution to force the elastic curve to be compatible with the natural constraints of the system. Used together, both of the basic requirements of equilibrium and compatibility are fulfilled.

Returning to Eq. (j), we now evaluate the vertical displacement under the load P. Substituting $\psi = 0$ into (j) gives

$$\delta = v(0) = \frac{\pi P R^3}{4EJ_z} \tag{k}$$

For the circular section of radius r, the last member of Eqs. (4.14) gives, after some lengthy integration,

$$J_z = \pi\left\{2R^4\left[1 - \sqrt{1 - \left(\frac{r}{R}\right)^2}\right] - R^2r^2\right\}$$

which we may introduce into Eq. (k). Before doing this, let us rewrite the expression for J_z in the more meaningful form

$$J_z = \frac{\pi r^4}{4}\,[1 + \phi(\lambda)] \tag{l}$$

in which $$\phi(\lambda) = \frac{1}{\lambda^4}\,[8(1 - \sqrt{1 - \lambda^2}) - \lambda^2(4 + \lambda^2)]$$

and $\lambda = r/R$, so that Eq. (k) becomes

$$\delta = \frac{PR^3/Er^4}{1 + \phi(\lambda)} \tag{m}$$

We recognize the first term of Eq. (l) as the moment of inertia I_z of the circular section. As we might have expected, it may be shown that

$$\lim_{\lambda \to 0} \phi(\lambda) = 0$$

so that when R is large compared with r we may replace J_z with I_z without a significant error.

To determine, for the structure under consideration, precisely when such a simplification is permissible, let us examine the curve shown in Fig. 6.4. The ordinates of points on this curve are the ratios of the tip deflections computed using I_z (δ_I) to those computed using J_z (δ_J), and the corresponding values of λ are the abscissas. We see that as λ decreases δ_J quickly approaches δ_I. For $\lambda = \frac{1}{3}$ an error of only 3.4 percent is introduced by using I_z instead of J_z. For $\lambda = \frac{1}{10}$ we find an error of only 0.8 percent. Although values of near unity are found in many curved machine parts (as we mentioned in Chap. 4), in the great majority of practical cases λ ranges from approximately $\frac{1}{10}$ to $\frac{1}{50}$, with values of $\frac{1}{100}$ not uncommon.

Calculations for other types of cross sections and generating curves yield results very similar to those obtained above. We conclude that in

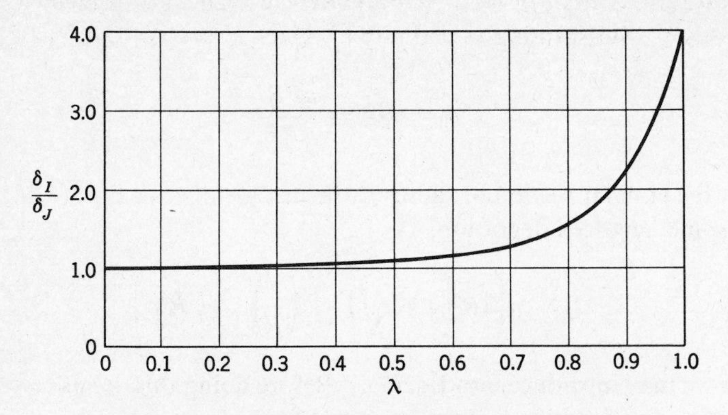

FIGURE 6.4 Curve showing that as $\lambda = r/R$ becomes smaller the deflection computed using J_z in Eq. (*k*) approaches that obtained by using I_z in place of J_z.

calculating deformations of bars of small curvature, it is often permissible to use the simpler section properties I_y, I_z, and I_{yz}. In addition, for such structures the strain of the centroidal axis due to bending, $-M_z/RAE$, is usually negligible compared with that due to N_s. Therefore, in the case of a bar of small curvature, Eqs. (6.4) reduce to

$$\frac{du_0}{ds} - \frac{v}{R} = \frac{N_s}{AE}$$

$$\frac{d^2v}{ds^2} + \frac{v}{R^2} = -\frac{M_zI_y - M_yI_{yz}}{E(I_yI_z - I_{yz}^2)} \tag{6.12}$$

$$\frac{d^2w}{ds^2} = -\frac{M_yI_z - M_zI_{yz}}{E(I_yI_z - I_{yz}^2)}$$

6.5 Deflection of unsymmetrical straight bars. In the case of straight bars, R is infinite, v becomes the transverse displacement of the bar, the coordinate s reduces to x, and Eqs. (6.4) become

$$\frac{du_0}{dx} = \frac{N_x}{AE}$$

$$\frac{d^2v}{dx^2} = -\frac{M_zI_y - M_yI_{yz}}{E(I_yI_z - I_{yz}^2)} \tag{6.13}$$

$$\frac{d^2w}{dx^2} = -\frac{M_yI_z - M_zI_{yz}}{E(I_yI_z - I_{yz}^2)}$$

If the dimensions of the cross sections do not vary with x, we can conveniently express the last two of these equations in terms of applied loads by using Eqs. (5.17):

$$\frac{d^4v}{dx^4} = \frac{p_yI_y - p_zI_{yz}}{E(I_yI_z - I_{yz}^2)}$$
$$\frac{d^4w}{dx^4} = \frac{p_zI_z - p_yI_{yz}}{E(I_yI_z - I_{yz}^2)} \tag{6.14}$$

Comparing Eqs. (6.14) with (6.7), we see that the order of the last equation has been reduced from five to four owing to the absence of an initial curvature of the bar. Also, there is no longer a coupling of the dependent variables v and w, so each equation can be solved independently.

We obtain solutions to either Eqs. (6.13) or (6.14) by direct integration. As a simple example, let us consider the cantilever beam shown in Fig. 6.5. In this case p_y, p_z, N_x, and M_y are zero and M_z is $-P(L - x)$. Equations (6.13) become

$$\frac{du_0}{dx} = 0$$
$$\frac{d^2v}{dx^2} = \frac{P(L - x)I_y}{E(I_yI_z - I_{yz}^2)} \tag{a}$$
$$\frac{d^2w}{dx^2} = -\frac{P(L - x)I_{yz}}{E(I_yI_z - I_{yz}^2)}$$

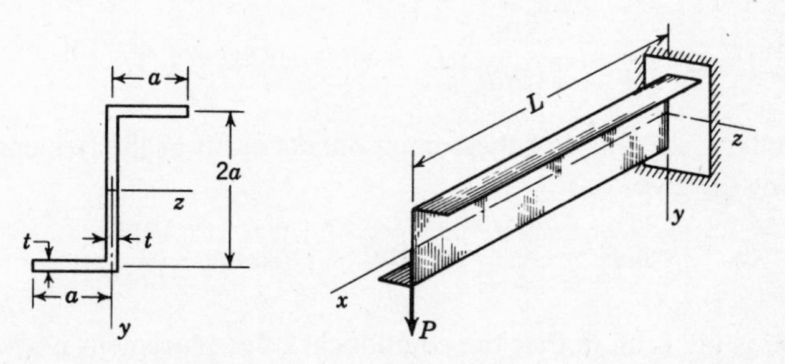

FIGURE 6.5 Cantilevered z section.

The stretching u_0 is zero, as is easily seen by integrating the first of Eqs. (*a*) and noting that u_0 is zero at the fixed end. Integrating each of the remaining equations twice, we find

$$v = \frac{P(L - x)^3 I_y}{6E(I_yI_z - I_{yz}^2)} + C_0x + C_1 \tag{b}$$

and

$$w = \frac{-P(L - x)^3 I_{yz}}{6E(I_yI_z - I_{yz}^2)} + C_2x + C_3 \tag{c}$$

where C_0, C_1, C_2, and C_3 are constants of integration. The kinematic boundary conditions are

$$v(0) = 0 \qquad \frac{dv(0)}{dx} = 0$$
$$w(0) = 0 \qquad \frac{dw(0)}{dx} = 0 \tag{d}$$

from which we obtain

$$C_0 = \frac{PL^2 I_y}{2E(I_y I_z - I_{yz}^2)} \qquad C_1 = -\frac{PL^3 I_y}{6E(I_y I_z - I_{yz}^2)}$$
$$C_2 = -\frac{PL^2 I_{yz}}{2E(I_y I_z - I_{yz}^2)} \qquad C_3 = \frac{PL^3 I_{yz}}{6E(I_y I_z - I_{yz}^2)} \tag{e}$$

For this section,

$$\frac{I_y}{I_y I_z - I_{yz}^2} = \frac{\frac{2}{3}ta^3}{\frac{2}{3}\cdot\frac{8}{3}\cdot t^2a^6 - t^2a^6} = \frac{6}{7ta^3}$$
$$\frac{I_{yz}}{I_y I_z - I_{yz}^2} = \frac{-ta^3}{\frac{2}{3}\cdot\frac{3}{8}t^2a^6 - t^2a^6} = \frac{-9}{7ta^3} \tag{f}$$

so that the final equations for the elastic curve are

$$v = \frac{P}{7Eta^3}[(L - x)^3 + 3L^2x - L^3]$$
$$w = \frac{3P}{14Eta^3}[(L - x)^3 + 3L^2x - L^3] \tag{g}$$

The maximum values of these components occur at the free end $x = L$. Equations (*g*) give

$$v_{\max} = \frac{2PL^3}{7Eta^3} \qquad \text{and} \qquad w_{\max} = \frac{3PL^3}{7Eta^3} \tag{h}$$

It is interesting to note that the component of displacement normal to the plane of loading is larger than that in the plane of loading.

Finally, the maximum displacement is given by

$$\delta_{\max} = \sqrt{v_{\max}^2 + w_{\max}^2} = 0.515\frac{PL^3}{Eta^3} \tag{i}$$

which occurs in a direction perpendicular to the neutral axis.

6.6 Deflections of symmetrical bars. It is clear from the above example that a lack of symmetry of the cross section of a bar with respect to the plane of loading can significantly influence the final shape of its elastic curve. Lack of symmetry further complicates the problem by making it necessary to evaluate two transverse displacements rather than one for loadings in the xy plane. Fortunately, in many structural systems an effort is made to assemble and load individual members so that bending occurs only about one principal axis. The majority of literature on the theory of bar structures concerns members loaded in this manner; the governing equations are somewhat simplified and a variety of special effects can be more conveniently considered. It is to this class of problems that we now direct our attention. The results may be altered so that they apply to beams with unsymmetrical cross sections by simply replacing I_z by $(I_y I_z - I_{yz}^2)/I_y$, assuming, of course, that bending occurs only about the z axis. Referring, then, to Eqs. (6.13) (w and I_{yz} now being zero), we find for the differential equations of the elastic curve

$$\frac{du_0}{dx} = \frac{N_x}{AE} \qquad \frac{d^2v}{dx^2} = -\frac{M_z}{EI_z} \tag{6.15}$$

or, since

$$\frac{d^2M_z}{dx^2} = \frac{dV_y}{dx} = -p_y$$

we may use the fourth-order equation

$$\frac{d^2}{dx^2}\left(EI_z \frac{d^2v}{dx^2}\right) = p_y \tag{6.16}$$

If EI_z is constant,

$$\frac{d^4v}{dx^4} = \frac{p_y}{EI_z} \tag{6.17}$$

The above simplified forms of the differential equations of the elastic curve constitute the governing equations for what is commonly referred to as the Bernoulli-Euler theory of bending. The complete solution to Eq. (6.17) is found from simple quadrature:

$$v = \iiiint \frac{p_y}{EI_z}(dx)^4 + C_1\frac{x^3}{6} + C_2\frac{x^2}{2} + C_3x + C_4 \tag{6.18}$$

in which the integral represents the particular solution which, as before, provides equilibrium for each element, and the remaining terms represent the complementary solution which adjusts the curve to fit given boundary conditions.

By inspection of Eqs. (5.15) to (5.17) and (6.15), it is clear that the following chain of relationships exist:

$$
\begin{aligned}
v &= \text{deflection} \\
\frac{dv}{dx} &= \theta = \text{slope} \\
-EI_z \frac{d^2v}{dx^2} &= M_z = \text{bending moment} \\
-EI_z \frac{d^3v}{dx^3} &= V_y = \text{shearing force} \\
EI_z \frac{d^4v}{dx^4} &= p_y = \text{load intensity}
\end{aligned}
\tag{6.19}
$$

Thus, for the standard idealized support conditions, we may use as boundary conditions

$$
\begin{aligned}
\frac{d^3v}{dx^3} = \frac{d^2v}{dx^2} &= 0 \qquad \text{for a free end} \\
v = \frac{d^2v}{dx^2} &= 0 \qquad \text{for a hinged end} \\
\text{and} \qquad v = \frac{dv}{dx} &= 0 \qquad \text{for a fixed end}
\end{aligned}
\tag{6.20}
$$

The free- and fixed-end conditions are also obtained from Eqs. (6.10) and (6.11) by setting w and J_{yz} equal to zero and noting that R is infinitely large. Obviously, other combinations are often possible.

The solution to any given problem of this class, then, follows from applying Eq. (6.18) or Eqs. (6.15) and conditions such as those in Eqs. (6.20). The general method of attack is well known and straightforward. In some cases, however, the evaluation of the constants of integration may prove quite laborious if some shortcuts are not taken.

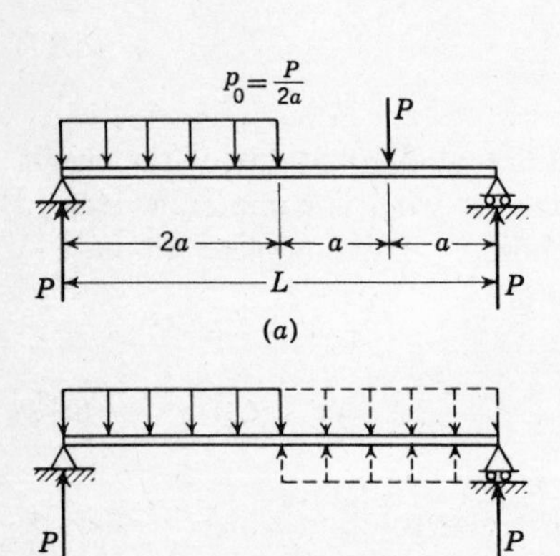

FIGURE 6.6 Simply supported straight beam.

Consider, for example, the beam shown in Fig. 6.6*a*. The flexural rigidity of the beam is constant, $EI_z = EI$, and the loading is discontinuous. In this case three different functions are needed to define the elastic curve: v_1, v_2, v_3, corresponding to the three intervals shown. We obtain the equations for the bending moments developed in each interval from statics:

$$M_z = \begin{cases} Px - \dfrac{p_0 x^2}{2} & 0 \le x \le 2a \\ Pa & 2a \le x \le 3a \\ P(L-x) & 3a \le x \le L \end{cases} \tag{a}$$

Thus, according to the second equation in Eqs. (6.15),

$$-EI\frac{d^2v_1}{dx^2} = Px - \frac{p_0x^2}{2}$$

$$-EI\frac{d^2v_2}{dx^2} = Pa$$

$$-EI\frac{d^2v_3}{dx^2} = P(L-x)$$

where v_1, v_2, and v_3 are the equations for v in the respective intervals $0 \le x \le 2a$, $2a \le x \le 3a$, and $3a \le x \le L$. Integrating each of these equations twice, we find

$$v_1 = -\frac{1}{EI}\left(P\frac{x^3}{6} - \frac{p_0x^4}{24}\right) + C_{11}x + C_{12}$$

$$v_2 = -\frac{Pax^2}{2EI} + C_{21}x + C_{22}$$

$$v_3 = -\frac{P(L-x)^3}{6EI} + C_{31}x + C_{32}$$

where $C_{11}, C_{12}, \ldots, C_{32}$ are constants. The boundary conditions for each segment $(v_1)_{x=0} = 0$; $(v_1)_{x=2a} = (v_2)_{x=2a}$; $(dv_1/dx)_{x=2a} = (dv_2/dx)_{x=2a}$; $(v_2)_{x=3a} = (v_3)_{x=3a}$; $(dv_2/dx)_{x=3a} = (dv_3/dx)_{x=3a}$; $(v_3)_{x=L} = 0$ lead us to the simultaneous equations

$$C_{12} = 0 \qquad 2a(C_{11} - C_{21}) - C_{22} = 0$$

$$2P\frac{a^2}{3}EI + C_{11} - C_{21} = 0$$

$$13P\frac{a^3}{3}EI + 3a(C_{31} - C_{21}) + C_{32} - C_{22} = 0$$

$$-5P\frac{a^2}{6}EI + C_{21} - C_{31} = 0 \qquad 4aC_{31} + C_{32} = 0$$

Hence, we are faced with the tedious task of solving five (six) simultaneous equations for the unknown constants of integration.

We can now appreciate a much easier way to solve the problem. Imagine that the uniform load p_0 is extended so that it acts throughout the length of the beam, as shown in Fig. 6.6*b*. In intervals for which $x > 2a$, a load of $-p_0$ is applied so that the result is the same as that of the beam in Fig. 6.6*a*. We now write the following relationships:

$$0 \le x \le 2a \qquad -EI\frac{d^2v_1}{dx^2} = Px - \frac{p_0x^2}{2}$$

$$-EI\theta_1 = \frac{Px^2}{2} - \frac{p_0x^3}{6} + \bar{C}_{11}$$

$$-EIv_1 = \frac{Px^3}{6} - \frac{p_0x^4}{24} + \bar{C}_{11}x + \bar{C}_{12}$$

$$2a \le x \le 3a \qquad -EI\frac{d^2v_2}{dx^2} = Px - \frac{p_0x^2}{2} + \frac{p_0}{2}(x-2a)^2$$

$$-EI\theta_2 = \frac{Px^2}{2} - \frac{p_0x^3}{6} + \frac{p_0}{6}(x-2a)^3 + \bar{C}_{21}$$

$$-EIv_2 = \frac{Px^3}{6} - \frac{p_0x^4}{24} + \frac{p_0}{24}(x-2a)^4 + \bar{C}_{21}x + \bar{C}_{22}$$

$$3a \le x \le L \qquad -EI\frac{d^2v_3}{dx^2} = Px - \frac{p_0x^2}{2} + \frac{p_0}{2}(x-2a)^2 - P(x-3a)$$

$$-EI\theta_3 = \frac{Px^2}{2} - \frac{p_0x^3}{6} + \frac{p_0}{6}(x-2a)^3 - \frac{P}{2}(x-3a)^2 + \bar{C}_{31}$$

$$-EIv_3 = \frac{Px^3}{6} - \frac{p_0x^4}{24} + \frac{p_0}{24}(x-2a)^4 - \frac{P}{6}(x-3a)^3 + \bar{C}_{31}x + \bar{C}_{32}$$

Now applying the boundary conditions, we find

$$x = 0 \qquad v_1 = 0 \qquad \bar{C}_{12} = 0$$

$$x = 2a \qquad \theta_1 = \theta_2 \qquad v_1 = v_2 \qquad \bar{C}_{21} = \bar{C}_{11} \qquad \bar{C}_{22} = 0$$

$$x = 3a \qquad \theta_2 = \theta_3 \qquad v_2 = v_3 \qquad \bar{C}_{31} = \bar{C}_{11} \qquad \bar{C}_{32} = 0$$

which leaves only one constant $\bar{C}_{11}$ to be determined. This is done quite simply by using the condition

$$v_3 = 0 \qquad \text{at } x = L$$

We find
$$\bar{C}_{11} = -\tfrac{33}{24}Pa^2$$

The above procedure is quite systematic and no simultaneous equations were solved. The constant terms in the expressions for v_1, v_2, and v_3 are automatically zero, and the constants in θ_1, θ_2, and θ_3 are equal. The most important consequence of solving the problem in this manner is that we may write in a single expression the equation for the entire elastic curve:

$$-EIv = \frac{P}{6}x^3 - \frac{33Pa^2}{24}x - \frac{p_0}{24}x^4 + \frac{p_0}{24}(x-2a)^4 - \frac{P}{6}(x-3a)^3 \qquad (b)$$

Only the first three terms in this equation are valid for $0 \leq x \leq 2a$. These, then, represent v_1. For $2a \leq x \leq 3a$, four terms (representing v_2) are used. For the interval $3a \leq x \leq L$ all terms must be used. Obviously,

$$v_1 = v_1$$

$$v_2 = v_1 + \frac{p_0}{24}(x - 2a)^4$$

$$v_3 = v_2 - \frac{P}{6}(x - 3a)^3$$

The idea of representing the elastic curve in this manner was introduced by Clebsch[1] in 1862. Some years later a similar procedure was presented by Macauley,[2] and in English literature it is often called *Macauley's method.*

The technique is more properly classified as a special application of the *Heaviside step function*[3] with which it is possible to obtain, in a single

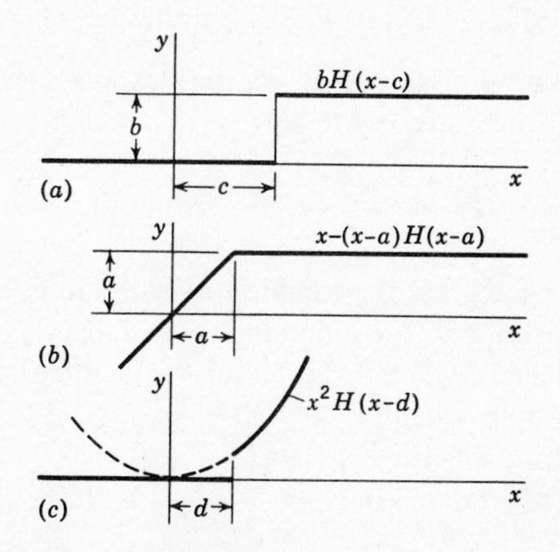

FIGURE 6.7 Examples of the representation of irregular curves by using Heaviside's unit step function.

equation, a mathematically correct representation of the elastic curve of the beam shown in Fig. 6.6. The Heaviside unit step function is defined as follows:

$$H(x - a) = \begin{cases} 0 & \text{if } x < a \\ 1 & \text{if } x > a \end{cases} \tag{6.21}$$

Some examples of the use of step functions to represent curves which are discontinuous or have discontinuous derivatives are given in Fig. 6.7.

[1] An account of Clebsch's method is given in Ref. 12.

[2] See Ref. 38. More detailed information on the subject is given in Refs. 22 and 73.

[3] Named after the British electrical engineer Oliver Heaviside (1850–1925).

Using this notation, Eq. (*b*) acquires the form

$$-EIv = \frac{Px^3}{6} + \frac{11Pa^2x}{12} - \frac{p_0x^4}{24} + \frac{p_0(x-2a)^4}{24} H(x-2a) - \frac{P(x-3a)^3}{6} H(x-3a) \quad (c)$$

The advantage in using this function lies in the fact that it can be integrated. For example, we can represent the bending moment [or the transverse loading if Eq. (6.17) is used] by a single expression valid throughout the length of the beam. We then obtain the elastic curve by successive integrations of a single function. It is merely necessary to keep in mind that each step is nonzero only in the interval indicated by its argument. Thus, the bending moment of the beam in Fig. 6.6*a* may be written

$$M_z = \left(Px - \frac{p_0x^2}{2}\right)[H(x) - H(x-2a)] + Pa[H(x-2a) - H(x-3a)] + P(L-x)[H(x-3a) - H(x-L)] \quad (d)$$

which can be further simplified; or we may express M_z in terms of the equivalent loading shown in Fig. 6.6*b*:

$$M_z = Px - \frac{p_0x^2}{2} + \frac{p_0(x-2a)^2}{2} H(x-2a) - P(x-3a)H(x-3a) \quad (e)$$

Since we intend to integrate the function over the interval $0 \leq x \leq L$, it is not necessary to "cut off" the function at $x = L$ [as was done in Eq. (*d*)]; we need only to define M_z for this interval. The integral of the last term in Eq. (*e*), for example, is simply

$$\int P(x-3a)H(x-3a)\,dx = \frac{P(x-3a)^2}{2} H(x-3a) + C$$

Integrating either Eq. (*d*) or (*e*) twice, we find that only *two* constants of integration appear, rather than the six found earlier. Again, these are evaluated from the conditions of zero deflection at the supports. The resulting equation of the elastic curve is given by Eq. (*c*).

6.7 Shear deformations. In the previous developments we have tacitly assumed that transverse beam deflections are produced solely by the bending moments on each section. No consideration was given to the fact that the shearing forces developed on each element are accompanied by a detrusion of the element, as shown in Fig. 6.8, which produces an additional deflection. Shear deformations are usually very small compared with those due to bending; but in some cases, such as short deep members subjected to high shearing forces, it is necessary to consider them in order to obtain a more accurate description of the elastic curve.

From previous developments, we recall that

$$\gamma_{xy} = \frac{\tau_{xy}}{G} = \frac{V_y Q_z}{I_z b G} = \frac{\partial u}{\partial y} + \frac{\partial v}{\partial x}$$

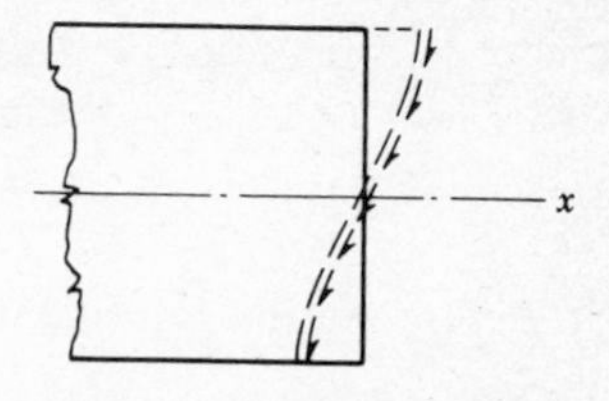

FIGURE 6.8 Warping of cross section due to shear deformation.

Note that $V_y Q_z / I_z b$ in this expression is an approximation of τ_{xy} derived from purely static considerations. Referring to the expression for u_2 in Art. 6.2, we find that

$$u = -y \frac{dv_b}{dx}$$

where v_b is the deflection due to bending. Thus,

$$\frac{V_y Q_z}{I_z b G} = -y \frac{\partial^2 v_b}{\partial y\, \partial x} - \frac{\partial v_b}{\partial x} + \frac{\partial v}{\partial x} \tag{6.22}$$

The first term on the right side of this equation represents a slight warping of the cross sections due to shear deformation, as shown in Fig. 6.8. In view of our assumption that planes remain plane (on the basis of which we obtained the equation for τ_{xy}), this term is small and can be neglected. The left side of Eq. (6.22) is a function of both x and y. This, again, is inconsistent with the assumption that planes remain plane since we found that v_b is a function of only x. To remedy this situation, we replace τ_{xy} by a uniform stress distribution and account for its variation with y by introducing a *shear correction factor* κ. The left side of Eq. (6.22) then becomes

$$\frac{\kappa V_y}{AG}$$

The factor κ, therefore, is a constant by which the average shearing strain must be multiplied in order to obtain an equivalent uniform shearing strain which leads to the same transverse (shear) displacement as the distributed shearing strains $V_y Q_z / I_z b G$. With this definition, we find values of κ of $\frac{6}{5}$ for rectangular sections[4] and near unity for I sections.

$$\frac{\kappa V_y}{AG} = -\frac{dv_b}{dx} + \frac{dv}{dx}$$

[4] The calculation of shear correction factors is discussed in Chap. 9. See Art. 9.15, particularly Eqs. (9.107) and (9.108).

It is clear that this equation gives the slope of the elastic curve due to shear deformation θ_s. That is,

$$\theta_s = \frac{dv}{dx} - \frac{dv_b}{dx} = \frac{\kappa V_y}{AG} \tag{6.23}$$

Differentiating Eq. (6.23) with respect to x, we find

$$\frac{d^2v}{dx^2} - \frac{d^2v_b}{dx^2} = \frac{\kappa}{AG}\frac{dV_y}{dx} = -\frac{\kappa p_y}{AG}.$$

$$\frac{d^2v}{dx^2} = -\left(\frac{M_z}{EI_z} + \frac{\kappa p_y}{AG}\right) \tag{6.24}$$

This is the final differential equation of the elastic curve accounting for shear deformation. Comparing this result with Eqs. (6.15), it is clear that shear deformations alter the curvature of the elastic curve by an amount $\kappa p_y/AG$.

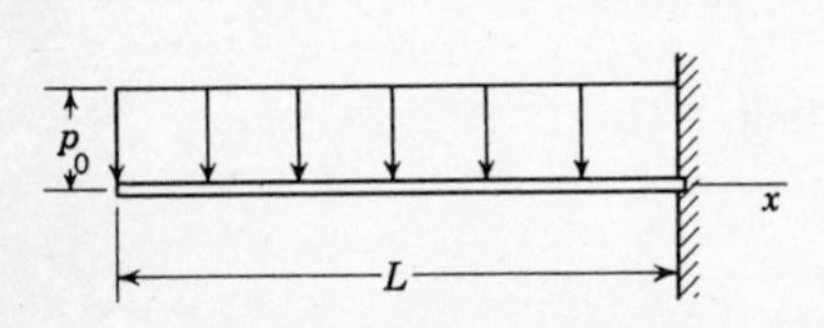

FIGURE 6.9 Uniformly loaded cantilever beam.

As a simple example, let us consider the influence of shear on the deflection of the beam shown in Fig. 6.9. Assuming a rectangular section of dimensions $b \times h$, $E = 2G$, and $\kappa = \frac{6}{5}$, Eq. (6.24) becomes[5]

$$\frac{d^2v}{dx^2} = -\frac{1}{EI}\left(-\frac{p_0x^2}{2} - \frac{1}{5}h^2p_0\right)$$

Integrating twice and evaluating the constants of integration from the conditions of zero deflection and slope at the fixed end, we find

$$v = \frac{p_0}{24EI}(x^4 - 4L^3x + 3L^4) + \frac{h^2p_0}{10EI}(x^2 - 2Lx + L^2)$$

The tip deflection $[\delta = v(0)]$ is

$$\delta = \frac{p_0L^4}{8EI}\left[1 + \frac{4}{5}\left(\frac{h}{L}\right)^2\right]$$

The last term in the parentheses represents the influence of shear deformation, which we see depends upon the square of the ratio of depth of the beam to length. It is interesting to note that even for a rather short deep beam for which $h/L = \frac{1}{5}$, the shear deformation contributes only 4 percent of the total tip deflection in this example.

For cases in which p_y is zero (for example, a cantilevered beam with a

[5] A minus p_0 is used in this case since V_y is negative.

concentrated force applied at its free end), the influence of shear deformation enters the results when boundary conditions are applied.

6.8 Anticlastic bending. Thus far we have assumed that longitudinal beam fibers deformed independently. In other words, we have effectively assumed that the beam is made of a material which has a Poisson's ratio of zero. Obviously, this is never true.

Consider, for example, the strip of flexible material in pure bending shown in Fig. 6.10*a*. The top fibers of the strip are clearly in compression, and the bottom fibers are in tension. We recall from Chap. 2 that every extensional strain ϵ is accompanied by a strain of $-\nu\epsilon$ perpendicular to ϵ, where ν is Poisson's ratio for the material. Thus, as the top fibers are compressed in the x direction they become slightly longer in the z direction. Bottom fibers elongate in the x direction but shorten in the z direction. The bar acquires a small curvature $1/\rho'$ in the yz plane which results in a distortion of the cross section, as is indicated in Fig. 6.10*b*. This phenomenon is called *anticlastic bending*.

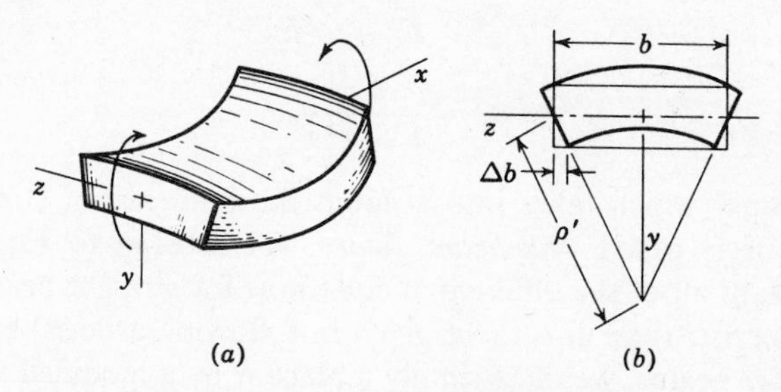

FIGURE 6.10 Illustration of anticlastic bending.

To evaluate the effects of anticlastic bending, we refer to the general stress-strain relations given by Eqs. (2.27). Since σ_y and σ_z are zero, we find

$$\epsilon_x = \frac{\sigma_x}{E}$$

$$\epsilon_y = \epsilon_z = -\nu\frac{\sigma_x}{E}$$

If we assume that plane sections normal to the z axis remain plane after (anticlastic) bending, an inspection of Fig. 6.10*b* reveals that

$$\frac{b}{\rho'} = \frac{b - \Delta b}{\rho' - y}$$

or

$$\frac{1}{\rho'} = \frac{d^2v}{dz^2} = \frac{\Delta b}{by} = -\frac{\epsilon_z}{y} = \nu\frac{\epsilon_x}{y}$$

in which $1/\rho'$ is the anticlastic curvature and b is the width of the section. Since $\epsilon_x = y/\rho = -y\, d^2v/dx^2$, we find[6]

$$\frac{1}{\rho'} = \nu \frac{1}{\rho}$$

where $1/\rho$ is the change of curvature in the xy plane. Thus,

$$\frac{d^2v}{dz^2} = -\nu \frac{d^2v}{dx^2} \tag{6.25}$$

If no anticlastic bending of the strip occurs, additional stresses must be developed in the z direction to eliminate curvature in the yz plane. In this case

$$\epsilon_z = 0 = \frac{\sigma_z}{E} - \nu \frac{\sigma_x}{E}$$

so that

$$\sigma_z = \nu \sigma_x$$

The extensional strain in the x direction is

$$\epsilon_x = \frac{\sigma_x}{E} - \nu^2 \frac{\sigma_x}{E}$$

Therefore

$$\sigma_x = \frac{E}{1 - \nu^2} \epsilon_x \tag{6.26}$$

Beam theory which takes into account the influence of stresses σ_z on deformations is called *wide-beam theory*. According to Eq. (6.26), in order to adapt all of the differential equations for straight beams that we developed earlier (and also those given in following articles) so that they apply to *wide* beams, we must simply replace E by a modified modulus of elasticity $\bar{E}$ where

$$\bar{E} = \frac{E}{1 - \nu^2} \tag{6.27}$$

6.9 Deflection of ties and beam columns. According to Eqs. (6.15) and (6.16), the presence of an axial force N_x does not influence the transverse displacements of a beam. The displacements u, v, and w may be calculated independently, and final displacements are proportional to the applied loads. Furthermore, the principle of superposition is valid.

In the case of slender flexible members subjected to large axial loads, however, experience has shown that axial forces may greatly alter the shape of the elastic curve. In such structures the transverse displacements are no longer so small that their influence on the equilibrium conditions for a deformed element can be neglected. As a result, deflections are no longer

[6] This is obtained from Eq. (6.3) by setting w and $1/R$ equal to zero and s equal to x.

proportional to the transverse loading, the principle of superposition is not directly applicable, and the equations that we developed earlier may lead to serious errors. Beams in which this effect is significant are called *ties* if the axial load is a tensile force and *beam columns* if it is a compressive force.

To begin our study of this effect, let us consider the beam shown in Fig. 6.11*a*. The beam is simultaneously subjected to a general system of

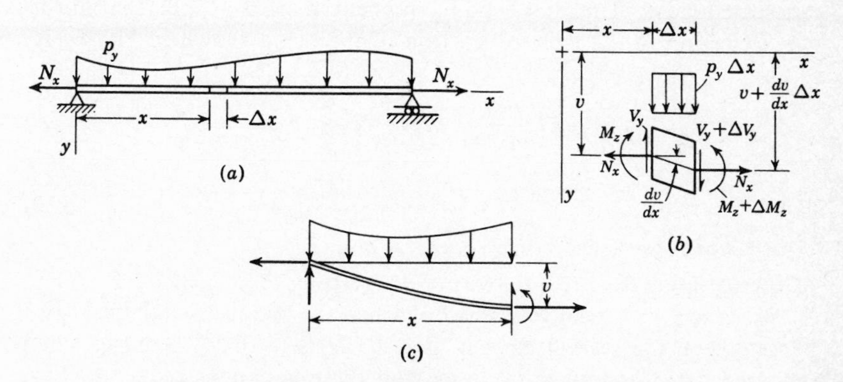

FIGURE 6.11 (*a*) Tie subjected to a general system of loads; (*b*) typical element of the tie; (*c*) segment of length *x*.

applied loads $p_y = p$ and an axial force N_x. In Fig. 6.11*b* we see a typical segment of the beam isolated as a free body. Applying statics to this element, we find for the conditions of equilibrium

$$\frac{dV_y}{dx} = -p_y$$

which is identical to the relationship in Eqs. (5.15), and

$$\Delta M_z - V_y \, \Delta x + N_x \frac{dv}{dx} \Delta x = 0$$

or

$$V_y = \frac{dM_z}{dx} + N_x \frac{dv}{dx} \tag{6.28}$$

which is noticeably different from the relationship obtained earlier.

We have found earlier that $-EI_z \, d^2v/dx^2$ equals M_z. Thus, if N_x is constant,

$$\frac{d^2}{dx^2}\left(EI_z \frac{d^2v}{dx^2}\right) = -\frac{d^2M_z}{dx^2} = -\frac{dV_y}{dx} + N_x \frac{d^2v}{dx^2}$$

so that on introducing Eq. (6.28) we arrive at the governing differential equation:

$$\frac{d^2}{dx^2}\left(EI_z \frac{d^2v}{dx^2}\right) - N_x \frac{d^2v}{dx^2} = p_y \tag{6.29}$$

We see that the presence of N_x has introduced an additional term in the differential equation of the elastic curve.

An alternative form of this equation is often more convenient to use. Let us denote by M_l the bending moment at x due to just the applied transverse loads. The total bending moment, then, is M_l plus that due to N_x. Referring to the isolated portion of the beam shown in Fig. 6.11c, we find

$$-EI_z \frac{d^2v}{dx^2} = M_z = M_l - N_x v$$

Thus

$$\frac{d^2v}{dx^2} - \frac{N_x}{EI_z} v = -\frac{M_l}{EI_z} \tag{6.30}$$

Differentiating Eq. (6.30) twice with respect to x yields Eq. (6.29). Clearly, when N_x is relatively small or when the member is relatively stiff (EI_z is large), the second term in Eqs. (6.29) and (6.30) becomes small compared with the remaining terms and these equations reduce to Eqs. (6.16) and (6.15), respectively.

We also note that the principle of superposition is not directly applicable, as is illustrated in Fig. 6.12. The elastic curve of the beam shown in Fig. 6.12a obviously cannot be obtained by adding the curves of the beams shown in Fig. 6.12b and c. The deflection of b is independent of N_x while that of c is zero. This is also obvious from a mathematical viewpoint: The elastic curves of the beams in Fig. 6.12a and c must

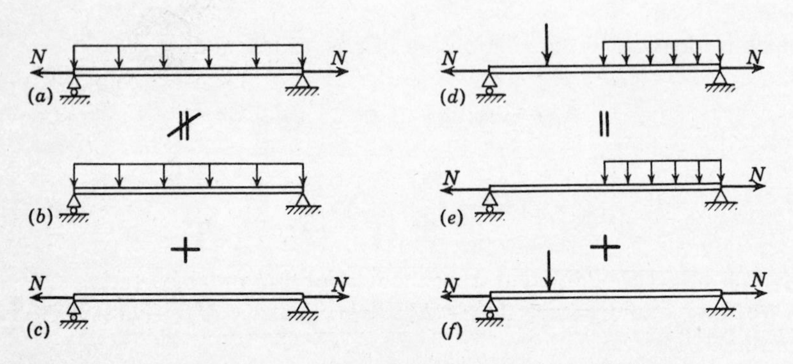

FIGURE 6.12 Flexible beams under axial load showing examples in which superposition is not legitimate, (a) ≠ (b) + (c), and is legitimate, (d) = (e) + (f).

satisfy a completely different differential equation from that of the beam in Fig. 6.12*b*. The sum of a number of solutions to the same linear differential equation, of course, is always a solution to that equation. Thus, the superposition of elastic curves shown in Fig. 6.12*e* and *f* is perfectly valid. If N_x of the beam in Fig. 6.12*e* were different from that of the beam in Fig. 6.12*f*, however, the corresponding differential equations would be different and superposition would be invalid. The equations describing the deflection of beams *d*, *e*, and *f* differ only in the "nonhomogeneous" part involving the independent variable x. Terms involving the dependent variable v are identical. By adding a number of loading cases, then, in effect we are adding to the complementary solution a number of particular solutions, each of which corresponds to a particular loading case.

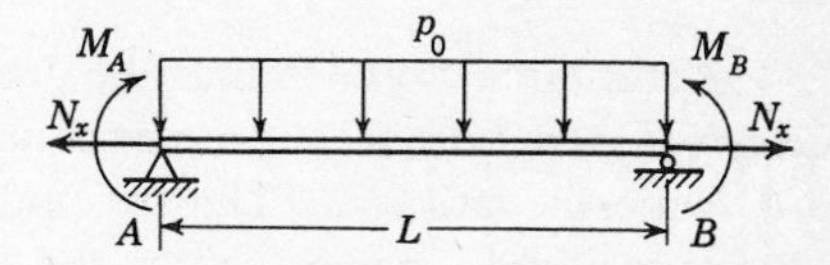

FIGURE 6.13 Simply supported tie.

To illustrate the application of the above equations, let us consider the simple beam of constant section shown in Fig. 6.13. The beam is subjected to a uniform load and to applied moments at its ends M_A and M_B in addition to an axial force N_x. The beam is a tie, since N_x is a tensile force. From simple statics we find that the bending moment due to the applied loads is

$$M_l = \frac{M_B - M_A}{L} x + M_A + \frac{p_0 L}{2} x - \frac{p_0 x^2}{2}$$

Thus, Eq. (6.30) becomes

$$\frac{d^2v}{dx^2} - \lambda^2 v = \frac{1}{EI_z}\left[\frac{p_0 x^2}{2} - \frac{p_0 L x}{2} - \frac{M_B x}{L} - \frac{M_A}{L}(L - x)\right] \qquad (a)$$

where

$$\lambda = \sqrt{\frac{N_x}{EI_z}} \qquad (b)$$

To find a particular solution to Eq. (*a*), we assume v_p to be of the form $A_0 + A_1 x + A_2 x^2$, in which A_0, A_1, and A_2 are constants evaluated by substituting the assumed function into Eq. (*a*) and equating like coefficients. Proceeding in this manner, we find for the particular solution

$$v_p = \frac{1}{N_x}\left[M_A - \frac{p_0}{\lambda^2} + \left(\frac{p_0 L}{2} + \frac{M_B - M_A}{L}\right)x - \frac{p_0 x^2}{2}\right] \qquad (c)$$

Since the equation has constant coefficients, we assume a complementary solution of the form Ce^{mx} which, on substitution into the homogeneous equation $d^2v/dx^2 - \lambda^2 v = 0$, yields the characteristic polynomial

$$m^2 - \lambda^2 = 0 \qquad (d)$$

Equation (*d*) has two real roots $m = \pm\lambda$. Hence, the complementary solution to Eq. (*a*) is

$$v_c = C_0 e^{\lambda x} + C_1 e^{-\lambda x} \tag{e}$$

which may be written in the equivalent form

$$v_c = A \cosh \lambda x + B \sinh \lambda x \tag{f}$$

A and B being arbitrary constants. Thus, the general solution to Eq. (*a*) is

$$v = A \cosh \lambda x + B \sinh \lambda x - \frac{p_0}{N_x}\left(\frac{1}{\lambda^2} - \frac{Lx}{2} + \frac{x^2}{2}\right) + M_B \frac{x}{N_x L} + M_A \frac{L - x}{N_x L} \tag{g}$$

The constants A and B must be determined from boundary conditions. Since we solved the second-order equation, static boundary conditions are automatically satisfied. The two independent kinematic boundary conditions required are provided by the conditions that the ends of the beam do not displace, that is, $v(0) = v(L) = 0$. From these conditions, Eq. (*g*) yields the following equations for A and B:

$$v(0) = 0 = A - \frac{p_0}{\lambda^2 N_x} + \frac{M_A}{N_x}$$

$$v(L) = 0 = A \cosh \lambda L + B \sinh \lambda L - \frac{p_0}{\lambda^2 N_x} + \frac{M_B}{N_x}$$

Solving these equations, we find

$$A = \frac{1}{N_x}\left(\frac{p_0}{\lambda^2} - M_A\right) \tag{h}$$

$$B = + \frac{(p_0/\lambda^2)(1 - \cosh \lambda L) - M_B + M_A \cosh \lambda L}{N_x \sinh \lambda L} \tag{i}$$

Finally, substituting these results into Eq. (*g*) and rearranging terms, we get the equation of the elastic curve:

$$v = \frac{M_A}{N_x}\left[\coth \lambda L \sinh \lambda x - \cosh \lambda x + \frac{L - x}{L}\right] + \frac{M_B}{N_x}\left[-\frac{\sinh \lambda x}{\sinh \lambda L} + \frac{x}{L}\right] + \frac{p_0}{\lambda^2 N_x}\left[\cosh \lambda x - 1 + (1 - \cosh \lambda L)\frac{\sinh \lambda x}{\sinh \lambda L} + \frac{\lambda^2 x}{2}(L - x)\right] \tag{j}$$

Equation (*j*) is written so that we may easily identify the influence of each part of the applied loading; the first term containing the brackets is obviously the elastic curve due to M_A, the second term is that due to M_B, and the last term represents the elastic curve due to the uniform loading. The influence of N_x on the shape of the elastic curve is not immediately apparent on inspecting Eq. (j), though, intuitively, we might expect an increase in N_x to decrease v. This, of course, is true; and it is easily demonstrated by considering as a typical example the center deflection (v at $x = L/2$) of a uniformly loaded tie ($M_A = M_B = 0$). From Eq. (j) we find after some transformations

$$\delta = v\left(\frac{L}{2}\right) = \frac{5p_0L^4}{384EI} \frac{384[\operatorname{sech}(\lambda L/2) - 1] + 48\lambda^2L^2}{5L^4\lambda^4} \tag{k}$$

The coefficient $5p_0L^4/384EI$ in Eq. (k) is the deflection due to p_0 acting alone. The quantity in the last fraction, then, represents the influence of N_x. Expanding sech $(\lambda L/2)$ in a series, we find for this term

$$\frac{384(-\frac{1}{2}\lambda L + \frac{5}{384}\lambda^4L^4 - \cdots) + 48\lambda^2L^2}{5\lambda^4L^4}$$

Dividing the numerator and denominator by λ, we see that as N_x (and consequently λ) approaches zero the last fraction approaches unity and Eq. (k) yields the center deflection of a beam with no axial load. However, if N_x (and λ) becomes infinitely large, this term approaches zero.

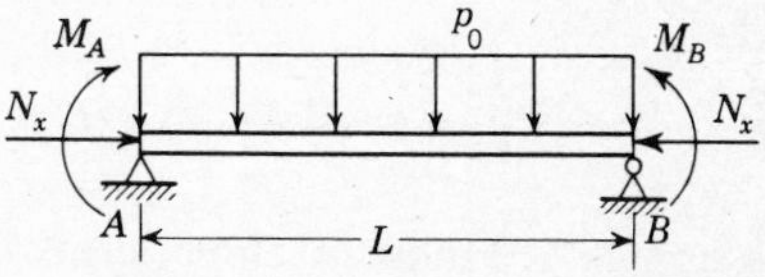

FIGURE 6.14 Simply supported beam column.

Quite different results are found when N_x is compressive. To analyze the beam column shown in Fig. 6.14, we must replace N_x by $-N_x$ in Eqs. (6.29) and (6.30), so that the governing differential equations become

$$\frac{d^2}{dx^2}\left(EI_z\frac{d^2v}{dx^2}\right) + N_x\frac{d^2v}{dx^2} = p_y \tag{6.31}$$

and

$$\frac{d^2v}{dx^2} + \frac{N_x}{EI_z}v = -\frac{M_l}{EI_z} \tag{6.32}$$

The procedure by which we apply Eq. (6.32) to evaluate the elastic curve of the beam in Fig. 6.14 is very similar to that which we used in studying the tie. M_l is unchanged and the particular solution acquires the slightly modified form

$$v_p = \frac{-1}{N_x}\left[M_A + \frac{p_0}{\lambda^2} + \left(\frac{p_0L}{2} + \frac{M_B - M_A}{L}\right)x - \frac{p_0x^2}{2}\right] \tag{l}$$

The complementary solution, however, is considerably different. Instead of Eq. (d), the characteristic polynomial for the beam column is

$$m^2 + \lambda^2 = 0 \tag{m}$$

which has two imaginary roots $m = \pm\lambda i$. Thus,

$$v_c = A \cos \lambda x + B \sin \lambda x \tag{n}$$

We obtain this same result by substituting $i\lambda$ for λ in Eq. (f), since $\cosh i\lambda x = \cos \lambda x$ and $\sinh i\lambda x = i \sin \lambda x$.

Adding Eqs. (l) and (n), we obtain the general solution to Eq. (6.32). From the conditions $v(0) = v(L) = 0$, we find that

$$A = \frac{1}{N_x}\left(\frac{p_0}{\lambda^2} + M_A\right) \tag{o}$$

and
$$B = \frac{(p_0/\lambda^2)(1 - \cos \lambda L) + M_B - M_A \cos \lambda L}{N_x \sin \lambda L} \tag{p}$$

Finally, introducing Eqs. (o) and (p) into the sum of Eqs. (l) and (n), we arrive at the final equation of the elastic curve:

$$v = -\frac{M_A}{N_x}\left(\cot \lambda L \sin \lambda x - \cos \lambda x + \frac{L - x}{L}\right) - \frac{M_B}{N_x}\left(-\frac{\sin \lambda x}{\sin \lambda L} + \frac{x}{L}\right)$$
$$+ \frac{p_0}{\lambda^2 N_x}\left[\cos \lambda x - 1 + (1 - \cos \lambda L)\frac{\sin \lambda x}{\sin \lambda L} - \frac{\lambda^2 x}{2}(L - x)\right] \tag{q}$$

Again, we see that the three terms on the right side of Eq. (q) represent the elastic curve due to M_A, M_B, and p_0, respectively. Following a procedure similar to that which we used in studying the deflections of the tie, we now consider the center-line deflection of a uniformly loaded beam column. M_A and M_B are again zero, and substituting $L/2$ for x in Eq. (q) gives

$$\delta = \frac{p_0 L^4}{384EI}\,\frac{384[\sec(\lambda L/2) - 1] - 48\lambda^2 L^2}{5\lambda^4 L^4} \tag{r}$$

It is interesting to compare Eq. (r) with Eq. (k). Again, as N_x and λ approach zero, the final fraction approaches unity and δ of Eq. (r) approaches the center-line deflection of a simple beam. In this case, however, when λ approaches a value of π/L the deflections increase indefinitely. Equation (b) shows that this value of λ corresponds to a value of the axial load of

$$N_x = \frac{\pi^2 EI}{L^2} \tag{6.33}$$

which we recognize as the *critical load*, or the *Euler buckling load* for a pinned-end column (usually denoted N_{cr}). Thus, for this value of N_x the structure becomes *elastically unstable* and buckles. We note that N_{cr} is independent of p_0 and is a function of only the elastic and geometric properties of the bar and of the boundary conditions. Calculations show[7]

[7] See, for example, Ref. 64.

that for the loading shown in Figs. 6.13 and 6.14 when N_x is $N_{cr}/10$, beam deflections are approximately 10 percent larger than those found by neglecting the influence of axial forces. For a value of N_x equal to $N_{cr}/2$, however, deflections computed by ignoring N_x may be more than 100 percent in error.

6.10 Theory of cables. Our study of flexible ties leads us to the consideration of the interesting limiting case of a tie with zero or negligible flexural rigidity—the structural cable. A cable is a slender flexible member capable of developing only tensile forces. Technically, a cable differs from a simple tension tie in that it has practically no flexural stiffness and it derives its capacity to resist transverse loads by undergoing significant changes in slope at points at which transverse loads are applied. This causes force components to be developed transverse to the path of the cable, thereby establishing equilibrium. Strings, ropes, wires, and chains are typical examples of cables. In the developments to follow, we are primarily concerned with evaluating the shape that the cable acquires in order to support the applied load.

To begin our study, let us consider the cable shown in Fig. 6.15*a* suspended between two points *A* and *B* and subjected to a general system of vertical loads p_y. The straight line *AB*, called the *chord* of the cable, is of length *L*. Since no horizontal forces act on the structure, the

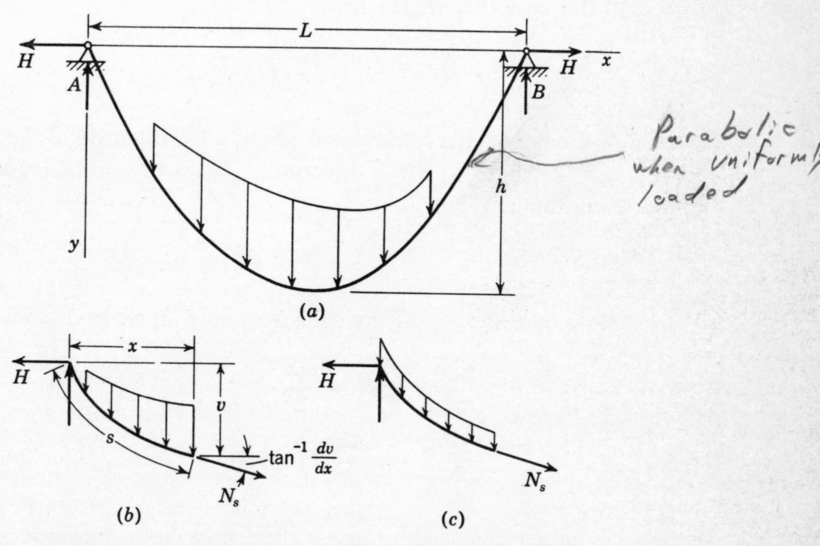

FIGURE 6.15 Geometry and statics of a simple cable.

horizontal component of the cable tension N_s is a constant, denoted H.[8] The loaded cable acquires some shape, $v = v(x)$, which provides equilibrium.

From the statics of the segment of the cable shown in Fig. 6.15*b*, we find that N_s is related to H by

$$H = N_s \frac{dx}{ds} \tag{6.34}$$

where s is the length of arc measured along the deformed axis of the cable. Since, by definition, the flexural rigidity of the cable is zero, the bending moment at each section must vanish. Thus,

$$Hv = M_l \tag{6.35}$$

where M_l, we recall, is the bending moment due to the applied loads. Equation (6.35) is a statement of the *theorem of simple cables: the product of the horizontal component of the tension in a cable under vertical loads and the vertical distance from a point P on the cable to the cable chord is equal to the bending moment at P of a simple beam of the same length under the same load.* A simple calculation shows that this theorem is also valid in cases in which AB is inclined with respect to the x axis.

It is interesting to compare Eq. (6.35) with the results obtained for the tie. For small deflections of a straight tie, dx/ds is unity so that according to Eq. (6.34) N_x, N_s, and H are equal. If we replace N_x with H in Eq. (6.35), multiply both sides by EI_z, and then set EI_z equal to zero, we get Eq. (6.35).

The case of a uniformly loaded cable is a classic example of cable theory. In this case, Eq. (6.35) yields

$$Hv = M_l = \frac{p_0 x}{2}(L - x)$$

where L is the length of the cable chord and p_0 is the intensity of the applied load per unit of length in the x direction. Thus, the cable acquires the shape of a parabola:

$$v = \frac{p_0 x}{2H}(L - x) \tag{a}$$

If we denote by h the sag of the cable at $x = L/2$, we find from Eq. (*a*) that

$$H = \frac{p_0 L^2}{8h} \tag{6.36}$$

Hence,

$$v = \frac{4hx}{L^2}(L - x) \tag{6.37}$$

[8] We return to the notation N_s rather than N_x here, since the cable tension is directed tangent to the cable curve, in the "s" direction.

The tension developed in the cable is now easily evaluated. Referring to Eq. (6.34), we note that

$$N_s = H\frac{ds}{dx} = H\left[1 + \left(\frac{dv}{dx}\right)^2\right]^{\frac{1}{2}} \tag{b}$$

Thus, introducing Eq. (6.37) gives

$$N_s = H\left[1 + 16\gamma^2\left(1 - \frac{2x}{L}\right)^2\right]^{\frac{1}{2}} \tag{6.38}$$

where $\gamma = h/L$ is called the *sag ratio* of the cable. We note that the maximum tension occurs at A and B, and that it is of magnitude $H(1 + 16\gamma^2)^{\frac{1}{2}}$.

A second loading case of practical interest is that of a cable loaded by its own weight. A free body of a portion of such a cable is shown in Fig. 6.15*c*. Referring to this figure and summing forces in the vertical direction, we find

$$p(s + S/2) + N_s \sin\left(\tan^{-1}\frac{dv}{dx}\right) = 0 \tag{c}$$

where p, in this case, is the weight of the cable per unit length and S is the total cable length. The second term in this equation can be written

$$N_s \sin\left(\tan^{-1}\frac{dv}{dx}\right) = N_s\frac{dv}{ds}\frac{dx}{dx} = H\frac{dv}{dx}$$

so that Eq. (*c*) becomes

$$\frac{dv}{dx} = -\frac{p}{H}(s + S/2) \tag{d}$$

Differentiating this equation with respect to s gives

$$\frac{dv'}{ds} = \frac{1}{\sqrt{1 + v'^2}}\frac{dv'}{dx} = -\frac{p}{H} \tag{e}$$

where $v' = dv/dx$.

It is interesting to compare Eq. (*e*) with that for the initially straight tie undergoing small displacements. For the tie, v'^2 is negligible in comparison with unity and N_s becomes N_x and is equal to H. In this case Eq. (*e*) becomes

$$-N_x\frac{d^2v}{dx^2} = p \tag{f}$$

This result is identical to Eq. (6.29) if EI_z is zero.

Returning to Eq. (*e*), we find on integrating that

$$\sinh^{-1} v' = -\frac{px}{H} + C$$

or

$$\frac{dv}{dx} = \sinh\left(-\frac{px}{H} + C\right) \tag{g}$$

C being the constant of integration. Since v' is zero at $x = L/2$, we find that C is $pL/2H$. Another integration yields

$$v = -\frac{H}{p}\cosh\frac{p(L-2x)}{2H} + C_1$$

From the condition that v is zero when x is zero, we find

$$C_1 = \frac{H}{p}\cosh\frac{pL}{2H}$$

so that the final shape of the loaded cable is

$$v = -\frac{H}{p}\left[\cosh\frac{p(L-2x)}{2H} - \cosh\frac{pL}{2H}\right] \tag{6.39}$$

This is the equation of a *catenary*, the geometric curve with the lowest center of gravity. H, in this case, is determined numerically from the condition that $v = h$ at $x = L/2$. We now evaluate N_s by simply substituting Eq. (6.39) into Eq. (*b*):

$$N_s = H\left[1 + \sinh^2\frac{p(L-2x)}{2H}\right]^{\frac{1}{2}}$$

which may be written

$$N_s = H\cosh\frac{p(L-2x)}{2H} \tag{6.40}$$

We see that the maximum value of N_s occurs at $x = 0$ and that it is equal to $H\cosh(pL/2H)$.

The total length of the cable, regardless of the loading, is given by

$$S = \int_0^S ds = \int_0^L \sqrt{1 + \left(\frac{dv}{dx}\right)^2}\,dx \tag{6.41}$$

which, according to Eq. (*b*), may be written

$$S = \frac{1}{H}\int_0^L N_s\,dx \tag{6.42}$$

Using Eq. (6.38), we find for the length of the uniformly loaded cable

$$S = \int_0^L [1 + 16\gamma^2(L-2x)^2]^{\frac{1}{2}}\,dx$$

$$= \frac{L}{2}\left[\sqrt{1+16\gamma^2} + \frac{1}{4\gamma}\ln\left(4\gamma + \sqrt{1+16\gamma^2}\right)\right] \tag{h}$$

Similarly, introducing Eq. (6.40) into Eq. (6.42), we find for the length of the cable supporting its own weight

$$S = \int_0^L \cosh\frac{p(L-2x)}{2H}\,dx = \frac{2p}{H}\sinh\frac{pL}{2H} \tag{i}$$

The change in length of the cable due to N_s is also of interest. If δ_s is the change in length, we know that for an element of length Δs, the incremental change in length $\Delta\delta_s$ is such that

$$\sigma_s = \frac{N_s}{A} = E\epsilon_s = E \lim_{s\to 0} \frac{\Delta\delta_s}{\Delta s} = E\frac{d\delta_s}{ds}$$

where A is the cross-sectional area of the cable. The total change in length, then, is

$$\delta_s = \int_0^S \frac{N_s\, ds}{AE}$$

or

$$\delta_s = \frac{H}{AE}\int_0^L \left[1 + \left(\frac{dv}{dx}\right)^2\right]^{\frac{1}{2}} dx \tag{6.43}$$

Thus, for the uniformly loaded parabolic cable, we find on substituting Eq. (6.37) into Eq. (6.43) that

$$\delta_s = \frac{HS(3 + 16\gamma^2)}{AE(3 + 8\gamma^2)} \tag{j}$$

where S is given by Eq. (h). The unstressed length of the cable is $S - \delta_s$. Similar results may be obtained for the catenary by simply substituting Eq. (6.39) into Eq. (6.43) and performing the indicated integration.

6.11 Beams on elastic foundations. We shall now investigate a class of problems in beam bending that leads us to a differential equation of the elastic curve which is considerably different from those studied previously—namely, that of a straight beam resting on a linearly elastic foundation. By an "elastic foundation," we mean that the beam is supported by some load-bearing medium which responds elastically to beam deformations by developing a resisting "load" distribution which is proportional to the transverse deflection of the beam. We may think of such a foundation as an infinite number of independent linear springs distributed continuously along the length of the beam. The spring stiffness, or modulus, k is assumed to be constant and has the units force per length squared.

Originally, the problem of a beam on an elastic foundation arose in the study of stresses and deflections of rails resting on "elastic" soils. The general theory, however, has since been applied to a wide variety of problems including the analysis of grids, floor systems of ships and bridges, underground framed structures, missile hulls, and cylindrical shells.

Let us consider the beam shown in Fig. 6.16a which is supported by a linearly elastic foundation and is subjected to a general system of applied loads. By definition, the foundation exerts an upward force per unit length of magnitude kv. Referring to the free body of a typical beam

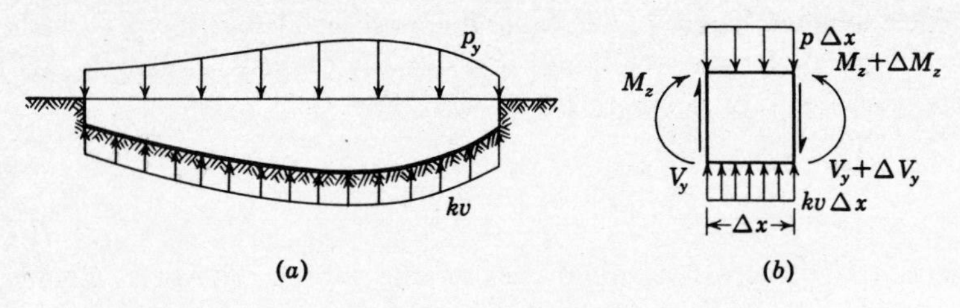

FIGURE 6.16 (*a*) Straight beam on an elastic foundation; (*b*) typical beam element.

element shown in Fig. 6.16*b*, we see that the resultant transverse force per unit length is now $p_y - kv$ rather than p_y, as before. Hence, from the statics of the element we find

$$\frac{dV_y}{dx} = kv - p_y \tag{6.44}$$

and

$$\frac{d^2M_z}{dx^2} + p_y = kv \tag{6.45}$$

Recalling that $-EI_z\, d^4v/dx = d^2M_z/dx^2$ [see Eqs. (6.19)], we rewrite Eq. (6.45) in the form

$$\frac{d^2}{dx^2}\left(EI_z \frac{d^2v}{dx^2}\right) + kv = p_y \tag{6.46}$$

This is the governing differential equation of the elastic curve of a straight beam on an elastic foundation. We obtain this same result, of course, by simply replacing p_y in Eq. (6.16) by $p_y - kv$. The elastic curve is necessarily defined by a fourth-order equation; no alternative second-order equation is available as was the case in our previous investigations. Hence, it is always necessary to have four independent boundary conditions.

Mz + Vy @ BOTH ENDS

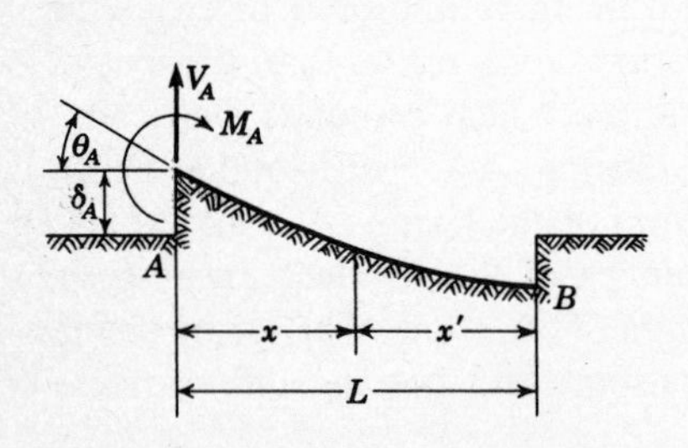

FIGURE 6.17 Beam on an elastic foundation subjected to end moment and shear.

Consider, for example, the beam *AB* shown in Fig. 6.17. The beam is subjected to an applied moment M_A and an applied shear V_A at end *A* where $x = 0$. The deflection and slope of the beam at $x = 0$ are denoted δ_A and θ_A, respectively. For this structure, Eq. (6.46) becomes

$$\frac{d^4v}{dx^4} + 4\beta^4 v = 0 \tag{6.47}$$

where

$$\beta = \sqrt[4]{\frac{k}{4EI_z}} \tag{6.48}$$

The characteristic polynomial corresponding to Eq. (6.47) is

$$m^4 + 4\beta^4 = 0$$

which has the four roots

$$m_{1,2,3,4} = \pm\beta(1 \pm i)$$

Thus, the solution is

$$v = C_1 e^{\beta(1+i)x} + C_2 e^{-\beta(1+i)x} + C_3 e^{\beta(1-i)x} + C_4 e^{-\beta(1-i)x}$$

which, after some simple transformations, may be put in the more convenient form

$$v = e^{\beta x}(A \cos \beta x + B \sin \beta x) + e^{-\beta x}(C \cos \beta x + D \sin \beta x) \qquad (a)$$

From the conditions

$$v(0) = \delta_A \qquad \frac{dv(0)}{dx} = \theta_A$$

$$-EI_z \frac{d^2v(0)}{dx^2} = M_A \qquad -EI_z \frac{d^3v(0)}{dx^3} = V_A$$

we arrive at the four equations for the constants A, B, C, and D:

$$\delta_A = A + C \qquad -\frac{M_A}{EI_z} = 2\beta^2(B - D)$$

$$\theta_A = \beta(A + B + D - C) \qquad -\frac{V_A}{EI_z} = 2\beta^3(-A + B + D + C)$$

Solving these, we find

$$A = \frac{\delta_A}{2} + \frac{\theta_A}{4\beta} + \frac{V_A}{8\beta^3 EI_z} \qquad B = \frac{\theta_A}{4\beta} - \frac{M_A}{4\beta^2 EI_z} - \frac{V_A}{8\beta^3 EI_z}$$

$$C = \frac{\delta_A}{2} - \frac{\theta_A}{4\beta} - \frac{V_A}{8\beta^3 EI_z} \qquad D = \frac{\theta_A}{4\beta} + \frac{M_A}{4\beta^2 EI} - \frac{V_A}{8\beta^3 EI_z} \qquad (b)$$

Finally, substituting these values into Eq. (*a*) and recalling that $\cosh \beta x = (e^{\beta x} + e^{-\beta x})/2$ and $\sinh \beta x = (e^{\beta x} - e^{-\beta x})/2$, we find for the equation of the elastic curve

$$v = \delta_A \psi_1(\beta x) + \frac{\theta_A}{\beta}\psi_2(\beta x) - \frac{M_A}{\beta^2 EI_z}\psi_3(\beta x) - \frac{V_A}{\beta^3 EI_z}\psi_4(\beta x) \qquad (c)$$

where

$$\begin{aligned} \psi_1(\beta x) &= \cosh \beta x \cos \beta x \\ \psi_2(\beta x) &= \tfrac{1}{2}(\sinh \beta x \cos \beta x + \cosh \beta x \sin \beta x) \\ \psi_3(\beta x) &= \tfrac{1}{2} \sinh \beta x \sin \beta x \\ \psi_4(\beta x) &= \tfrac{1}{4}(\cosh \beta x \sin \beta x - \sinh \beta x \cos \beta x) \end{aligned} \qquad (d)$$

δ_A and θ_A of Eq. (c) may be expressed in terms of M_A and V_A by introducing the conditions that the moment and shear at B,

$$-EI_z \frac{d^2v(L)}{dx^2} \qquad \text{and} \qquad -EI_z \frac{d^3v(L)}{dx^3}$$

are zero. Such calculations give

$$\delta_A = -\frac{2M_A\beta^2}{k} \frac{\sinh^2 \beta L + \sin^2 \beta L}{\sinh^2 \beta L - \sin^2 \beta L} - \frac{2V_A\beta}{k} \times \frac{\sinh \beta L \cosh \beta L - \sin \beta L \cos \beta L}{\sinh^2 \beta L - \sin^2 \beta L} \tag{e}$$

and

$$\theta_A = \frac{4M_A\beta^3}{k} \frac{\cosh \beta L \sinh \beta L - \cos \beta L \sin \beta L}{\sinh^2 \beta L - \sin^2 \beta L} + \frac{2V_A\beta^2}{k} \frac{\sinh^2 \beta L + \sin^2 \beta L}{\sinh^2 \beta L - \sin^2 \beta L} \tag{f}$$

Denoting $x' = L - x$ and substituting Eqs. (e) and (f) into Eq. (c) and simplifying, we get

$$\begin{aligned} v = M_A\Big\{&\frac{2\beta^2/k}{\sinh^2 \beta L - \sin^2 \beta L} [\sinh \beta L (\cosh \beta x' \sin \beta x - \sinh \beta x' \cos \beta x) \\ &+ \sin \beta L (\sinh \beta x' \cos \beta x - \cosh \beta x' \sin \beta x)]\Big\} \\ - V_A\Big\{&\frac{2\beta/k}{\sinh^2 \beta L - \sin^2 \beta L} [\sinh \beta L \cos \beta x \cosh \beta x' \\ &- \sin \beta L \cosh \beta x \cos \beta x']\Big\} \end{aligned} \tag{g}$$

It is clear from this example that the determination of the arbitrary constants in studying beams on elastic foundations is often quite a laborious job, even for relatively simple boundary conditions. For this reason, it is common practice to take full advantage of the principle of superposition. Superposition is valid provided that the foundation modulus k is the same for each case to be superimposed. This conclusion should follow directly from an argument similar to that given earlier for ties and beam columns; the "complementary" portion of the equation for each elastic curve must satisfy Eq. (6.47). As a simple example, consider the beam shown in Fig. 6.18a whose elastic curve is the sum of the elastic curves of the beams shown in Fig. 6.18b and c. The elastic curve of the beam in Fig. 6.18b is given by Eq. (c) and may be written in the abbreviated form

$$v_b = M_A\Phi_1(\beta x) - V_A\Phi_2(\beta x)$$

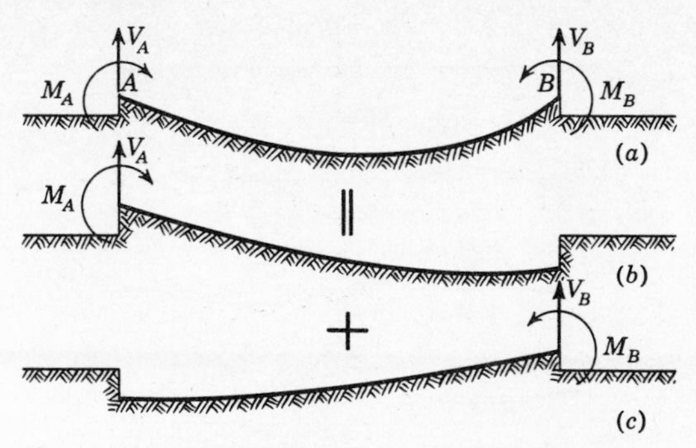

FIGURE 6.18 Superposition of elastic curves.

where $\Phi_1(\beta x)$ and $\Phi_2(\beta x)$ are the complicated functions following M_A and V_A, respectively, in Eq. (*g*). Obviously, the elastic curve of the beam in Fig. 6.18*c* is

$$v_c = M_B\Phi_1(\beta x') - V_B\Phi_2(\beta x')$$

where, as before, $x' = L - x$. Finally,

$$v_a = v_b + v_c = M_A\Phi_1(\beta x) - V_A\Phi_2(\beta x) + M_B\Phi_1(\beta x') - V_B\Phi_2(\beta x') \quad (h)$$

in which v_a is the elastic curve of the beam in Fig. 6.18*a*. Numerous other cases can be found in a similar manner.

We now investigate a class of problems less complicated than those just studied but of significant practical interest—namely, the deflections of beams of infinite or semi-infinite length resting on an elastic foundation. Infinite beams extend indefinitely in both the positive and negative directions from a given origin, while semi-infinite beams have one end, usually taken at $x = 0$, and extend indefinitely in the positive direction. We discuss only a few examples of each here, since other cases may be easily obtained by following similar procedures.

Let us first consider the beam of infinite length shown in Fig. 6.19 acted upon by a single concentrated load P at the origin. Equation (*a*) is still valid for all values of x; but we see that the first term (the term containing $e^{\beta x}$) leads to infinite deflections for infinite x. This, obviously, is impossible, and we must conclude that A and B are zero. This leaves only two constants C and D to be determined.

By inspection, we see that the slope of the elastic curve is zero at the origin. Also, by considering the equilibrium of the segment under the load P, we find that the shear is $-P/2$ as x approaches zero from the right. Therefore, from Eq. (*a*) we find the relations

$$\frac{dv(0)}{dx} = 0 = \beta(D - C)$$

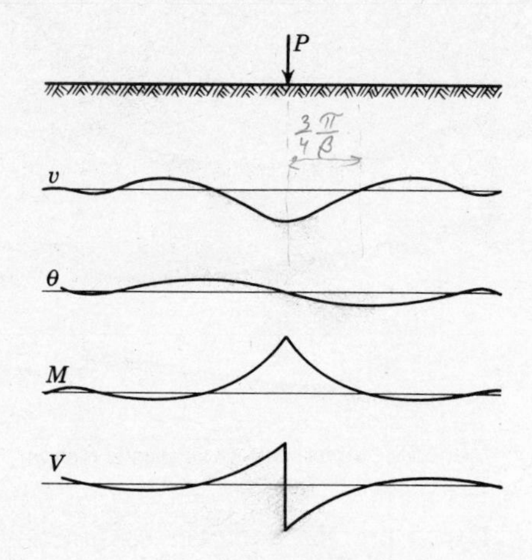

FIGURE 6.19 Infinite beam subjected to a single force P.

and
$$-EI_z \frac{d^3v(0)}{dx^3} = -\frac{P}{2} = -2\beta^3 EI_z(D + C)$$

Hence,
$$C = D = \frac{P}{8\beta^3 EI_z} = \frac{P\beta}{2k}$$

It follows that the equation of the elastic curve of the infinite beam is

$$v = \frac{P\beta}{2k} e^{-\beta x}(\cos \beta x + \sin \beta x) \tag{6.49}$$

Following a standard procedure,[9] we introduce the notation

$$\begin{aligned} f_1(\beta x) &= e^{-\beta x}(\cos \beta x + \sin \beta x) \\ f_2(\beta x) &= e^{-\beta x} \sin \beta x \\ f_3(\beta x) &= e^{-\beta x}(\cos \beta x - \sin \beta x) \\ f_4(\beta x) &= e^{-\beta x} \cos \beta x \end{aligned} \tag{6.50}$$

so that
$$\begin{aligned} v &= \frac{P\beta}{2k} f_1(\beta x) \\ \theta &= \frac{dv}{dx} = -\frac{P\beta^2}{k} f_2(\beta x) \\ M_z &= -EI_z \frac{d^2v}{dx^2} = \frac{P}{4\beta} f_3(\beta x) \\ V_y &= -EI_z \frac{d^3v}{dx^3} = -\frac{P}{2} f_4(\beta x) \end{aligned} \tag{6.51}$$

[9] See, for example, Ref. 26, pp. 10–37.

Numerical values of the functions f_1, f_2, f_3, f_4 are given in Table 6.1. The shape of the elastic curve, the variation in slope, the bending-moment diagram, and the shear diagram are shown in Fig. 6.19. Note that Eqs. (6.51) are applicable only for $x \geq 0$; the curves v and M_z are symmetrical with respect to a vertical axis through the origin, while θ and V_y are antisymmetrical.

Table 6.1 Beam functions

βx	$f_1(\beta x)$	$f_2(\beta x)$	$f_3(\beta x)$	$f_4(\beta x)$
0.0	1.0000	0.0000	1.0000	1.0000
0.2	0.9651	0.1627	0.6398	0.8024
0.4	0.8784	0.2610	0.3564	0.6174
0.6	0.7628	0.3099	0.1430	0.4529
0.8	0.6353	0.3223	−0.0093	0.3131
1.0	0.5083	0.3096	−0.1109	0.1987
1.2	0.3898	0.2807	−0.1716	0.1091
1.4	0.2849	0.2430	−0.2011	0.0419
1.6	0.1960	0.2018	−0.2077	−0.0059
1.8	0.1234	0.1610	−0.1985	−0.0376
2.0	0.0667	0.1230	−0.1793	−0.0563
2.2	0.0244	0.0895	−0.1547	−0.0652
2.4	−0.0056	0.0613	−0.1282	−0.0669
2.6	−0.0254	0.0383	−0.1020	−0.0637
2.8	−0.0369	0.0204	−0.0777	−0.0573
3.0	−0.0422	0.0071	−0.0563	−0.0493
3.2	−0.0431	−0.0024	−0.0383	−0.0407
3.4	−0.0408	−0.0085	−0.0238	−0.0323
3.6	−0.0366	−0.0121	−0.0124	−0.0245
3.8	−0.0314	−0.0137	−0.0040	−0.0177
4.0	−0.0258	−0.0139	0.0019	−0.0120
4.2	−0.0204	−0.0131	0.0057	−0.0096
4.4	−0.0155	−0.0117	0.0079	−0.0038
4.6	−0.0111	−0.0100	0.0089	−0.0012
4.8	−0.0075	−0.0082	0.0089	0.0007
5.0	−0.0046	−0.0065	0.0084	0.0019

In the case of the semi-infinite beam shown in Fig. 6.20, the elastic curve must satisfy the conditions

$$\frac{d^2v(0)}{dx^2} = 0 \qquad -EI_z\frac{d^3v(0)}{dx^3} = -P$$

Again, A and B of Eq. (a) must be zero. Substitution of the remaining terms into the above conditions gives

$$D = 0 \qquad C = \frac{P}{2\beta^3 EI_z}$$

Therefore,

$$v = \frac{2\beta P}{k} e^{-\beta x} \cos \beta x$$

Semiinfinite beam may be analyzed as infinite beam with contractive forces + moments @ point of cut

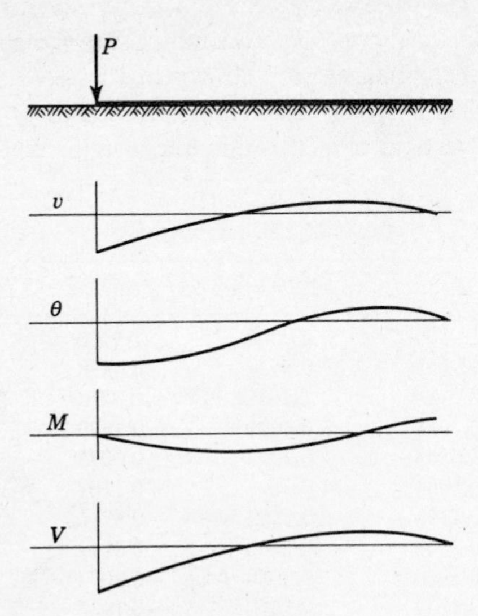

FIGURE 6.20 Semi-infinite beam subjected to a force P at its end.

Through successive differentiations of this equation and the use of the notation in Eqs. (6.50), we find

$$
\begin{aligned}
v &= \frac{2\beta P}{k} f_4(\beta x) \\
\theta &= -\frac{2\beta^2 P}{k} f_1(\beta x) \\
M_z &= -\frac{P}{\beta} f_2(\beta x) \\
V_y &= -P f_3(\beta x)
\end{aligned}
\tag{6.52}
$$

These functions are shown in Fig. 6.20.

Finally, for the semi-infinite beam in Fig. 6.21 which is subjected to an end moment M_0 a similar series of calculations gives

$$
-EI_z \frac{d^2v(0)}{dx^2} = EI\beta^2 2D = M_0
$$

$$
-EI_z \frac{d^3v(0)}{dx^3} = C + D = 0
$$

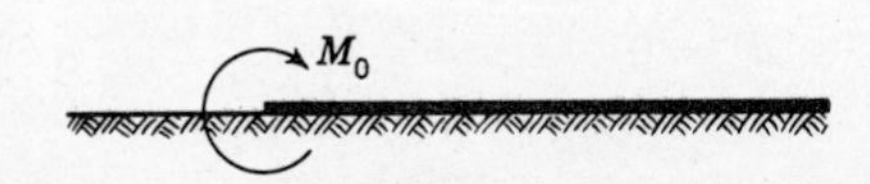

FIGURE 6.21 Semi-infinite beam with end moment.

so that

$$v = -\frac{2\beta^2 M_0}{k} f_3(\beta x)$$

$$\theta = \frac{M_0}{EI\beta} f_4(\beta x) \tag{6.53}$$

$$M_z = M_0 f_1(\beta x)$$

$$V_y = -2\beta M_0 f_2(\beta x)$$

It is clear from these examples that f_1, f_2, f_3, and f_4 are quite useful functions. What is also important to note, however, is that each of these functions "dampens out" very rapidly; hence, end loads such as P and M_0 influence only a small portion of the beam. In fact, on inspecting Table 6.1 we see that for $\beta x = 3.0$, each function is only about 5 percent of its maximum value. For $\beta x = 5.0$, they amount to only a fraction of 1 percent of their maximum values.

This fact is quite useful in analyzing beams of finite length greater than $\beta/5$. Because of this rapid dampening of end effects, deformations at one end of beams of finite length are not affected by moments and shears existing at the opposite end. For example, if the length of the beam in Fig. 6.18*a* is greater than $\beta/5$ (which is often the case), the end slope and the deflection at A are not affected by V_B and M_B and vice versa.

6.12 Beam columns and ties on elastic foundations. If the beam on an elastic foundation is flexible and is subjected to large axial forces, it is often necessary to account for the influence of the axial forces on deformations. In order to do this, we must solve a differential equation of a different form from those considered previously. Referring to Eq. (6.29), we recall that the reaction of the foundation diminishes the magnitude of the total transverse load by an amount kv, so that the right-hand side of this equation becomes $p_y - kv$. Thus, the governing differential equation of flexible beams on elastic foundations is

$$\frac{d^2}{dx^2}\left(EI_z \frac{d^2v}{dx^2}\right) - N_x \frac{d^2v}{dx^2} + kv = p_y \tag{6.54}$$

The sign in front of the second term becomes positive if N_x is a compressive force.

As a simple example, let us consider the semi-infinite tie shown in Fig. 6.22 subjected to an end moment and an end shear M_A and V_A.

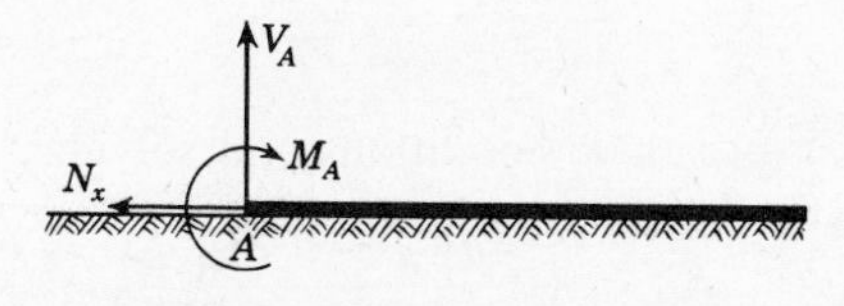

FIGURE 6.22 Semi-infinite tie on an elastic foundation.

EI_z is assumed to be constant. The characteristic polynomial is

$$m^4 - \lambda^2 m^2 + 4\beta^4 = 0 \tag{a}$$

where, as before, $\lambda = (N_x/EI_z)^{\frac{1}{2}}$ and $\beta = (k/4EI_z)^{\frac{1}{4}}$. The four roots are

$$m_{1,2,3,4} = \pm\sqrt{\tfrac{1}{2}\lambda^2 \pm i\sqrt{4\beta^4 - \tfrac{1}{4}\lambda^4}} \tag{b}$$

To proceed, we must know the relative magnitude of λ and β in order to determine whether the quantity $4\beta^2 - \frac{1}{4}\lambda^4$ is positive, zero, or negative. A completely different form of the solution exists for each case, excluding, of course, the three other cases pertaining to problems in which N_x is compressive. We consider only one of these cases here, since the general procedure is the same for each.

Let us assume that N_x is a tensile force of such a magnitude that $4\beta^2 - \frac{1}{4}\lambda^4$ is negative. Then $N_x < 2\sqrt{2kEI_z}$, which is often the case in practical problems. We then write

$$m_{1,2,3,4} = \pm(a \pm ib) \tag{c}$$

where

$$a = (\beta^2 + \tfrac{1}{4}\lambda^2)^{\frac{1}{2}} \qquad b = (\beta^2 - \tfrac{1}{4}\lambda^2)^{\frac{1}{2}} \tag{d}$$

Thus, the solution to Eq. (6.54) is

$$v = (C_0 e^{ax} + C_1 e^{-ax}) \cos bx + (C_2 e^{ax} + C_3 e^{-ax}) \sin bx \tag{e}$$

Since the beam extends infinitely in the x direction, it follows from our earlier arguments that C_0 and C_2 must be zero. The remaining constants must be determined from the conditions that the bending moment and the shear at $x = 0$ are M_A and V_A. In this case, however, a component of the axial force N_x contributes to the transverse shear at every point. According to Eq. (6.28),

$$V_y = \frac{dM_z}{dx} + N_x \frac{dv}{dx}$$

Thus, the shear boundary condition at $x = 0$ is

$$\frac{d^3v(0)}{dx^3} - \lambda^2 \frac{dv(0)}{dx} = -\frac{V_A}{EI_z} \tag{f}$$

The second condition to be satisfied is

$$\frac{d^2v(0)}{dx^2} = -\frac{M_A}{EI_z} \tag{g}$$

Substituting Eq. (e) into these conditions and solving for the constants, we find

$$C_1 = -\frac{aV_A + \beta^2 M_A}{\beta^2(2\beta^2 + \lambda^2)EI_z} \tag{h}$$

and
$$C_3 = \frac{-\lambda^2 V_A + 4a\beta^2 M_A}{4b\beta^2(2\beta^2 + \lambda^2)EI_z} \tag{i}$$

Finally, introducing C_1 and C_3 into Eq. (*e*) and simplifying, we arrive at the equation for the elastic curve:

$$v = -V_A \frac{4ab \cos bx + \lambda^2 \sin bx}{4b\beta^2(2\beta^2 + \lambda^2)EI_z} e^{-ax} + M_A \frac{-b \cos bx + a \sin bx}{b(2\beta^2 + \lambda^2)EI_z} e^{-ax} \tag{j}$$

If N_x is zero, a and b are equal to β. If, for this case, V_A and M_A are replaced by $-P$ and M_0, respectively, the first term of Eq. (*f*) reduces to the expression for v in Eqs. (6.52) and the second term reduces to the expression for v in Eqs. (6.53).

PROBLEMS

6.1. The semicircular curved bar shown is subjected to vertical and horizontal loads P and H. Evaluate the displacement of support A in the direction of H in terms of R, J_z, P, and H.

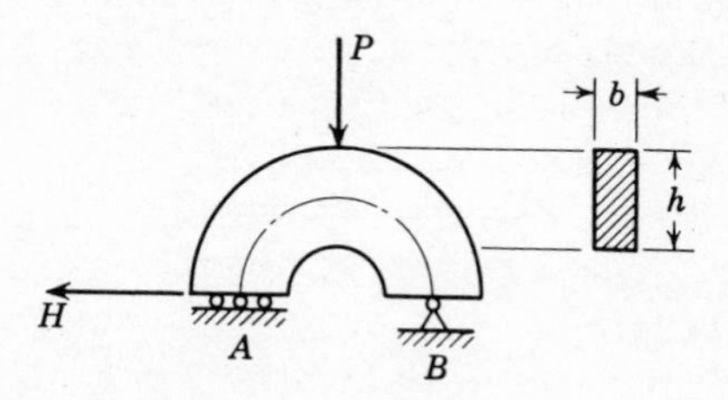

6.2. Derive the equation(s) for the elastic curve of the structure shown and evaluate the maximum displacement of point A. Take $E = 6 \times 10^6$ psi.

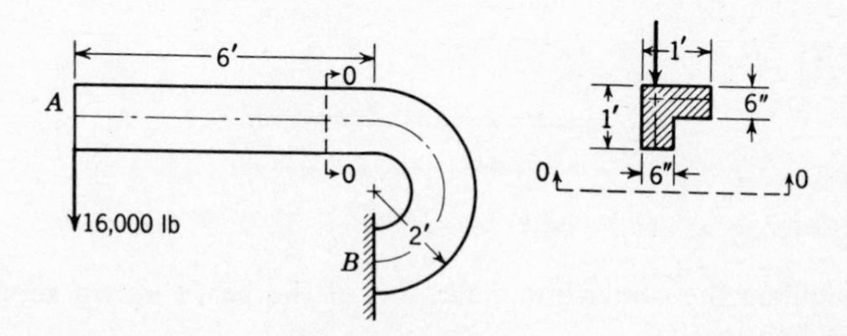

6.3. Find the equation(s) for the elastic curve of the beam shown and calculate the radial displacement of the free end.

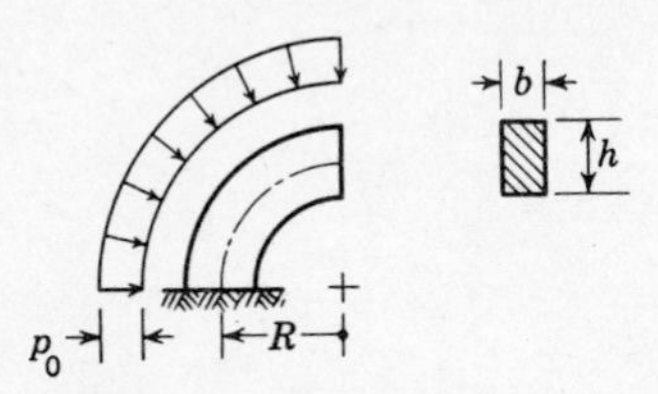

6.4. For what value of h/R will the end displacement of the beam in Prob. 6.3 differ 5 percent from that obtained by using I_z rather than J_z in the calculations?

6.5. Obtain the equations for the elastic curve of the beam in Prob. 4.10 and evaluate the deflection of the free end for the case in which $\beta = 30°$.

6.6. Obtain the equations for the elastic curve of the beam in Prob. 5.4 and evaluate the deflection of the free end.

6.7–6.9. Find the equations for the elastic curves of the beams shown below. Each beam is loaded in a plane of symmetry and I_z is constant.

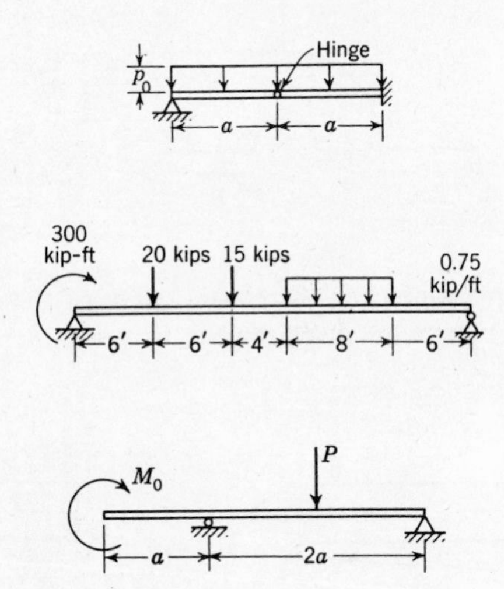

6.10. Calculate the center-line deflection of the beam shown accounting for shear deformation.

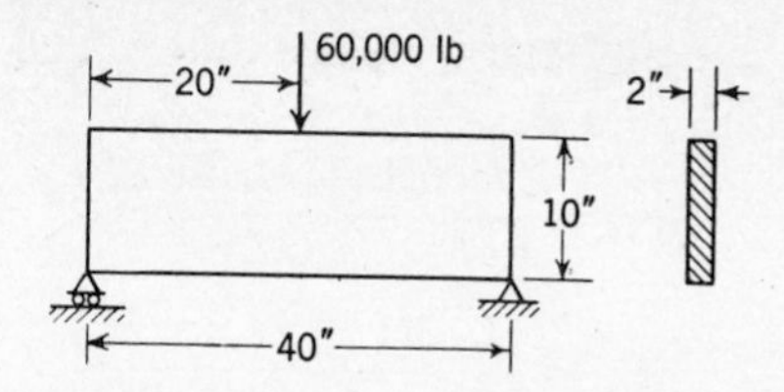

6.11. Show that the moment developed at the fixed end of the beam shown is given by

$$M_B = -\frac{p_0L^2}{2}\frac{\tan \lambda L[2 \tan (\lambda L/2) - \lambda L]}{\lambda L(\tan \lambda L - \lambda L)}$$

where $\lambda = \sqrt{N_x/EI}$.

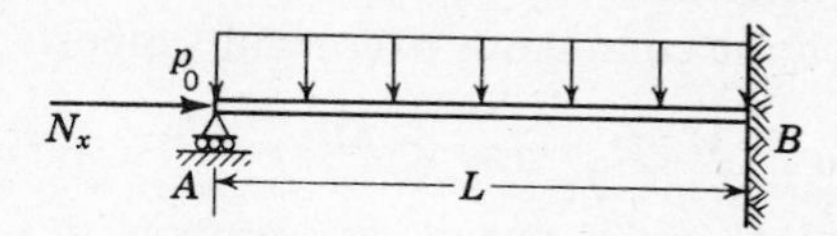

6.12. Evaluate the end slopes of the beam shown assuming that N_x is

(*a*) a tensile force
(*b*) a compressive force

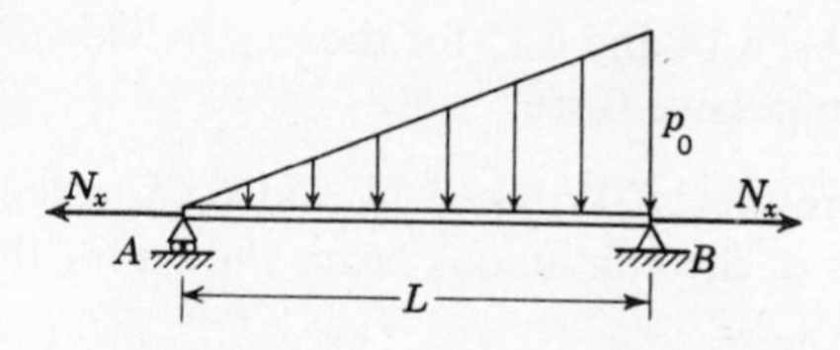

6.13. Find the elastic curve of the beam shown assuming that N_x is

(*a*) a tensile force
(*b*) a compressive force

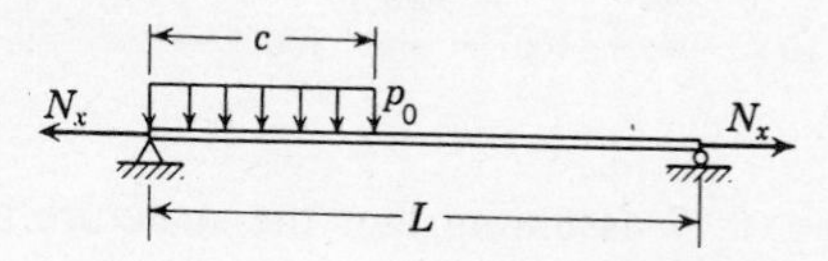

6.14. Show that the end slope at j of a simply supported beam column ij which is subjected to transverse loads and end moments is

$$\theta_{ji} = M_i g_{ij} + M_j f_{ji} + \tau_{ji}$$

where τ_{ji} is the slope at j due to loads and g_{ij} and f_{ji} are the angular flexibilities defined by

$$g_{ij} = \frac{L}{6EI} \frac{6(1 - \lambda L \csc \lambda L)}{(\lambda L)^2}$$

$$f_{ji} = \frac{L}{3EI} \frac{3(1 - \lambda L \cot \lambda L)}{(\lambda L)^2}$$

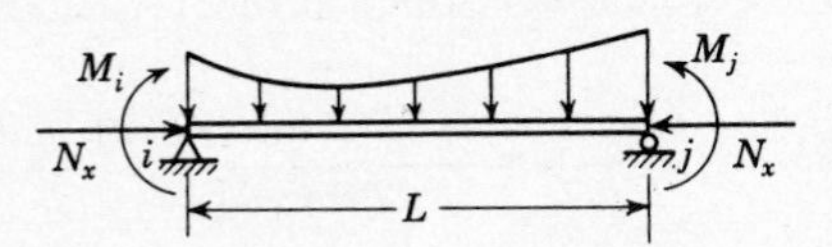

6.15. Derive the three-moment equation for beam columns of constant cross section which are continuous over rigid supports:

$$M_i g_{ij} + M_j(f_{ji} + f_{jk}) + M_k g_{kj} + \tau_{ji} + \tau_{jk} = 0$$

where M_i, M_j, and M_k are moments over supports i, j, and k; g_{ij}, g_{kj}, f_{ji}, and f_{jk} are end slopes at j of spans ij and jk due to unit moments at i, j, and k (angular flexibilities, see Prob. 6.14); and τ_{ji} and τ_{jk} are end slopes of spans ij and jk at j due to applied loads. (*Hint:* use the continuity condition that the slope of span ij at j must be minus the slope of span jk at j.)

6.16. Repeat Probs. 6.14 and 6.15 for the case in which N_x is a tensile force rather than a compressive force.

6.17. Using the results of Probs. 6.14 and 6.15, calculate the moments over the supports of the continuous beam shown for the following cases:

(*a*) $N_x = 0$
(*b*) $N_x = 45$ kips
(*c*) $N_x = 400$ kips

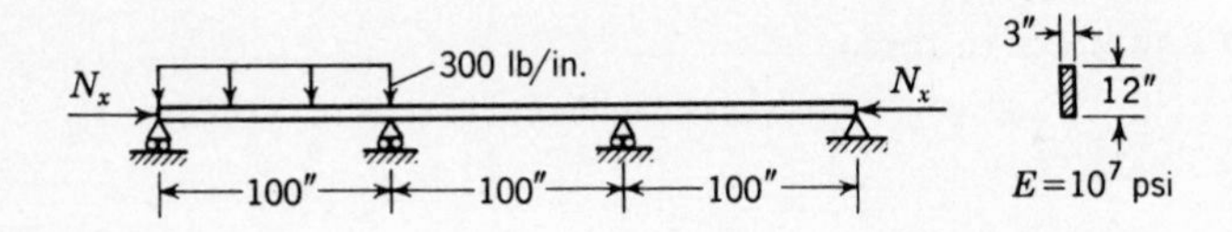

6.18. Rework Prob. 6.17 assuming that the structure is a continuous tie rather than a continuous beam column, that is, assuming that N_x is a tensile force.

6.19. The ends of the beam column shown are elastically restrained by rotational springs of stiffness k_θ which develop end moments proportional to the end slopes.

(*a*) Set up the equation for finding the critical buckling load.
(*b*) Solve for N_{cr} by trial and error for the case in which $k_\theta = \sqrt{4EI/L}$.

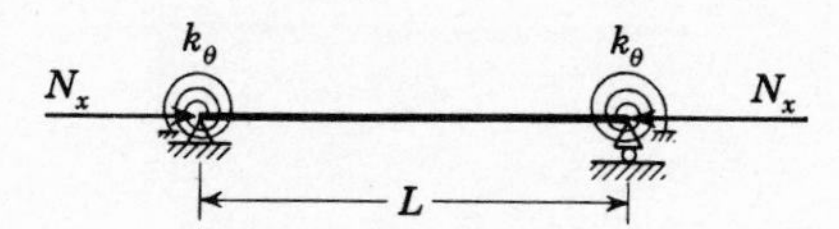

6.20. The chord of a uniformly loaded cable is inclined an angle α with respect to the x axis. Show that the static shape of the cable is given by

$$v = \frac{4hx}{L^2}(L - x) - x \tan \alpha$$

where h is the distance from the cable to the chord at $x = L/2$.

6.21. A cable supporting a uniform load of 100 lb/horizontal ft is suspended between two equally elevated points 400 ft apart. The sag is adjusted so that $h = 20$ ft. What is the maximum tension in the cable?

6.22. A cable is suspended between two points of equal elevation which are 1,800 ft apart. The horizontal thrust produced by a uniform load of 1,000 lb/horizontal ft is 3,000 lb.

(*a*) What is the maximum tension in the cable?
(*b*) What is the length of the loaded cable?
(*c*) If $E = 29 \times 10^6$ psi and $A = 40$ sq in., what is the unstressed length of the cable?

6.23. Calculate the horizontal thrust H developed in the cable shown.

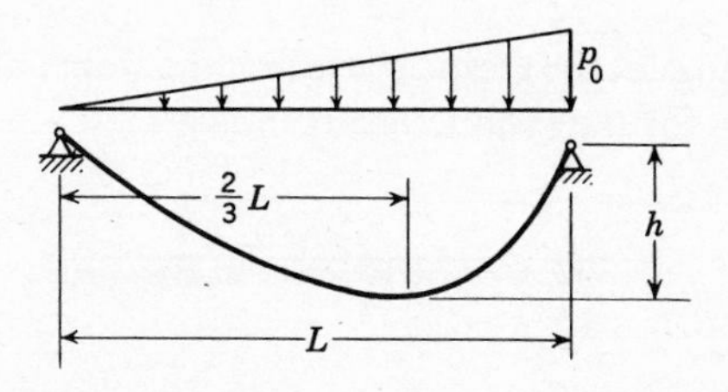

6.24. A cable weighing 100 lb/ft is suspended between two equally elevated points 1,000 ft apart.

(*a*) Calculate the maximum tension in the cable if the maximum sag h is 20 ft.
(*b*) Calculate the maximum sag of the cable if $H = 12{,}000$ lb.
(*c*) What is the length of the cable in part (*a*)?

6.25–6.28. Find the equations for the elastic curves of the beams on elastic foundations shown.

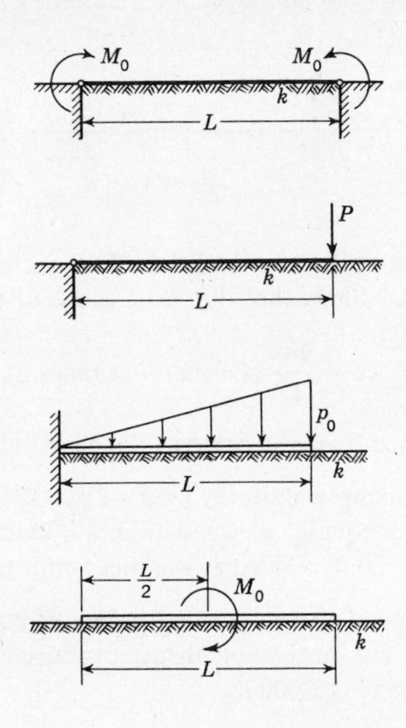

6.29. Compute the end moments developed in the beam shown.

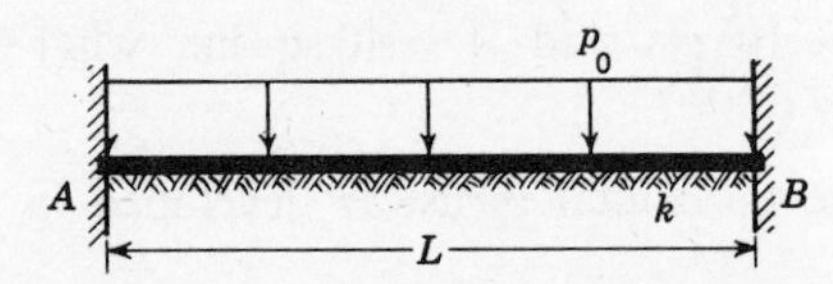

6.30. Evaluate and plot the elastic curve and the slope, shear, and bending-moment diagrams for the infinite beam shown.

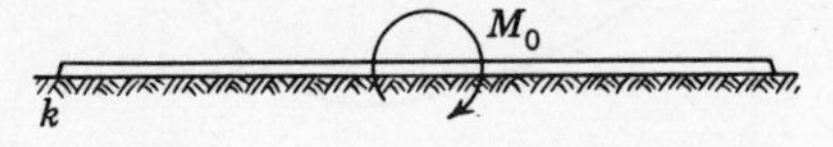

6.31. Calculate the slope and the deflection of points A, B, and C of the beam shown.

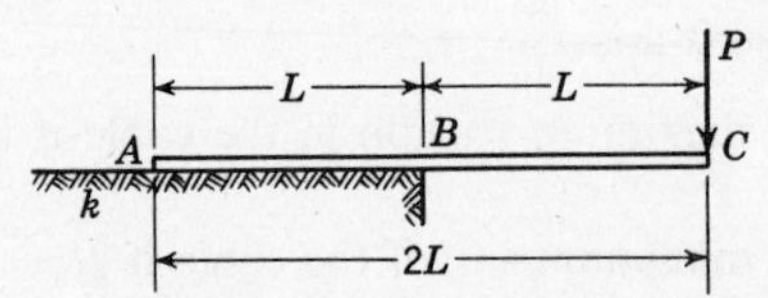

6.32. The radial displacement v of an elemental strip of a cylindrical shell under axially symmetrical loading behaves as a beam on an elastic foundation with modulus

$$k = \frac{\bar{E}t}{R^2}$$

where t is the thickness, R is the radius of the shell, and $\bar{E} = E/(1 - \nu^2)$. The longitudinal bending moment at any point in the strip is

$$-\left(\frac{Et^3}{12}\right)\frac{d^2v}{dx^2}$$

Compute and plot the bending-moment diagram for the cylindrical tank shown. Take $\nu = 0.3$, $E = 30 \times 10^6$ psi, and assume that the unit weight of the fluid in the tank is 60 lb/cu ft.

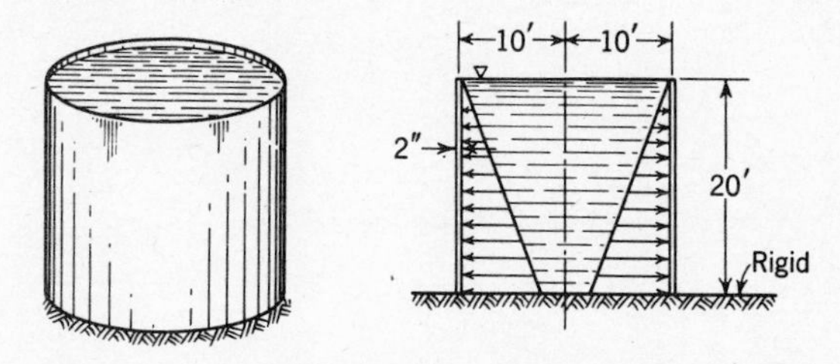

6.33–6.36. Find the equations for the deflection, slope, shear, and bending moments of the semi-infinite beams shown below.

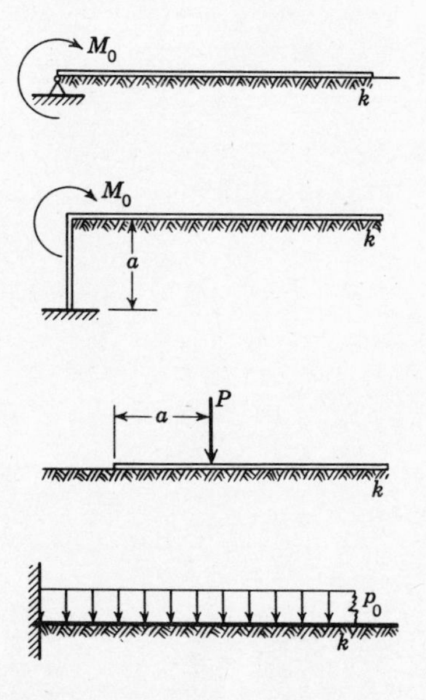

6.37. Find the equation of the elastic curve and the tip deflection of the cantilevered beam column on an elastic foundation shown below. Assume that $4\beta^2 - \frac{1}{4}\lambda^4$ is

(*a*) positive
(*b*) negative
(*c*) zero

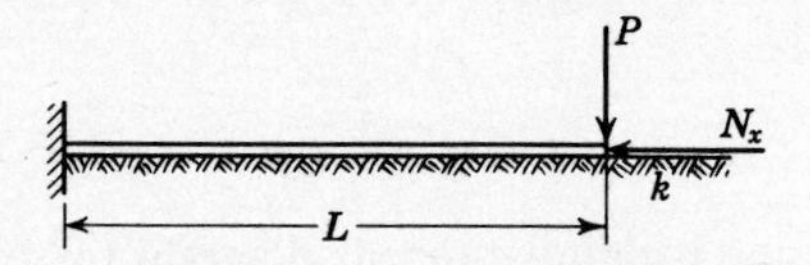

seven

BENDING AND TWISTING OF THIN-WALLED BEAMS

7.1 Introduction. When a thin-walled beam having one or more cross sections constrained against warping is subjected to a general system of external loads, a complex distribution of longitudinal stresses is developed that cannot be evaluated using the elementary theories. The assumption that plane sections remain plane during deformation is no longer valid, and applications of Saint-Venant's principle[1] may lead to serious errors.

The existence of these restrained-warping stresses is easy to verify from a physical viewpoint. We recall that a twisting moment applied to an unrestrained open section produces a relative longitudinal sliding (a warping) of the free ends of the section such as that indicated in Fig. 5.23. Obviously, if one section is constrained in such a way that it cannot undergo out-of-plane displacements, a certain system of normal stresses must be developed to eliminate this warping. In general, these normal stresses vary from point to point along the member; hence, they are accompanied by a nonuniform shearing-stress distribution which, in turn, alters the twist of the section. As a result, the twisting moment developed on each section is no longer proportional to the rate of twist and final

[1] See Art. 2.4.

shearing stresses cannot be obtained by superimposing those produced by unrestrained torsion and bending.

To complicate matters, such "warping" stresses are also developed by subjecting thin-walled members to eccentric axial loads. Consider, for example, the beam in Fig. 7.1*a*, which is composed of four stringers of equal area connected to thin skins. For simplicity, we assume that the

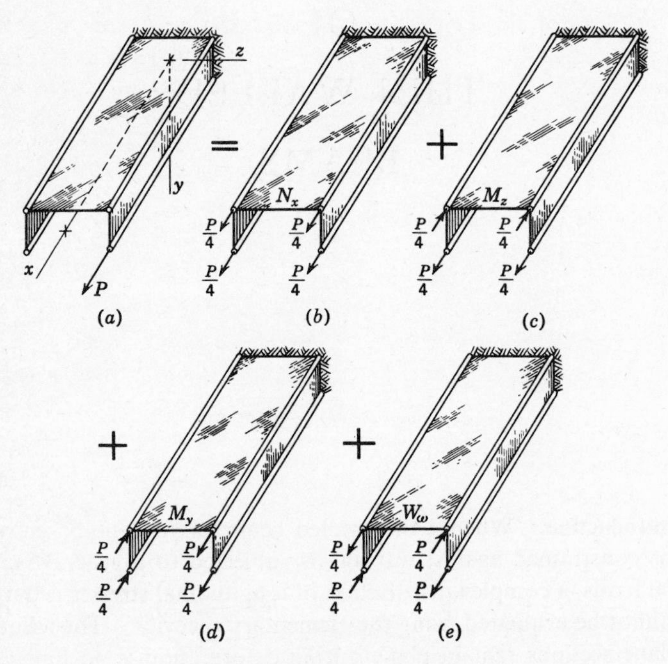

FIGURE 7.1 Thin-walled beam subjected to an eccentric axial load.

stringers develop only normal stresses and the skins develop only shearing stresses. According to the Bernoulli-Euler theory of bending, the eccentric load P produces the axial force N_x and the bending moments M_z and M_y shown in Fig. 7.1*b*, *c*, and *d*. No additional stress resultants are given by the elementary theory because it is based on the assumption that planes remain plane. This, however, is impossible. The superposition of cases (*b*), (*c*), and (*d*) in Fig. 7.1 does not lead to a force P in the bottom stringer; the additional force system shown in Fig. 7.1*e* is also necessary. This system is equivalent to two equal and opposite couples, denoted W_ω, developed in parallel planes. Clearly, these couples are statically zero; that is, they are self-equilibrating. We refer to such self-equilibrating normal stress resultants as *bimoments*;[2] they produce an out-of-plane warping of the section very similar to that produced by torsion. Conversely, if a torque, instead of an axial load, is applied to the free end of the beam, the constraints which eliminate warping at the fixed end also produce bimoments.

According to Saint-Venant's principle, the effects of such statically zero stress systems should decay rapidly from their points of application. For closed sections this is usually true; but in the case of thin-walled open sections applications of Saint-Venant's principle may lead to significant errors. In thin-walled open sections, stresses produced by restrained warping diminish very slowly from their points of application and may constitute the primary stress system developed in the structure.

7.2 State of stress in open sections. To describe stress distributions in thin-walled beams, we establish in addition to the usual x, y, z coordinates an orthogonal curvilinear coordinate system (s,n), as shown in Fig. 7.2*a*. The tangential coordinate s is measured counterclockwise along the center line of the tube wall, and n is in the direction of an outward normal to s. In the following, we assume that cross-sectional dimensions are constant along the longitudinal axis of the beam so that s and n are independent of x and x, s, and n form an orthogonal system.

Now the wall thickness t of a thin-walled beam is, by definition, very small compared with the other cross-sectional dimensions. Consequently

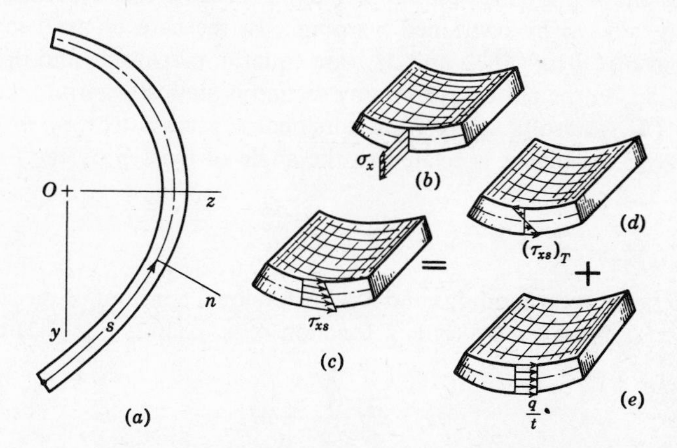

FIGURE 7.2 (*a*) Cross-sectional coordinate systems. Normal-stress (*b*) and shearing-stress (*c–e*) variations over wall thickness of a thin-walled open section.

[2] In general, a bimoment can consist of more than two self-equilibrating couples.

we are justified in assuming that the normal stress σ_x is essentially uniform over the thickness (Fig. 7.2*b*) and that the stress and strain components normal to the wall are negligible. That is,

$$\sigma_n = \tau_{xn} = \tau_{sn} = \epsilon_n = \gamma_{xn} = \gamma_{sn} = 0 \tag{7.1}$$

This means that only one shearing-stress component τ_{xs} need be evaluated.

In the case of thin-walled open sections, these shearing stresses are developed owing to two distinct modes of deformation—a pure twisting of the beam during which all sections are free to warp and a transverse bending of the beam coupled with nonuniform axial deformations. The first mode of deformation, we recall, leads to shearing stresses which vary linearly over the thickness. We denote these stresses by the symbol $(\tau_{xs})_T$. The additional shearing stresses developed during the second mode of deformation are essentially uniform over t. These stresses produce a shear flow q which acts along the contour of the section. Therefore, the total shearing stress at any point is given by the formula

$$\tau_{xs} = (\tau_{xs})_T + \frac{q}{t} \tag{7.2}$$

The superposition of these stresses is indicated in Fig. 7.2*c*, *d*, and *e*.

Similarly, we may regard the total twisting moment developed on any section as the sum of two parts—a moment (M_t) due to $(\tau_{xs})_T$ and a moment $M_{t\omega}$ due to q:

$$M_t = (M_t)_T + M_{t\omega} \tag{7.3}$$

The moment $M_{t\omega}$ is called the *warping torque* because it is the resultant of stresses produced by restrained warping. In the case of axial loading, such as in Fig. 7.1*a*, $(M_t)_T$ and $M_{t\omega}$ are equal in magnitude and opposite in direction, so the net torque on any section is statically zero.

Since $(M_t)_T$ results solely from torsional shearing stresses, we recall from Eq. (3.18) that it is related to the angle of twist ϕ by the formula

$$(M_t)_T = GJ\frac{d\phi}{dx} \tag{7.4}$$

where G is the shear modulus and J is the torsional constant of the section. In this case, however, $d\phi/dx$ is a function of x. Thus, Eq. (7.3) can be written

$$M_t = GJ\frac{d\phi}{dx} + M_{t\omega} \tag{7.5}$$

Furthermore, according to Eq. (3.29),

$$(\tau_{xs})_T = \frac{2(M_t)_T n}{J} = 2G\frac{d\phi}{dx}n \tag{7.6}$$

where n is the normal coordinate indicated in Fig. 7.2*a*. Therefore, the component $(\tau_{xs})_T$ is independent of s, and the variation in shearing stress along the contour of the section is due exclusively to the shear flow produced by bending and restrained warping.

7.3 Strains in open sections. To continue our development, we now extend some of the ideas of Saint-Venant's theory of torsion and assume that, regardless of the loading, the original shape of every cross section is unaltered during deformation. Thus, the geometric dimensions of every plane normal to the bar's longitudinal axis remain unchanged, although the section may undergo any type of out-of-plane deformation. The derivation of Eq. (7.4), we recall, is also based on this assumption. It follows that the in-plane strains ϵ_s and γ_{sn} are zero and that ϕ and the y and z components of displacement v and w are functions of only x. The longitudinal displacement u is a function of s and x but not of n due to the small wall thickness. Furthermore, owing to the great flexibility of thin-walled open sections, the effect of the shearing strain γ_{xs} on final deformations is extremely small and can be neglected. Final deformations are due almost exclusively to longitudinal strains produced by bending, twisting, and warping of the structure. The corresponding shearing stress τ_{xs}, however, cannot be neglected; it must be determined from static considerations in a manner similar to elementary beam theory.

In view of the above assumptions and Eqs. (7.1), we refer to Eqs. (2.18) and (2.27) and obtain the following relations:

$$\gamma_{xs} \cong 0 = \frac{\partial u}{\partial s} + \frac{\partial \eta}{\partial x} \tag{7.7a}$$

$$\epsilon_x = \frac{\partial u}{\partial x} = \frac{1}{E}(\sigma_x - \nu\sigma_s) \tag{7.7b}$$

$$\epsilon_s = 0 = \frac{1}{E}(\sigma_s - \nu\sigma_x) \tag{7.7c}$$

where η is the component of displacement in the s direction. Note that Eq. (7.7*a*) applies only to open sections.

From the last two of these equations, we find

$$\sigma_s = \nu\sigma_x \tag{7.8}$$

and

$$\sigma_x = \bar{E}\frac{\partial u}{\partial x} \tag{7.9}$$

where

$$\bar{E} = \frac{E}{1 - \nu^2} \tag{7.10}$$

We recognize $\bar{E}$ as the modified elastic modulus given by Eq. (6.27). Since σ_s is usually very small compared with σ_x, we neglect it in the developments to follow and use E rather than $\bar{E}$ in Eq. (7.9).

With $(\tau_{xs})_T$ given in terms of ϕ by Eq. (7.6), the problem of stress analysis reduces to one involving two unknowns σ_x and q. We have not, as yet, used Eq. (7.7*a*); to proceed with our investigation we now turn to a more detailed study of beam displacements.

7.4 Kinematics of deformation. In view of our assumption that the geometry of every cross section is unaltered during deformation, we see that the deformation of any section in its own plane can be described by the translations of points on the section plus the displacements of these points due to a rotation of the section about some point in its plane. This point in the cross-sectional plane about which the section rotates is called the *center of twist;* it is the only point on the cross section that remains at rest during deformation.

Let us now consider the thin-walled cross section of general shape shown in Fig. 7.3*a*. The origin O of the xyz coordinate system is located at the centroid of the section. Let A denote an arbitrary point on the cross section having coordinates a_y and a_z and let v and w denote the y and z components of displacement of A. Further, let c_y and c_z be the coordinates of C, another point on the section, and let v_C and w_C denote the displacement components of C. We now investigate the kinematic relations between the displacements of A and C by examining the geometry of the displaced cross-sectional elements shown in Fig. 7.3*b*.

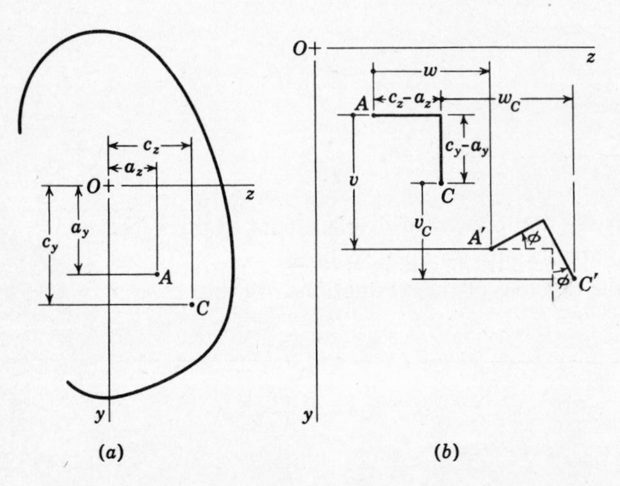

FIGURE 7.3 (*a*) Geometry of a thin-walled section of general shape; (*b*) geometry of deformation.

The displacements of A and C due to translations of the section are equal. If the cross section also rotates through a small counterclockwise angle ϕ, then the displacements of C in terms of those of A are

$$\begin{aligned} v_C &= v - (c_z - a_z)\phi \\ w_C &= w + (c_y - a_y)\phi \end{aligned} \tag{7.11}$$

Note that these relationships are independent of the location of the center of twist.

Let us identify point C as the center of twist of the section. Then, by definition, v_C and w_C are zero. Therefore, the coordinates of the center of twist are

$$\begin{aligned} c_y &= a_y - \frac{w}{\phi} \\ c_z &= a_z + \frac{v}{\phi} \end{aligned} \tag{7.12}$$

We see that in general the location of the center of twist varies from point to point along the member. Its location depends upon the deformation of the beam and, consequently, upon the boundary conditions and the loading.

Now let S denote a point located on the center line of the tube wall which has the coordinates (y,z). According to Eqs. (7.11), the displacements of this point are

$$\begin{aligned} v_S &= v - (z - a_z)\phi \\ w_S &= w + (y - a_y)\phi \end{aligned} \tag{7.13}$$

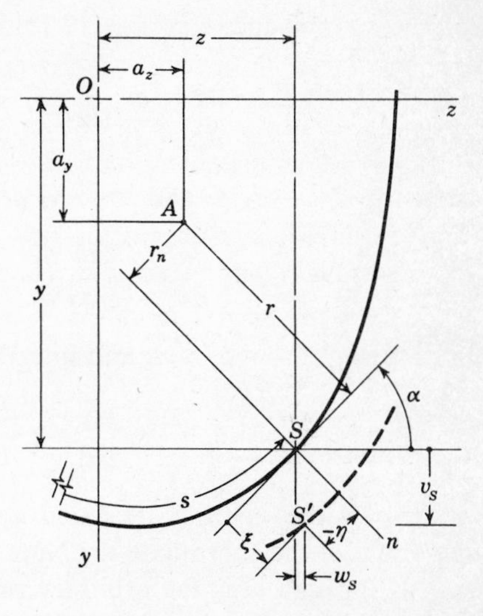

FIGURE 7.4 Normal and tangential components of displacement.

Further, let η denote the displacement of S in a direction tangent to the tube wall and ξ denote the displacement of S normal to this tangent (Fig. 7.4). Then, from simple geometry, we find

$$\begin{aligned} \xi &= v_S \cos\alpha + w_S \sin\alpha \\ \eta &= -v_S \sin\alpha + w_S \cos\alpha \end{aligned}$$

Substituting Eqs. (7.13) into this result gives

$$\xi = v\cos\alpha + w\sin\alpha + [-(z - a_z)\cos\alpha + (y - a_y)\sin\alpha]\phi \tag{7.14a}$$

$$\eta = v\sin\alpha - w\cos\alpha + [(z - a_z)\sin\alpha + (y - a_y)\cos\alpha]\phi \tag{7.14b}$$

Referring to Fig. 7.4, we find that the quantities within the brackets in Eqs. (7.14) are, respectively, the lengths of lines drawn from A perpendicular to the normal and the tangent to the center line of the tube wall at S. We denote these quantities by r_n and r, respectively, and Eqs. (7.14) become

$$\begin{aligned} \xi &= v \cos \alpha + w \sin \alpha + r_n \phi \\ \eta &= -v \sin \alpha + w \cos \alpha + r\phi \end{aligned} \tag{7.15}$$

Clearly, the last terms in these equations represent displacements due to a rotation of the section with respect to A. If no translations occur and the rate of twist $d\phi/dx$ is a constant, the above formula for η reduces to Eq. (3.42), which we obtained for pure torsion of unrestrained sections.

Returning to the strain-displacement relations which we established earlier, we introduce η of Eqs. (7.15) into Eq. (7.7*a*) and obtain

$$\frac{\partial u}{\partial s} = \frac{dv}{dx} \sin \alpha - \frac{dw}{dx} \cos \alpha - \frac{d\phi}{dx} r \tag{7.16}$$

Since $\sin \alpha = -dy/ds$ and $\cos \alpha = dz/ds$,

$$\frac{\partial u}{\partial s} = -\frac{dv}{dx}\frac{dy}{ds} - \frac{dw}{dx}\frac{dz}{ds} - \frac{d\phi}{dx} r$$

Multiplying through by ds and integrating, we find

$$u = -y\frac{dv}{dx} - z\frac{dw}{dx} - 2\omega\frac{d\phi}{dx} + u_0 \tag{7.17}$$

where ω is the sectorial area[3] and u_0 is the longitudinal displacement of the origin of the coordinate s. Note that the pole about which the arm r sweeps through ω is the arbitrary point A on the cross section. Comparing Eq. (7.17) with Eq. (3.53), we see that the term $-2\omega(d\phi/dx)$ represents a relative longitudinal displacement due to twisting of the section. The displacement u_0 is independent of s; for a given section it represents a uniform displacement of all points in the x direction. The first and second terms on the right side of Eq. (7.17) represent longitudinal displacements due to rotations of the cross-sectional plane through angles dv/dx and dw/dx about the z and y axes. These terms, along with u_0, satisfy the Navier hypothesis and thus lead to a deformed shape in which all initially plane sections remain plane. The departure from this shape is contributed by the term $-2\omega(d\phi/dx)$. We see that longitudinal displacements given by the ordinary Bernoulli-Euler theory of bending and Saint-Venant's theory of torsion may be regarded as special cases of Eq. (7.17).

[3] See Art. 3.8, Eq. (3.39).

7.5 Stress resultants in open sections. We can now obtain the normal stress in terms of the displacements by introducing Eq. (7.17) into Eq. (7.9). This gives

$$\sigma_x = -E\left(y\frac{d^2v}{dx^2} + z\frac{d^2w}{dx^2} - \frac{du_0}{dx} + 2\omega\frac{d^2\phi}{dx^2}\right) \tag{7.18}$$

Clearly, the last term on the right side of this equation represents the effects of restrained warping. The remaining terms lead to the linear stress distribution given by elementary beam theory. The linear variation, we recall, is statically equivalent to a normal force N_x and bending moments M_y and M_z which are defined by the formulas

$$N_x = \int_A \sigma_x\,dA \qquad M_z = \int_A \sigma_x y\,dA \qquad M_y = \int_A \sigma_x z\,dA \tag{7.19}$$

Introducing Eq. (7.18) into these expressions, we obtain the following equations:

$$N_x = -E\left(\frac{d^2v}{dx^2}\int_A y\,dA + \frac{d^2w}{dx^2}\int_A z\,dA - \frac{du_0}{dx}\int_A dA + \frac{d^2\phi}{dx^2}\int_A 2\omega\,dA\right) \tag{7.20a}$$

$$M_z = -E\left(\frac{d^2v}{dx^2}\int_A y^2\,dA + \frac{d^2w}{dx^2}\int_A yz\,dA - \frac{du_0}{dx}\int_A y\,dA + \frac{d^2\phi}{dx^2}\int_A 2\omega y\,dA\right) \tag{7.20b}$$

$$M_y = -E\left(\frac{d^2v}{dx^2}\int_A zy\,dA + \frac{d^2w}{dx^2}\int_A z^2\,dA - \frac{du_0}{dx}\int_A z\,dA + \frac{d^2\phi}{dx^2}\int_A 2\omega z\,dA\right) \tag{7.20c}$$

We note that

$$\int_A y\,dA = \int_A z\,dA = 0$$

since y and z pass through the centroid of the section, and that

$$A = \int_A dA \qquad I_y = \int_A z^2\,dA \qquad I_{yz} = \int_A yz\,dA \qquad I_z = \int_A y^2\,dA$$

The remaining integrals are referred to as *sectorial area properties* of the cross section. In particular, the *first sectorial moment* is defined by the equation

$$S_\omega = \int_A 2\omega\,dA \tag{7.21}$$

Its magnitude obviously depends upon the location of the pole A and the origin of the coordinate s. To identify the other two integrals, let us rewrite the last integral in Eq. (7.20c) in the following form:

$$\int_A 2\omega z\,dA = 2\int_A \omega\,dQ_y$$

where Q_y is the familiar first moment of area with respect to the y axis. Integrating by parts, we get

$$\int_A 2\omega z\, dA = 2\omega Q_y \Big|_A - 2\int_A Q_y\, d\omega$$

The first term on the right side is zero because y is a centroidal axis. Referring to Eqs. (5.40), we recognize the remaining integral as the *sectorial product of inertia* with respect to the y axis. Thus

$$\int_A 2\omega z\, dA = -2I_{\omega y}$$

Similarly,

$$\int_A 2\omega y\, dA = -2I_{\omega z}$$

Therefore, Eqs. (7.20) can be written in the more concise form

$$N_x = EA\frac{du_0}{dx} - ES_\omega \frac{d^2\phi}{dx^2} \qquad (7.22a)$$

$$M_z = -EI_z\frac{d^2v}{dx^2} - EI_{yz}\frac{d^2w}{dx^2} + 2EI_{\omega z}\frac{d^2\phi}{dx^2} \qquad (7.22b)$$

$$M_y = -EI_{yz}\frac{d^2v}{dx^2} - EI_y\frac{d^2w}{dx^2} + 2EI_{\omega y}\frac{d^2\phi}{dx^2} \qquad (7.22c)$$

Solving these equations, we find

$$\frac{du_0}{dx} = \frac{N_x}{AE} + \frac{S_\omega}{A}\frac{d^2\phi}{dx^2} \qquad (7.23a)$$

$$\frac{d^2v}{dx^2} = -\frac{M_zI_y - M_yI_{yz}}{E(I_yI_z - I_{yz}^2)} + 2\frac{I_{\omega z}I_y - I_{\omega y}I_{yz}}{I_yI_z - I_{yz}^2}\frac{d^2\phi}{dx^2} \qquad (7.23b)$$

$$\frac{d^2w}{dx^2} = -\frac{M_yI_z - M_zI_{yz}}{E(I_yI_z - I_{yz}^2)} + 2\frac{I_{\omega y}I_z - I_{\omega z}I_{yz}}{I_yI_z - I_{yz}^2}\frac{d^2\phi}{dx^2} \qquad (7.23c)$$

The last terms in each of these equations represent, respectively, an axial strain at $s = 0$ and changes in curvature in the xy and xz planes due to twisting of the section.

Comparing the coefficients of $d^2\phi/dx^2$ in Eq. (7.23b and c) with Eqs. (5.41), we see that they are precisely the coordinates $(e_z - a_z)$ and $(a_y - e_y)$ of the shear center of the section with respect to point A. Hence, in the absence of transverse loads, nonuniform twisting of the section produces changes of curvature of magnitude

$$\frac{d^2v}{dx^2} = -(a_z - e_z)\frac{d^2\phi}{dx^2} \quad \text{and} \quad \frac{d^2w}{dx^2} = (a_y - e_y)\frac{d^2\phi}{dx^2} \qquad (7.24)$$

If the section is fixed at $x = 0$, integration of Eqs. (7.24) shows that

$$\frac{v}{\phi} = -(a_z - e_z) \qquad \text{and} \qquad \frac{w}{\phi} = (a_y - e_y)$$

Introducing these expressions into Eqs. (7.12), we find that in this case the shear center and the center of twist coincide. For any other boundary conditions, however, the shear center and the center of twist do not coincide and the location of the latter varies from section to section.

Now the location of the shear center is strictly a geometric property of the cross section. Consequently, regardless of the actual location of the center of twist, the terms containing $d^2\phi/dx^2$ in Eqs. (7.23*b* and *c*) vanish if we set $a_y = e_y$ and $a_z = e_z$. Thus, unless noted otherwise, in the developments to follow we assume that the sectorial pole A is coincident with the shear center of the section. This means that $I_{\omega y}$ and $I_{\omega z}$ vanish and that Eqs. (7.23*b* and *c*) reduce to the more familiar form

$$\begin{aligned}\frac{d^2v}{dx^2} &= -\frac{M_zI_y - M_yI_{yz}}{E(I_yI_z - I_{yz}^2)} \\ \frac{d^2w}{dx^2} &= -\frac{M_yI_z - M_zI_{yz}}{E(I_yI_z - I_{yz}^2)}\end{aligned} \tag{7.25}$$

Let us now interpret geometrically the constant S_ω/A in Eq. (7.23*a*). We noted earlier that the magnitude of S_ω depends upon the location of the origin of the coordinate s. There exist points on the contour for which, when they are used for the origin of s, the quantity S_ω vanishes. These points are called *sectorial centroids;* there may be several or even an infinite number of such points for a given section. To locate a sectorial centroid, we note that two sectorial areas with common poles but with different origins are related as follows:

$$\omega_a(s) - \omega_b(s) = \omega_0 \tag{7.26}$$

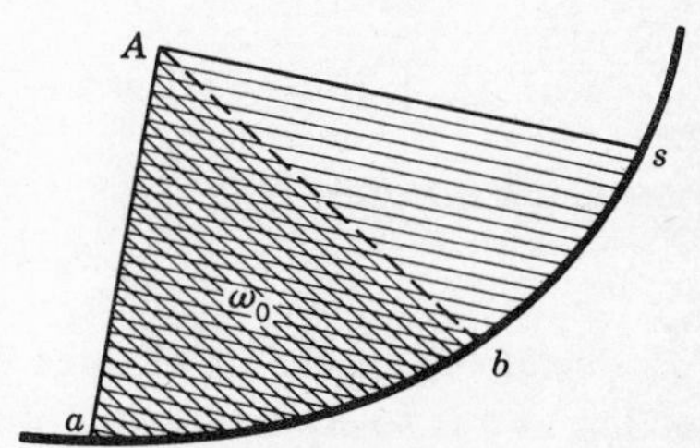

FIGURE 7.5 Relationship between sectorial areas with common pole A but different origins.

where $\omega_a(s)$ and $\omega_b(s)$ are sectorial areas measured from points a and b on the contour and ω_0 is a constant equal to the sectorial area between a and b (Fig. 7.5). If point b is a sectorial centroid,

$$(S_\omega)_b = \int_A 2\omega_b\,dA = 0 = \int_A 2\omega_a\,dA - 2\omega_0 A$$

Hence,

$$2\omega_0 = \frac{1}{A}\int_A 2\omega_a\,dA = \frac{(S_\omega)_a}{A} \tag{7.27}$$

Therefore, S_ω/A in Eq. (7.23*a*) is equal to twice the sectorial area between the origin of the coordinate s and a sectorial centroid of the section. We could, of course, choose the sectorial centroid for the s origin and thereby eliminate the second term in Eq. (7.23*a*). We shall find it more convenient, however, to select a free end of the open section for this origin and to retain the term containing $d^2\phi/dx^2$. Thus, if ω_0 denotes the sectorial area between an outside edge of the section and the sectorial centroid, Eq. (7.23*a*) becomes

$$\frac{du_0}{dx} = \frac{N_x}{AE} + 2\omega_0 \frac{d^2\phi}{dx^2} \tag{7.28}$$

Substituting Eqs. (7.25) and (7.28) into Eq. (7.18), we get

$$\sigma_x = \frac{N_x}{A} + \frac{M_z I_y - M_y I_{yz}}{I_y I_z - I_{yz}^2} y + \frac{M_y I_z - M_z I_{yz}}{I_y I_z - I_{yz}^2} z + 2E(\omega_0 - \omega)\frac{d^2\phi}{dx^2} \tag{7.29}$$

To eliminate $d^2\phi/dx^2$ from this equation, we must evaluate a fourth stress resultant W_ω which leads to a self-equilibrating stress system. This stress resultant must also be independent of N_x, M_y, and M_z, since it can be developed by applying external twisting moments to a restrained section. We notice that the quantity $2(\omega_0 - \omega)$ in Eq. (7.29) has the following properties:

$$\begin{aligned} \int_A 2(\omega_0 - \omega)\, dA &= 2\omega_0 A - S_\omega = 0 \\ \int_A 2(\omega_0 - \omega) y\, dA &= 2\omega_0 \int_A y\, dA - 2I_{\omega z} = 0 \\ \int_A 2(\omega_0 - \omega) z\, dA &= 2\omega_0 \int_A z\, dA - 2I_{\omega y} = 0 \end{aligned} \tag{7.30}$$

Hence, the integral

$$\int_A \sigma_x 2(\omega_0 - \omega)\, dA$$

is independent of the axial force and bending moments developed at a section and is solely a function of $d^2\phi/dx^2$ and the geometry of the cross section. With this in mind, we define the bimoment W_ω as follows:

$$W_\omega = \int_A 2\sigma_x(\omega_0 - \omega)\, dA \tag{7.31}$$

Note that W_ω has units of force times length squared (for example, lb-ft²). Substituting Eq. (7.29) into this formula and introducing the sectorial area properties given in Eqs. (7.30), we find

$$W_\omega = E\Gamma \frac{d^2\phi}{dx^2} \tag{7.32}$$

where

$$\Gamma = \int_A 4(\omega_0 - \omega)^2\, dA \tag{7.33}$$

The constant Γ is called the *warping constant* of the section; it has units of length to the sixth power. The quantity $E\Gamma$ is called the *warping rigidity*. Note that if a sectorial centroid is chosen for the s origin, ω_0 is zero and

$$\Gamma = \int_A (2\omega)^2 \, dA \tag{7.34a}$$

If the pole A is not coincident with the shear center, it can be shown that[4]

$$\Gamma_A = \Gamma + (a_y - e_y)^2 I_{0y} - 2(a_y - e_y)(a_z - e_z) I_{0yz} + (a_z - e_z)^2 I_{0z} \tag{7.34b}$$

where I_{0y}, I_{0z}, and I_{0yz} are the moments and the product of inertia with respect to a set of axes parallel to y and z through the point $s = 0$, and Γ is the warping constant with respect to the shear center [Eq. (7.34*a*)].

Finally, substituting Eq. (7.32) into Eq. (7.29) gives

$$\sigma_x = \frac{N_x}{A} + \frac{M_z I_y - M_y I_{yz}}{I_y I_z - I_{yz}^2} y + \frac{M_y I_z - M_z I_{yz}}{I_y I_z - I_{yz}^2} z + 2 \frac{W_\omega}{\Gamma} (\omega_0 - \omega) \tag{7.35}$$

When a thin-walled beam is subjected to transverse and axial loads, the normal force and the bending moments in Eq. (7.35) can be evaluated from simple statics. The bimoment, however, is statically zero and, in general, is a statically indeterminate quantity. Since both W_ω and ϕ in Eq. (7.32) are as yet unknown, additional information must be obtained before σ_x can be evaluated. To this end, we now investigate shearing stresses in thin-walled beams and the effect of restrained warping on the net twisting moment developed at a section.

7.6 Shear flow in open sections. Consider the thin-walled beam of Fig. 7.6*a*, which is subjected to a general system of transverse loads and a distributed external torque of intensity m_t per unit length. Since the bimoment W_ω is statically zero, the familiar transverse equilibrium equations for the beam segment in Fig. 7.6*b* are unaffected:

$$\frac{dV_y}{dx} = -p_y \qquad \frac{dV_z}{dx} = -p_z \tag{7.36a}$$

$$\frac{dM_z}{dx} = V_y \qquad \frac{dM_y}{dx} = V_z \tag{7.36b}$$

[4] See, for example, Ref. 25, pp. 126–134, or Ref. 68, p. 46. Warping constants for some common sections are given in Table 7.1.

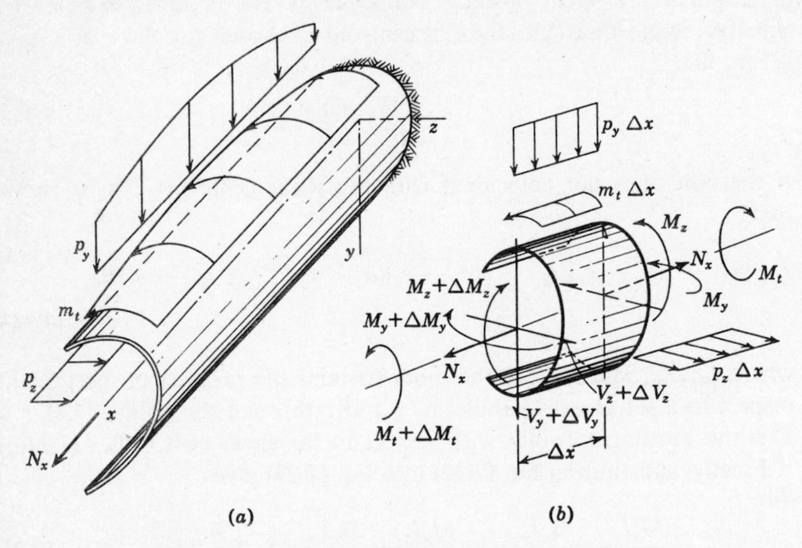

FIGURE 7.6 (*a*) Thin-walled open section under a general system of loads; (*b*) typical segment of the beam.

In this case, we obtain an additional independent equilibrium condition by summing moments about the longitudinal axis:

$$\frac{dM_t}{dx} = -m_t \tag{7.37}$$

or, in view of Eq. (7.5),

$$m_t = -GJ\frac{d^2\phi}{dx^2} - \frac{dM_{t\omega}}{dx} \tag{7.38}$$

Now let us examine the equilibrium conditions of a typical wall element, such as that shown in Fig. 7.7. Note that the linear shearing stresses $(\tau_{x_s})_T$ do not influence the equilibrium conditions in the longitudinal direction. Summing forces in the x direction and taking the limit as Δx and Δs approach zero, we get

$$\frac{\partial q}{\partial s} + t\frac{\partial \sigma_x}{\partial x} = 0 \tag{7.39}$$

Solving this equation for q, we find

$$q = \bar{q} - \int \frac{\partial \sigma_x}{\partial x}\,dA \tag{7.40}$$

where $\bar{q}$ is the shear flow at $s = 0$ and $dA = t\,ds$. Since we have agreed to choose a free edge of the section for the origin of the coordinate s, $\bar{q}$ is zero.

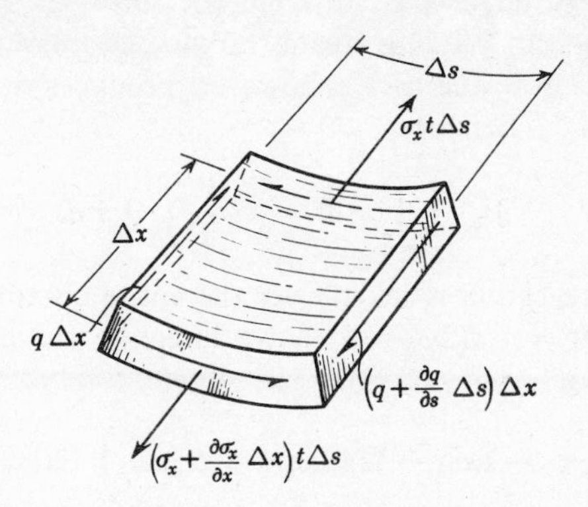

FIGURE 7.7 Forces on a typical wall element of a thin-walled beam.

We now introduce Eqs. (7.36*b*) into the first derivative of Eq. (7.35) with respect to x. Then substituting the result into the integral in Eq. (7.40), we find

$$q = -\left(\frac{I_y Q_z - I_{yz} Q_y}{I_y I_z - I_{yz}^2} V_y + \frac{I_z Q_y - I_{yz} Q_z}{I_y I_z - I_{yz}^2} V_z\right) - \frac{Q_\omega}{\Gamma} V_\omega \tag{7.41}$$

where Q_y and Q_z are the familiar first moments of area with respect to the y and z axes, Q_ω is the *first sectorial moment* with respect to the s origin,

$$Q_\omega = \int_{A'} 2(\omega_0 - \omega)\, dA \tag{7.42a}$$

and V_ω is the *warping shear*, defined by the equation

$$V_\omega = \frac{dW_\omega}{dx} \tag{7.42b}$$

We recognize the term within parentheses in Eq. (7.41) as the shear flow q_0 due to bending [see Eqs. (5.20) and (5.72)]. The minus sign in front of this term is due to our choice of counterclockwise shear flows as positive. Clearly, Q_y, Q_z, Q_ω, and consequently q vanish when the integration is carried out over the entire cross-sectional area. This is in agreement with the fact that q must vanish at the free edges of the cross section. We obtain the equation for the total shearing stress in an open section by substituting Eq. (7.41) and Eq. (7.6) into Eq. (7.2):

$$\tau_{xs} = \frac{2n}{J}(M_t)_T - \frac{I_y Q_z - I_{yz} Q_y}{t(I_y I_z - I_{yz}^2)} V_y - \frac{I_z Q_y - I_{yz} Q_z}{t(I_y I_z - I_{yz}^2)} V_z - \frac{Q_\omega}{t\Gamma} V_\omega \tag{7.43}$$

where $(M_t)_T$ is given by Eq. (7.4).

If the transverse loads pass through the shear center of the section, the shear flow due to bending creates no moment about the longitudinal axis. The shear flow due to V_ω, however, produces the warping torque $M_{t\omega}$ of Eq. (7.3). Therefore,

$$M_{t\omega} = \int_S qr\,ds = -\frac{V_\omega}{\Gamma}\int_S Q_\omega r\,ds \tag{7.44}$$

in which the integration is taken over the entire contour of the section. Noting that $r\,ds = -d(2\omega_0 - 2\omega)$, we integrate the above integral by parts and find

$$\int_S Q_\omega r\,ds = -2(\omega_0 - \Omega)(2\omega_0 A - S_\omega) + \int_S [2(\omega_0 - \omega)]^2\,dA$$

The first term on the right side of this expression vanishes because of the definition of ω_0 [see Eq. (7.27)]. We recognize the remaining integral as the warping constant Γ defined in Eq. (7.33). Therefore,

$$M_{t\omega} = -V_\omega = -\frac{dW_\omega}{dx} \tag{7.45a}$$

or, in view of Eq. (7.32),

$$M_{t\omega} = -E\Gamma\frac{d^3\phi}{dx^3} \tag{7.45b}$$

Finally, substituting this result into Eq. (7.38), we arrive at the governing differential equation for the twist of thin-walled open sections:

$$E\Gamma\frac{d^4\phi}{dx^4} - GJ\frac{d^2\phi}{dx^2} = m_t \tag{7.46a}$$

If M_t is constant, m_t is zero and we may use the third-order equation

$$E\Gamma\frac{d^3\phi}{dx^3} - GJ\frac{d\phi}{dx} = -M_t \tag{7.46b}$$

Equations (7.25), (7.28), and (7.46) represent the governing differential equations for the elastic curve of thin-walled open sections. Once these equations have been solved, W_ω and V_ω can be evaluated from Eqs. (7.32) and (7.42*b*) and the final normal stresses and shearing stresses can then be computed using Eqs. (7.35) and (7.43). It is emphasized that in these equations we assumed that the pole of the sectorial areas is the shear center of the section and that a free edge of the section is used for the origin of the coordinate s.

7.7 Physical interpretation of warping stresses. Let us now apply the above theory to some common structural shapes. Consider, for example, the cantilevered z section in Fig. 7.8 subjected to an end torque T. The shear center coincides with the centroid, and the origin of s is taken as the outside corner of the bottom flange. The thickness t is assumed to be constant, and the cross-sectional area equals $(2b + h)t$. From the definition of ω, we see that

$$
\begin{aligned}
2\omega &= \frac{sh}{2} && \text{for } 0 \leq s \leq b \\
2\omega &= \frac{bh}{2} && \text{for } b \leq s \leq b + h \qquad (a) \\
2\omega &= (2b + h - s)\frac{h}{2} && \text{for } b + h \leq s \leq 2b + h
\end{aligned}
$$

Thus, from Eq. (7.27),

$$2\omega_0 = \frac{1}{A}\int_0^{2b+h} 2\omega t\, ds = \frac{bh(b + h)}{2(2b + h)} \qquad (b)$$

Substituting these quantities into Eq. (7.33) and performing some lengthy integration, we find

$$\Gamma = \int_0^{2b+h} (2\omega_0 - 2\omega)^2 t\, ds = \frac{tb^3h^2(b + 2h)}{12(2b + h)} \qquad (c)$$

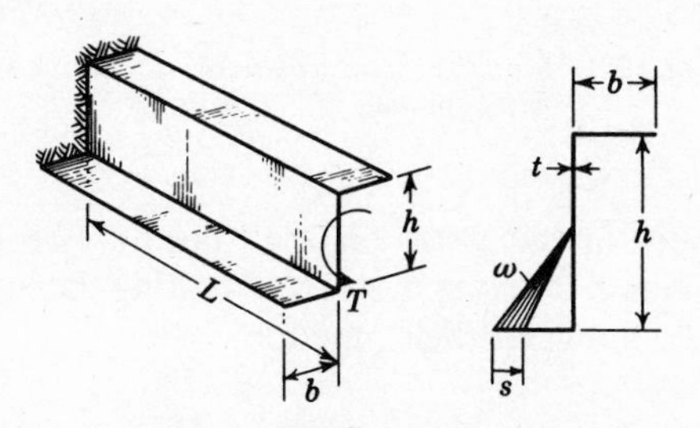

FIGURE 7.8 Cantilevered z section subjected to an end torque.

The above procedure for calculating Γ, though correct, is quite involved and mechanical. It is often more rewarding to evaluate the warping constant by investigating physically the process of warping. To do this, we note that if the member were perfectly free to warp, the center lines of the top and bottom flanges would remain straight during deformation, as is indicated by the dashed lines in Fig. 7.9a. When one section is restrained as in Fig. 7.9b, however, the flanges displace owing to twisting an amount $\phi h/2$ and acquire a certain amount of curvature. The web, on the other hand, merely rotates and remains plane. Since each flange has undergone a change in curvature, it follows that a flange bending moment M_f of magnitude $EI_f(d^2/dx^2)(\phi h/2)$ is developed, where I_f is the moment of inertia of the flange. Thus

$$M_f = E\frac{tb^3h}{24}\frac{d^2\phi}{dx^2} \qquad (d)$$

Now M_f varies from point to point, and in order that each flange element be in equilibrium a shearing force V_f must also be developed (Fig. 7.9c). Furthermore, since the longitudinal strains of the flange must be compatible with those of the web at their junction, we can also expect a normal force N_f to be developed. This, in turn, requires a normal force of $2N_f$

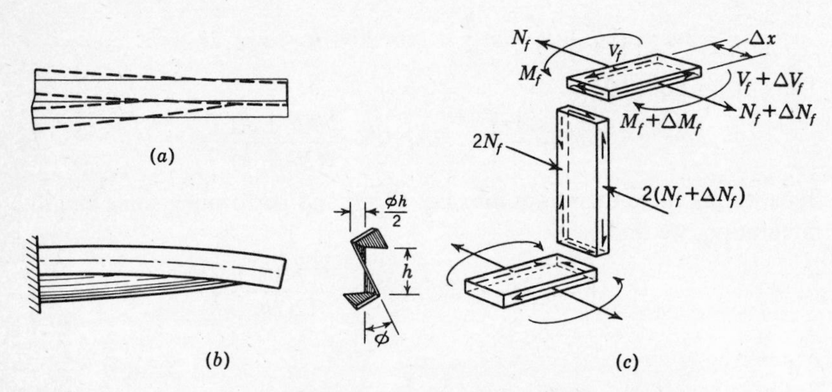

FIGURE 7.9 Physical interpretation of the warping stress system in a z section.

in the web because the net stress distribution must be self-equilibrating. N_f is easily determined in terms of M_f by equating the longitudinal strains of the web and the flange at their junction:

$$-E\frac{2N_f}{th} = E\left(\frac{N_f}{bt} - \frac{M_f b}{2I_f}\right)$$

Thus

$$N_f = \frac{tb^2hM_f}{2(2b+h)I_f} = \frac{Etb^2h^2}{4(2b+h)}\frac{d^2\phi}{dx^2} \qquad (e)$$

By considering the equilibrium of an element of the flange, we find

$$V_f = \frac{dM_f}{dx} + \frac{b}{2}\frac{dN_f}{dx}$$

or, in view of Eqs. (b) and (c),

$$V_f = \frac{Etb^3h}{12}\frac{b+2h}{2b+h}\frac{d^3\phi}{dx^3} \qquad (f)$$

A shearing force of equal magnitude but opposite direction is developed in the lower flange. Hence, it is the couple produced by these shearing forces that develops the warping torque $M_{t\omega}$. Referring to Eq. (7.45*b*), we see that .

$$M_{t\omega} = -E\Gamma\frac{d^3\phi}{dx^3} = -V_f h = -E\frac{tb^3h^2(b + 2h)}{12(2b + h)}\frac{d^3\phi}{dx^3} \tag{g}$$

Therefore,

$$\Gamma = \frac{tb^3h^2(b + 2h)}{12(2b + h)} \tag{h}$$

which agrees with Eqs. (*a*).

The normal stress at any point in the flange can now be evaluated:

$$(\sigma_x)_f = \frac{N_f}{bt} + \frac{M_f\bar{s}}{I_f}$$

where $\bar{s}$ is the distance measured from the centroid of the flange. For the bottom flange, $\bar{s} = b/2 - s$ and $(\sigma_x)_f$ becomes

$$(\sigma_x)_f = E\frac{d^2\phi}{dx^2}\left[\frac{bh(b + h)}{2(2b + h)} - \frac{hs}{2}\right] \tag{i}$$

Comparing the term within the brackets with Eqs. (*a*) and (*b*), we find that it is equal to $2(\omega_0 - \omega)$. Hence, Eq. (*i*) gives a normal stress distribution which agrees exactly with the general expression for warping stress in Eq. (7.29).

Similarly, the stress in the web is

$$(\sigma_x)_w = -\frac{2N_f}{ht} = -\frac{d^2\phi}{dx^2}\frac{b^2h}{2(2b + h)} \tag{j}$$

which is also verified by Eq. (7.29).

The shear flow in the flanges due to warping is the sum of a parabolic variation due to dM_f/dx and a linear variation due to dN_f/dx:

$$(q)_f = \frac{dN_f}{dx}\frac{s}{b} + \frac{dM_f}{dx}\frac{st(b - s)/2}{I_f} \tag{k}$$

The quantity $st(b - s)/2$ is the first moment of an outside portion of the flange area with respect to the centroid of the flange. Substituting Eqs. (*d*) and (*e*) into this expression gives

$$(q)_f = E\frac{d^3\phi}{dx^3}\left\{\frac{ths}{4(2b + h)}[bh + (2b + h)(b - s)]\right\} \tag{l}$$

By examining the equilibrium of an element of the web we find

$$(q)_w = \frac{dN_f}{dx}\left[1 - \frac{2(s - b)}{h}\right] \tag{m}$$

where $(q)_w$ is the shear flow in the web and $(s - b)$ is the distance measured from the bottom of the web to any point in the web. Note that $(q)_w$ is zero at the centroid of the section and that the net vertical shearing force developed in the web is zero. We now introduce Eq. (*e*) into this formula and obtain

$$(q)_w = E\frac{d^3\phi}{dx^3}\left\{\frac{tb^2h}{4(2b+h)}[h-2(s-b)]\right\} \tag{n}$$

It is easily shown that the quantities within braces in Eqs. (*l*) and (*n*) are precisely the first sectorial moments Q_ω defined in Eq. (7.42*a*). Hence, we obtain exactly the same equations for the shear flow by direct application of Eq. (7.41). It is important to note that in order to obtain final shearing stresses we must superimpose on those produced by warping

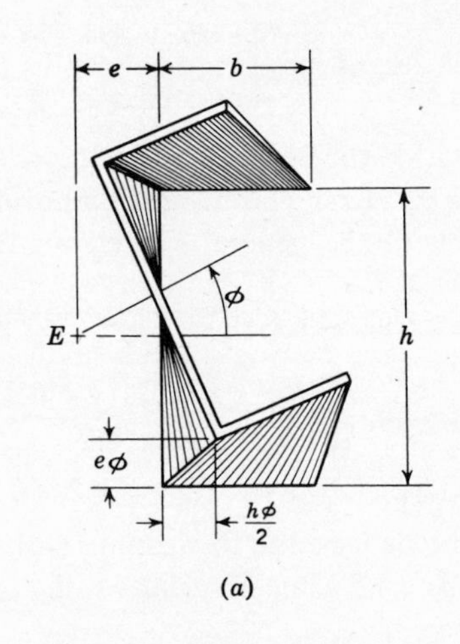

(*a*)

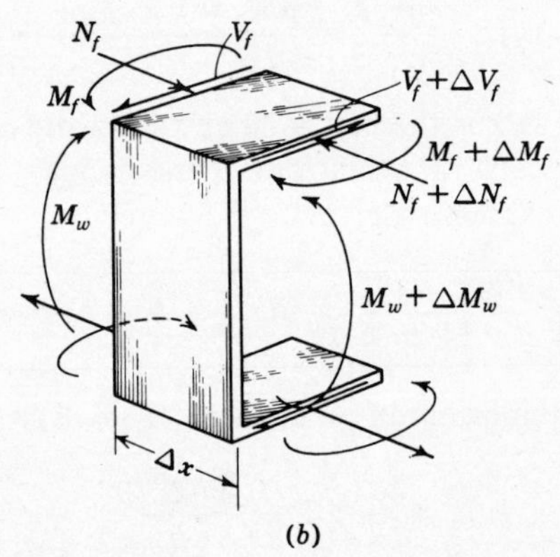

(*b*)

FIGURE 7.10 Warping stress system in a channel.

the linear torsional stresses of Eq. (7.6):

$$(\tau_{xs})_T = 2Gt\frac{d\phi}{dx}n = \frac{2Gt^3n}{3}(2b + h)\frac{d\phi}{dx} \tag{o}$$

where n is the coordinate measured normal to the center line of the wall (Fig. 7.2*a*).

A similar method of analysis can be used in the case of the channel shown in Fig. 7.10. In this case the analysis is complicated by the fact that the shear center is located a distance e from the web. Because of this, a twist ϕ of the section about the shear center not only displaces both flanges by $h\phi/2$ but also displaces the web by an amount $e\phi$. Therefore web moments M_w are developed proportional to the curvature of the web in addition to the flange moments M_f proportional to the flange curvatures. Since these curvatures are known in terms of the twist, these moments are easily obtained from elementary bending theory:

$$M_f = EI_f\frac{h}{2}\frac{d^2\phi}{dx^2} \qquad \text{and} \qquad M_w = EI_w e\frac{d^2\phi}{dx^2} \tag{p}$$

where I_f and I_w are the moments of inertia of the flange and the web. The distance e is given by Eq. (5.45).

Now the warping torque $M_{t\omega}$ is again $-V_fh$, where V_f is the flange shear. We proceed as before and isolate a typical element of the channel, as shown in Fig. 7.10*b*, and obtain the following equilibrium conditions:

$$V_f = \frac{dM_f}{dx} - \frac{b}{2}\frac{dN_f}{dx} \qquad \text{and} \qquad \frac{dM_w}{dx} = -h\frac{dN_f}{dx}$$

Thus

$$V_f = \frac{dM_f}{dx} + \frac{b}{2h}\frac{dM_w}{dx} = E\frac{d^3\phi}{dx^3}\left(\frac{I_fh}{2} + \frac{bI_we}{2h}\right) \tag{q}$$

If we denote by t_f and t_w the flange and web thicknesses, $I_f = t_fb^3/12$, $I_w = t_wh^3/12$, and $e = t_fh^2b^2/4I_z$, where I_z is the moment of inertia of the channel with respect to the z axis. Incorporating these values into Eq. (q) and referring to Eq. (7.45*b*), we find

$$M_t = -E\Gamma\frac{d^3\phi}{dx^3} = -V_fh = -E\frac{b^3t_fh^2}{24}\left(1 + \frac{t_wh^3}{4I_z}\right)\frac{d^3\phi}{dx^3}$$

Therefore,

$$\Gamma = \frac{b^3t_fh^2}{24}\left(1 + \frac{t_wh^3}{4I_z}\right) \tag{r}$$

Again, normal stresses are calculated by the elementary formulas

$$(\sigma_x)_f = \frac{N_f}{bt} + \frac{M_f\bar{s}}{I_f} \qquad (\sigma_x)_w = \frac{M_wy}{I_w} \tag{s}$$

Table 7.1 Warping constants

	Section	Γ	e (E = shear center)
①	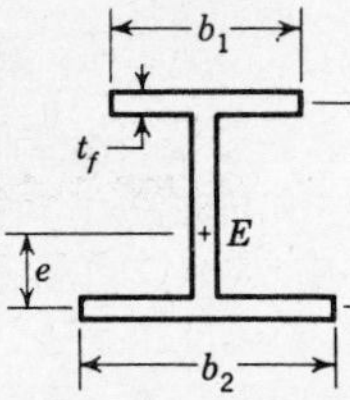 b_1, t_f, h, E, e, b_2	$\dfrac{t_f h^2}{12}\left(\dfrac{b_1^3 b_2^3}{b_1^3 + b_2^3}\right)$	$\dfrac{h b_2^3}{b_1^3 + b_2^3}$
②	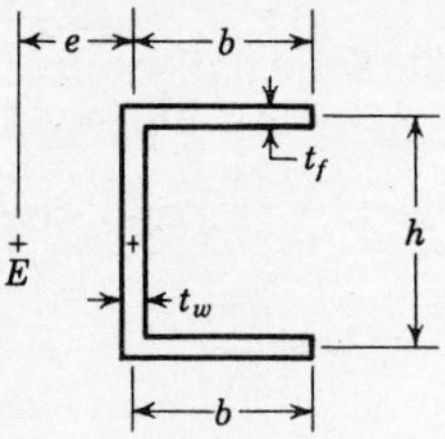 e, b, t_f, h, E, t_w, b	$\dfrac{t_f b^3 h^2}{12} \dfrac{2ht_w + 3bt_f}{ht_w + 6bt_f}$	$\dfrac{3t_f b^2}{6bt_f + ht_w}$
③	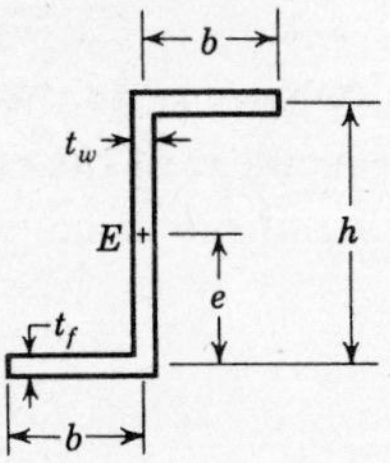 b, t_w, E, h, e, t_f, b	$\dfrac{t_f b^3 h^2}{12(2b+h)^2}\left[2(b+h)^2 - bh\left(2 - \dfrac{t_w}{t_f}\right)\right]$	$\dfrac{h}{2}$
④	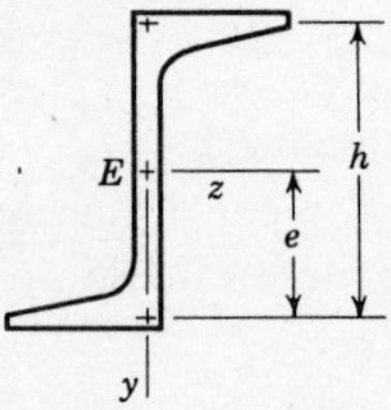 E, z, h, e, y	$\dfrac{h^2 I_y}{4}$	$\dfrac{h}{2}$
⑤	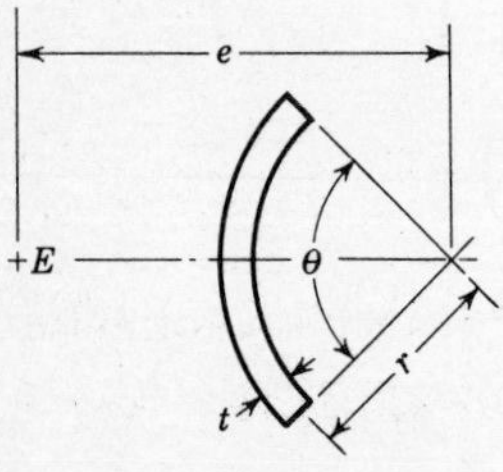 e, E, θ, r, t	$\dfrac{tr^5}{12}\left[\theta^3 - 6\dfrac{e^2}{r^2}(\theta - \sin\theta)\right]$	$2r\,\dfrac{2\sin(\theta/2) - \cos(\theta/2)}{\theta - \sin\theta}$

where $\bar{s}$ is the distance from the centroid of the flange. Note that N_f is negative (compressive) for the top flange. Similarly, the shear flows are

$$(q)_f = \frac{dM_f}{dx}\frac{Q_f}{I_f} - \frac{dN_f}{dx}\frac{s}{b} \qquad (q)_w = \frac{dM_w}{dx}\frac{Q_w}{I_w} - \frac{dN_f}{dx} \tag{t}$$

where Q_f and Q_w are the usual first moments of area with respect to the centroids of the flange and the web. The coordinate s is measured from the outside edge of the lower flange. We can easily express the warping stresses in terms of ϕ by substituting Eqs. (p) and (q) into Eqs. (s) and (t). Again, the resulting stress formulas are identical to those obtained by direct application of Eqs. (7.29) and (7.43).

The above procedures can be used for any open section composed of thin rectangular elements. The general formulas given by Eqs. (7.29) and (7.43) are more convenient to use when dealing with curved or irregular shaped sections, although they do not always permit a simple physical interpretation of the results. Table 7.1 contains some warping constants of some common thin-walled sections.[5]

7.8 Secondary warping. According to our theory, the sections shown in Fig. 7.11 twist without warping. The shear centers of these sections lie at the intersection of the center lines of the thin rectangular elements. Because of this, the resultants of shear flows produced by any type of

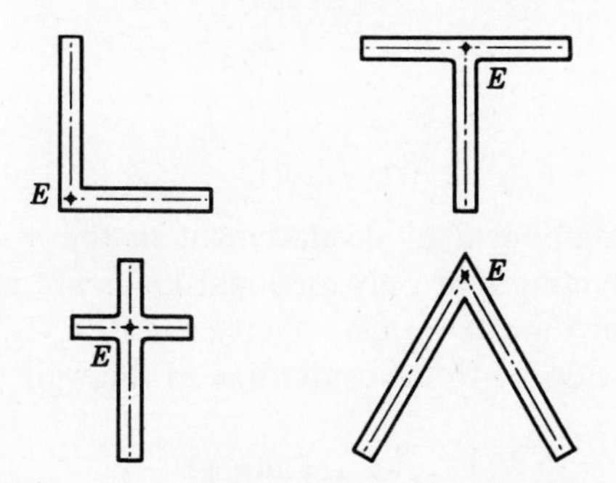

FIGURE 7.11 Thin-walled sections which bend and twist without primary warping.

loading must pass through these points; hence no warping torque $M_{t\omega}$ can be developed and no longitudinal stresses are produced by torsional loads. The warping constant can be taken equal to zero.

However, when the wall thickness t is not extremely small compared with the other dimensions, a secondary stress system can be developed

[5] Warping constants for other shapes are given in Refs. 8 and 25.

perpendicular to the contour line of the section. Normal stresses then vary linearly over t, the shearing stress τ_{nx} can no longer be neglected, and a secondary warping torque $M_{t\omega}^*$ is developed (Fig. 7.12). In such cases the sections in Fig. 7.11 undergo "secondary" warping and, consequently, possess a secondary warping rigidity $E\Gamma^*$.

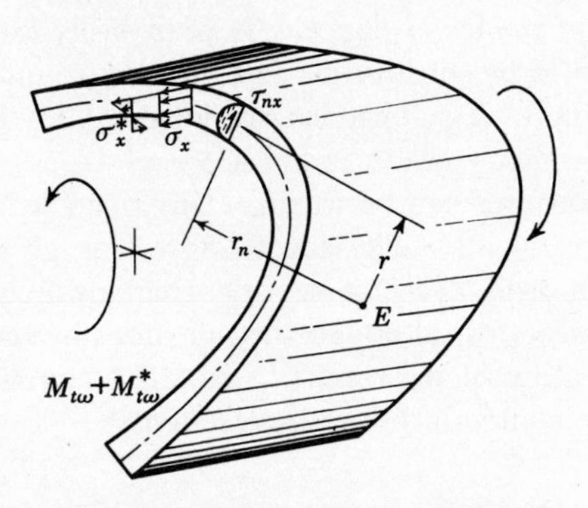

FIGURE 7.12 Secondary warping stresses.

To investigate these effects, let us denote the secondary longitudinal stresses and displacements by σ_x^* and u^*. Then

$$\gamma_{xn} \cong 0 = \frac{\partial u^*}{\partial n} + \frac{\partial \xi}{\partial x} \tag{7.47}$$

and

$$\frac{\partial \tau_{nx}}{\partial n} + \frac{\partial \sigma_x{}^*}{\partial x} = 0 \tag{7.48}$$

where ξ is the component of displacement in the n direction given by Eqs. (7.15). Assuming that only torsional loads are applied, v and w in Eqs. (7.15) are zero and $\xi = r_n\phi$, where, again, r_n is an arm from the shear center to a normal to the center line of the wall. Hence Eq. (7.47) gives

$$u^* = -\int \frac{\partial(r_n\phi)}{\partial x}\, dn$$

Since r_n is independent of n and u^* is zero at $n = 0$,

$$u^* = -r_n n \frac{d\phi}{dx} \tag{7.49}$$

Now $\sigma_x^* = E(\partial u^*/\partial x)$; therefore from Eq. (7.48) we find

$$\tau_{nx} = Er_n \frac{d^3\phi}{dx^3}\int n\, dn + \tau_0$$

where τ_0 is τ_{nx} at $n = 0$. From the condition that τ_{nx} must vanish at $n = \pm t/2$ we find

$$\tau_{nx} = E \frac{d^3\phi}{dx^3} \frac{r_n}{2} \left(\frac{t^2}{4} - n^2 \right) \tag{7.50}$$

Let V^* denote the shearing force per unit length of contour s produced by τ_{nx}. Then

$$V^* = \int_{-t/2}^{t/2} \tau_{xn} \, dn = E \frac{d^3\phi}{dx^3} \frac{t^3 r_n}{12} \tag{7.51}$$

These forces produce the secondary warping torque

$$M^*_{t\omega} = E\Gamma^* \frac{d^3\phi}{dx^3} = \int V^* r_n \, ds \tag{7.52}$$

Finally, substituting Eq. (7.51) into (7.52) and equating like coefficients, we find for the secondary warping constant

$$\Gamma^* = \frac{1}{12} \int_0^S t^3 r_n^2 \, ds \tag{7.53}$$

where S is the total length of the contour of the section. The total warping constant is $\Gamma + \Gamma^*$, where Γ is defined in Eq. (7.33).

For sections such as those in Fig. 7.11, the only warping stresses developed due to twisting are the secondary stresses given by the above formulas. In these cases secondary warping may lead to significant stresses. However, in the case of thin-walled sections which also possess a primary warping rigidity, Γ^* is usually very small compared with Γ and can be neglected. For example, for the z section with $b = h/2 = 10t$, Γ^* is less than 1 percent of Γ. Secondary warping constants for three common shapes are given in Table 7.2.[6]

Table 7.2 Secondary warping constants

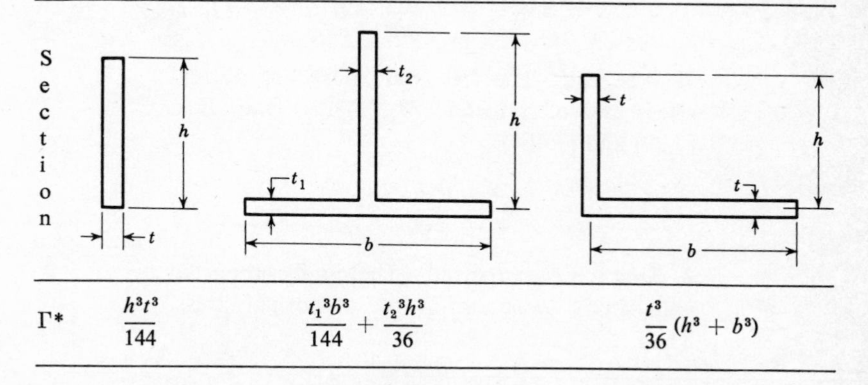

Γ^*	$\dfrac{h^3 t^3}{144}$	$\dfrac{t_1^3 b^3}{144} + \dfrac{t_2^3 h^3}{36}$	$\dfrac{t^3}{36}(h^3 + b^3)$

[6] For values of Γ^* for other sections, see Ref. 25, pp. 124–136.

7.9 Evaluation of twist of open sections. It is clear that the quantitative evaluation of warping stresses can be accomplished only if the twist ϕ is known as a function of x. This, of course, requires that we solve a differential equation of the form given by Eq. (7.46*a*) or (7.46*b*), whichever applies. In this section we investigate the procedure to be used in determining ϕ for a given problem.

Let us refer to the more general form of the twist equation given by (7.46*a*) and begin by rewriting it as follows:

$$\frac{d^4\phi}{dx^4} - c^2 \frac{d^2\phi}{dx^2} = \frac{m_t}{E\Gamma} \tag{7.54}$$

where
$$c = \sqrt{\frac{GJ}{E\Gamma}} \tag{7.55}$$

Given m_t as a function of x, we first obtain a particular solution ϕ_p of Eq. (7.54) and then assume a homogeneous solution of the form $Ae^{\lambda x}$. Following the usual procedure, we find, for the four roots of the characteristic equation, $\lambda_1 = \lambda_2 = 0$ and $\lambda_3 = -\lambda_4 = c$. Thus, the general solution to Eq. (7.54) is

$$\phi = A_0 + A_1 x + A_2 e^{cx} + A_3 e^{-cx} + \phi_p \tag{7.56}$$

where A_0, A_1, A_2, and A_3 are the arbitrary constants. These constants, of course, must be determined from boundary conditions.

We may arrive at the appropriate condition for a given type of end restraint by referring to the equations for u, σ_x, and τ_{xs} derived previously. Some of the more common conditions are as follows:

1. Fixed end—The built-in section can neither twist nor warp. According to Eq. (7.17), u is proportional to $d\phi/dx$. Thus,

$$\phi = 0 \qquad \frac{d\phi}{dx} = 0 \tag{7.57}$$

2. Free end—The free end is free of normal stress and of a torque M_t. Thus, from Eqs. (7.29) and (7.46*b*)

$$\frac{d^2\phi}{dx^2} = 0 \qquad \frac{d^3\phi}{dx^3} - c^2 \frac{d\phi}{dx} = 0 \tag{7.58}$$

3. Simply supported end—The simply supported end cannot twist and is free of normal stress. Thus

$$\phi = 0 \qquad \frac{d^2\phi}{dx^2} = 0 \tag{7.59}$$

Consider as a simple example the cantilevered z section in Fig. 7.8 and, for simplicity, let us take for the dimensions of the cross section $b = h/2 = a$.

In this case N_x, M_y, M_z, and m_t are zero and Eq. (7.46b) is applicable. We obtain the general solution to the third-order equation from Eq. (7.56) by setting the constant A_1 equal to zero. Further, from Eq. (c) of Art. 7.7, we have

$$\Gamma = \frac{tb^3h^2(b + 2h)}{12(2b + h)} = \frac{5ta^5}{12} \tag{a}$$

and the torsional constant

$$J = \frac{4t^3a}{3} \tag{b}$$

Since $M_t = T$ in this case, we can easily verify that a particular solution to Eq. (7.46b) is

$$\phi_p = \frac{Tx}{GJ}$$

We recognize ϕ_p as the twist of an unrestrained section under pure torsion. Again we observe that the particular solution leads to stresses which satisfy the equilibrium requirements but not the compatibility requirements. The complementary solution leads to the self-equilibrating warping stresses and provides compatibility. For the present problem Eq. (7.56) becomes

$$\phi = A_0 + A_2e^{cx} + A_3e^{-cx} + \frac{Tx}{GJ}$$

which we rewrite in the more convenient form

$$\phi = C_0 + C_1 \cosh cx + C_2 \sinh cx + \frac{Tx}{GJ} \tag{c}$$

In view of Eqs. (7.57) and (7.58), this function must satisfy the conditions

$$\phi = \frac{d\phi}{dx} = 0 \quad \text{at } x = 0 \qquad \text{and} \qquad \frac{d^2\phi}{dx^2} = 0 \quad \text{at } x = L$$

Substituting Eq. (c) into these conditions and solving for the constants, we find

$$C_0 = -C_1 = -\frac{T}{cGJ}\tanh cL \qquad C_2 = -\frac{T}{cGJ}$$

Therefore $$\phi = \frac{T}{cGJ}[\tanh cL(\cosh cx - 1) - \sinh cx + cx] \tag{d}$$

We can now evaluate stresses. Substituting the first derivative of ϕ into Eq. (7.6), we find

$$(\tau_{xs})_T = \frac{2Tn}{J} - \frac{2Tn}{J}\frac{\cosh c(L - x)}{\cosh cL} \tag{e}$$

This stress varies linearly over t. We see that $(\tau_{xs})_T$ vanishes at the fixed end of the section and is a maximum at the free end. We recognize the first term in this equation as the shearing stress in a completely unrestrained section in pure torsion. Hence, the remaining term represents the effects of restrained warping.

Introducing ϕ into Eq. (7.29), we find

$$\sigma_x = \frac{2ETc}{GJ}(\omega_0 - \omega)\frac{\sinh c(L-x)}{\cosh cL} \tag{f}$$

The maximum value of σ_x occurs at the fixed end where the warping restraints are applied. Note that the rate of decay of σ_x depends upon the magnitude of the constant cL.

We obtain the shear flow due to warping by substituting ϕ into Eq. (7.41):

$$q = \frac{TQ_\omega}{\Gamma}\frac{\cosh c(L-x)}{\cosh cL} \tag{g}$$

Note that the value of q at the free end is, in general, much less than that at the fixed end of the beam. It is clear that cross sections near the free end of the beam are essentially in pure torsion, while those near the fixed end are subjected primarily to restrained-warping stresses. We obtain the total shearing stress by substituting Eqs. (e) and (g) into Eq. (7.2).

For the section in Fig. 7.13, Eqs. (7.42a) and Eqs. (a) and (b) of Art. 7.7 give the following values of $2(\omega_0 - \omega)$ and Q_ω:

$$\begin{aligned}
&0 \le s \le a \qquad && 2(\omega_0-\omega) = \frac{a}{4}(3a-4s) \qquad && Q_\omega = \frac{ats}{4}(3a-2s)\\
&a \le s \le 3a \qquad && 2(\omega_0-\omega) = -\frac{a^2}{4} \qquad && Q_\omega = \frac{a^2t}{4}(2a-s)\\
&3a \le s \le 4a \qquad && 2(\omega_0-\omega) = \frac{a}{4}(4s-13a) \qquad && Q_\omega = \frac{at}{4}(2s^2-13as + 20a^2)
\end{aligned} \tag{h}$$

When we introduce these functions into Eqs. (f) and (g), the restrained-warping stresses are completely defined in terms of coordinates.

Let us now consider the more complex example of a cantilevered z section subjected to a vertical end force P that acts at the outside edge of the flange rather than at the shear center of the section (Fig. 7.14a). This loading is statically equivalent to an end torque $T = Pa$ plus a load P acting through the shear center.

Taking for illustration purposes $L/a = 10$, $a/t = 8$, and $\nu = \frac{1}{4}$, we find that the force through the shear center produces the unsymmetrical

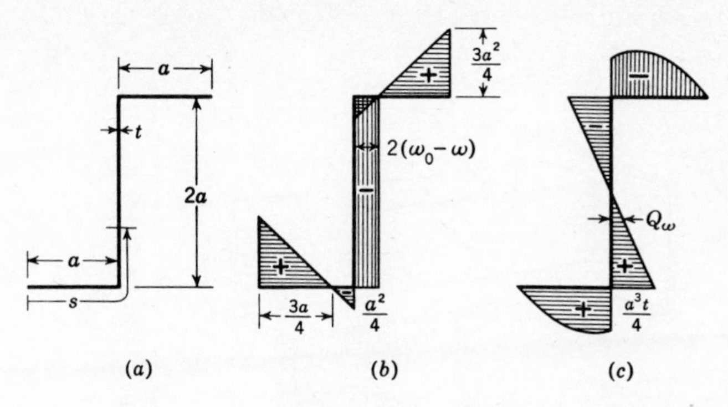

FIGURE 7.13 (*a*) Geometry of a *z* section; (*b*, *c*) variations in sectorial functions $2(\omega_0 - \omega)$ and Q_ω for the section.

bending stresses[7] shown in Fig. 7.14*b* and *c*. The end torque develops the torsional shearing stresses $(\tau_{xs})_T$ given by Eq. (*e*) and shown in Fig. 7.14*d*. Again note that these stresses are zero at the fixed end, since they are proportional to $d\phi/dx$, and that they are a maximum at the free end of the beam. Stresses due to restrained warping are obtained by substituting Eqs. (*h*) into Eqs. (*f*) and (*g*). These are shown in Fig. 7.14*e* and *f* for the section at $x = 0$. We see that the maximum normal stress due to restrained warping is 12.3 percent greater than the maximum due to bending. The largest stress produced by the shear flow q is 5 percent greater than the maximum shearing stress due to bending. We find that the maximum total shearing stress occurs at the free end and is of magnitude $35.08P/a^2$; of this, 85 percent is due to torsion. The total normal-stress and shearing-stress distributions at $x = 0$ are obtained by superimposing those in Fig. 7.14*b* and *e* and Fig. 7.14*c* and *f*. The resulting profiles are indicated in Fig. 7.14*g* and *h*.

7.10 Effects of axial loads. Thus far we have assumed that the presence of axial loads does not influence the twist or the transverse displacements of the structure. As a result, Eqs. (7.25) and (7.46) can be solved independently and the solutions can be superimposed to obtain the shape of the elastic curve. However, in the case of flexible thin-walled open sections subjected to large axial loads, this assumption can lead to serious errors; the twist and the transverse displacements are no longer so small that they can be ignored in equilibrium considerations and the principle of superposition is not directly applicable. In the following we develop

[7] See Arts. 4.6 and 5.3. The normal- and shearing-stress distributions due to this loading are also shown in Figs. 4.10 and 5.4.

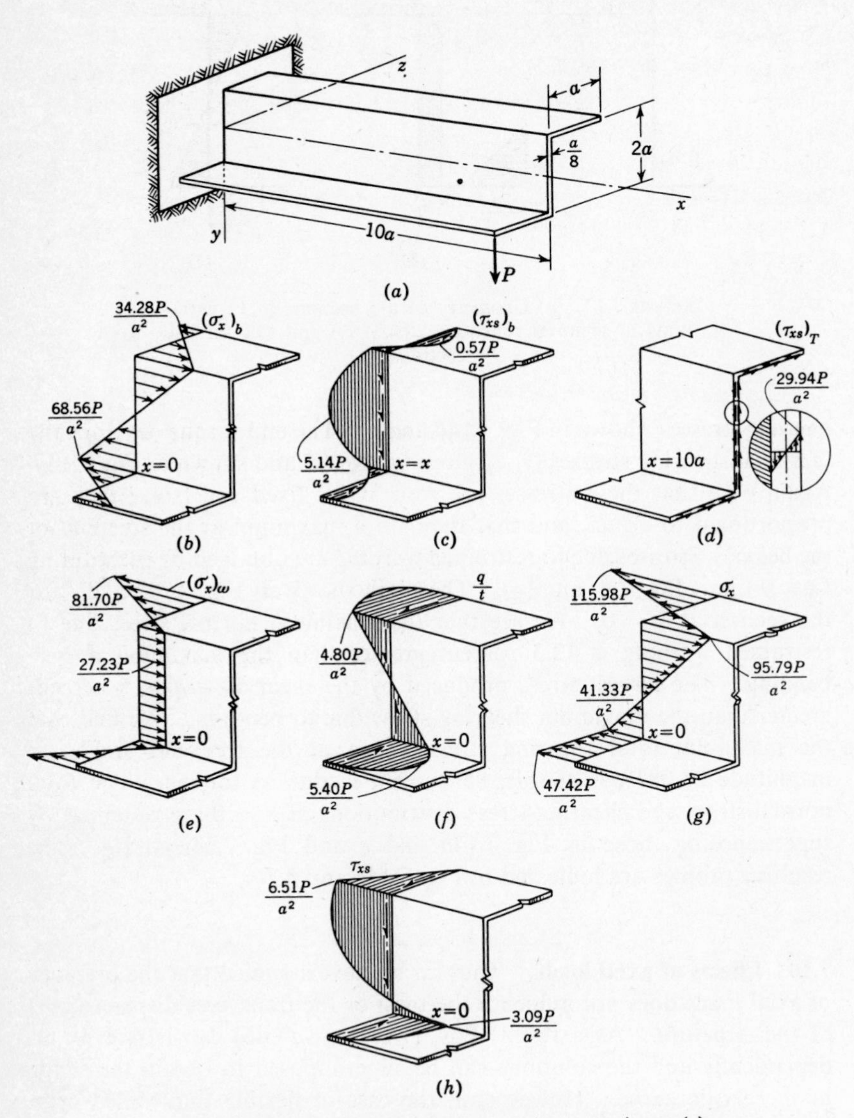

FIGURE 7.14 Combined stress system in a z section. (a) Cantilevered z section with off-center transverse load; (b, c) stresses due to bending; (d) torsional shearing stresses; (e, f) restrained-warping stresses; (g, h) total normal and shearing stresses developed at the fixed end.

a linear theory to account for these effects; it will serve as a generalization of the restrained-warping theory described in the preceding articles and the theory of ties and beam columns discussed in Chap. 6.[8]

To begin our investigation, we assume that the normal stress σ_x is still given by Eq. (7.35). In the present case, however, the static relations used to determine the moments in this equation involve the displacements of the structure. We now assume that displacements and curvatures are of such magnitude that components of the internal forces change the net transverse load and twisting moment per unit length.

To visualize these effects, let us consider the longitudinal fiber of length Δx and area ΔA located at an arbitrary point S on the contour.

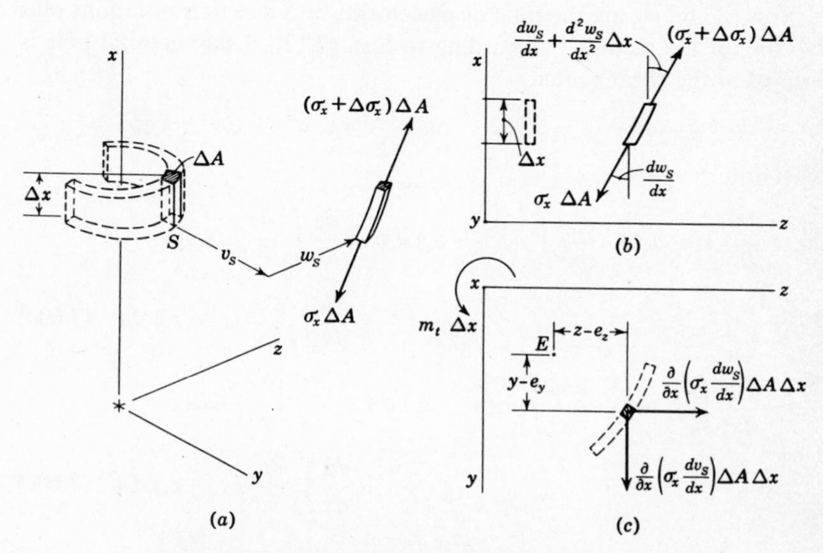

FIGURE 7.15 (*a*) Displacements of a fiber of area A; (*b*) projection of deformed fiber on xz plane; (*c*) in-plane forces produced by normal stresses.

Upon the applications of loads, the fiber displaces v_S and w_S in the y and z directions. When it reaches an equilibrium configuration, internal forces $\sigma_x \, \Delta A$ are developed which act along the deformed axis of the fiber (Fig. 7.15*a*). Referring to Fig. 7.15*b*, we see that the net projection of the internal force on the z axis is

$$-\sigma_x \, \Delta A \, \frac{dw_S}{dx} \, \Delta x + (\sigma_x + \Delta\sigma_x) \, \Delta A \left(\frac{dw_S}{dx} + \frac{d^2 w_S}{dx^2} \, \Delta x \right)$$

[8] See Art. 6.9.

Let $\bar{p}_y$ and $\bar{p}_z$ denote the total transverse force per unit length due to the projections of these internal forces. Then, taking the limit as ΔA and Δx approach zero and integrating over the contour of the section, we find

$$\bar{p}_z = \int_A \frac{\partial}{\partial x}\left(\sigma_x \frac{dw_S}{dx}\right) dA$$

Similarly
$$\bar{p}_y = \int_A \frac{\partial}{\partial x}\left(\sigma_x \frac{dv_S}{dx}\right) dA$$

Further, if $\bar{m}_t$ denotes the torque produced by these internal forces, we find from Fig. 7.15*c* that

$$\bar{m}_t = \int_A \left[(y - e_y)\frac{\partial}{\partial x}\left(\sigma_x \frac{dw_S}{dx}\right) - (z - e_z)\frac{\partial}{\partial x}\left(\sigma_x \frac{dv_S}{dx}\right)\right] dA$$

Now v_S and w_S are the total displacements of S due to translations plus twisting of the section. According to Eqs. (7.13), if the sectorial pole is located at the shear center,

$$v_S = v - (z - e_z)\phi \qquad \text{and} \qquad w_S = w + (y - e_y)\phi$$

Therefore

$$\bar{p}_y = \frac{d^2v}{dx^2}\int_A \sigma_x \, dA - \frac{d^2\phi}{dx^2}\int_A \sigma_x(z - e_z)\, dA + \frac{dv}{dx}\int_A \frac{\partial \sigma_x}{\partial x}\, dA - \frac{d\phi}{dx}\int_A \frac{\partial \sigma_x}{\partial x}(z - e_z)\, dA \tag{7.60a}$$

$$\bar{p}_z = \frac{d^2w}{dx^2}\int_A \sigma_x \, dA + \frac{d^2\phi}{dx^2}\int_A \sigma_x(y - e_y)\, dA + \frac{dw}{dx}\int_A \frac{\partial \sigma_x}{\partial x}\, dA + \frac{d\phi}{dx}\int_A \frac{\partial \sigma_x}{\partial x}(y - e_y)\, dA \tag{7.60b}$$

$$\begin{aligned}\bar{m}_t = &-\frac{d^2v}{dx^2}\int_A \sigma_x(z - e_z)\, dA + \frac{d^2w}{dx^2}\int_A \sigma_x(y - e_y)\, dA + \frac{d^2\phi}{dx^2}\int_A \sigma_x[(y - e_y)^2 \\ &+ (z - e_z)^2]\, dA - \frac{dv}{dx}\int_A \frac{\partial \sigma_x}{\partial x}(z - e_z)\, dA + \frac{dw}{dx}\int_A \frac{\partial \sigma_x}{\partial x}(y - e_y)\, dA \\ &+ \frac{d\phi}{dx}\int_A \frac{\partial \sigma_x}{\partial x}[(y - e_y)^2 + (z - e_z)^2]\, dA\end{aligned} \tag{7.60c}$$

We assume that N_x is applied at the centroid of the end sections. Then, from statics we find that the moments in the equation for σ_x are given by the formulas

$$\begin{aligned} M_y &= M_{ly} - N_x w_S = M_{ly} - N_x(w - e_y\phi) \\ M_z &= M_{lz} - N_x v_S = M_{lz} - N_x(v + e_z\phi) \end{aligned} \tag{7.61}$$

where M_{ly} and M_{lz} are the moments due solely to transverse loading and N_x is assumed to be a tensile force. Also from statics we find

$$\frac{dM_z}{dx} = V_y' \qquad \text{and} \qquad \frac{dM_y}{dx} = V_z' \tag{7.62}$$

where V_y' and V_z' are the modified shears

$$\begin{aligned} V_y' &= V_y - N_x\left[\frac{dv}{dx} - (z - e_z)\frac{d\phi}{dx}\right] \\ V_z' &= V_z - N_x\left[\frac{dw}{dx} + (y - e_y)\frac{d\phi}{dx}\right] \end{aligned} \tag{7.63}$$

The quantities V_y and V_z are the shearing forces due to the applied loads.

Finally, substituting Eq. (7.35) into Eqs. (7.60) and performing the indicated integration, we get

$$\bar{p}_y = N_x\frac{d^2v}{dx^2} - (M_y - e_zN_x)\frac{d^2\phi}{dx^2} - V_z'\frac{d\phi}{dx} \tag{7.64a}$$

$$\bar{p}_z = N_x\frac{d^2w}{dx^2} + (M_z - e_yN_x)\frac{d^2\phi}{dx^2} - V_y'\frac{d\phi}{dx} \tag{7.64b}$$

$$\begin{aligned} \bar{m}_t = {} & -(M_y - e_zN_x)\frac{d^2v}{dx^2} + (M_z - e_yN_x)\frac{d^2w}{dx^2} \\ & + \left(\frac{I_E}{A}N_x + C_yM_z + C_zM_y + \frac{H_\omega}{\Gamma}W_\omega\right)\frac{d^2\phi}{dx^2} \\ & - V_z'\frac{dv}{dx} + V_y'\frac{dw}{dx} + \left(C_zV_y' + C_yV_z' + \frac{H_\omega}{\Gamma}V_\omega\right)\frac{d\phi}{dx} \end{aligned} \tag{7.64c}$$

where I_E is the polar moment of inertia with respect to the shear center,

$$I_E = I_y + I_z + (e_y^2 + e_z^2)A \tag{7.65a}$$

and

$$C_z = \frac{I_yH_z - I_{yz}H_y}{I_yI_z - I_{yz}^2} - 2e_y \tag{7.65b}$$

$$C_y = \frac{I_zH_y - I_{yz}H_z}{I_yI_z - I_{yz}^2} - 2e_z \tag{7.65c}$$

The quantities H_y, H_z, and H_ω are new geometrical properties of the cross section defined as follows:

$$H_y = \int_A z(y^2 + z^2)\,dA \tag{7.66a}$$

$$H_z = \int_A y(y^2 + z^2)\,dA \tag{7.66b}$$

$$H_\omega = \int_A 2(\omega_0 - \omega)(y^2 + z^2)\,dA \tag{7.66c}$$

Now the net transverse and torsional loadings per unit length are $p_y + \bar{p}_y$, $p_z + \bar{p}_z$, and $m_t + \bar{m}_t$, where p_y, p_z, and m_t represent the applied loadings. Therefore, referring to Eqs. (6.14) and (7.46*a*), we obtain the following equations for the elastic curve:

$$
\begin{aligned}
E(I_yI_z - I_{yz}^2)\frac{d^4v}{dx^4} - (\bar{p}_yI_y - \bar{p}_zI_{yz}) &= p_yI_y - p_zI_{yz} \\
E(I_yI_z - I_{yz}^2)\frac{d^4w}{dx^4} - (\bar{p}_zI_z - \bar{p}_yI_{yz}) &= p_zI_z - p_yI_{yz} \\
E\Gamma\frac{d^4\phi}{dx^4} - GJ\frac{d^2\phi}{dx^2} - \bar{m}_t &= m_t
\end{aligned}
\tag{7.67}
$$

In view of Eqs. (7.61) and (7.63), when Eqs. (7.64) are introduced into Eqs. (7.67) we arrive at a complex system of nonlinear differential equations with variable coefficients. Since we set out to develop a linear theory, we now neglect products and squares of the displacements and their derivatives. Then Eqs. (7.67) become

$$
\begin{aligned}
E(I_yI_z - I_{yz}^2)\frac{d^4v}{dx^4} - I_yN_x\frac{d^2v}{dx^2} + I_{yz}N_x\frac{d^2w}{dx^2} + [I_yM_{ly} + I_{yz}M_{lz} \\
- N_x(e_zI_y + e_yI_{yz})]\frac{d^2\phi}{dx^2} + (I_yV_z + I_{yz}V_y)\frac{d\phi}{dx} = p_yI_y - p_zI_{yz}
\end{aligned}
\tag{7.68a}
$$

$$
\begin{aligned}
E(I_yI_z - I_{yz}^2)\frac{d^4w}{dx^4} - I_zN_x\frac{d^2w}{dx^2} + I_{yz}N_x\frac{d^2v}{dx^2} + [- I_zM_{lz} - I_{yz}M_{ly} \\
+ N_x(e_yI_z + e_zI_{yz})]\frac{d^2\phi}{dx^2} - (I_zV_y - I_{yz}V_z)\frac{d\phi}{dx} = p_zI_z - p_yI_{yz}
\end{aligned}
\tag{7.68b}
$$

$$
\begin{aligned}
E\Gamma\frac{d^4\phi}{dx^4} - \left(GJ + \frac{I_E}{A}N_x + C_yM_{lz} + C_zM_{ly} + \frac{H_\omega}{\Gamma}W_\omega\right)\frac{d^2\phi}{dx^2} \\
+ (M_{ly} - e_zN_x)\frac{d^2v}{dx^2} - (M_{lz} - e_yN_x)\frac{d^2w}{dx^2} + V_z\frac{dv}{dx} \\
- V_y\frac{dw}{dx} - \left(C_zV_y + C_yV_z + \frac{H_\omega}{\Gamma}V_\omega\right)\frac{d\phi}{dx} = m_t
\end{aligned}
\tag{7.68c}
$$

These equations are three simultaneous linear differential equations with variable coefficients. Since each is of fourth order, their solution will contain twelve arbitrary constants which must be evaluated from twelve independent boundary conditions. The precise form of the complementary solutions to these equations depends upon the functions M_{ly}, M_{lz}, V_y, and V_z; the relative magnitudes of the cross-sectional constants; and the sign and relative magnitude of N_x. We are again reminded that N_x is assumed to be a tensile force in these equations; otherwise the signs of all terms containing N_x must be changed. Note that if ϕ and I_{yz} are zero, Eq. (7.68*a*) reduces to Eq. (6.29), which we derived in our study of ties and beam columns.

To simplify Eqs. (7.68), let us assume that the transverse loads p_y and p_z are zero and that the moments M_{ly} and M_{lz} do not vary with x. The transverse shears V_y and V_z are zero, and all terms containing first derivatives of the displacements disappear in Eqs. (7.68*a*) and (7.68*b*). This is realized when the beam is bent by equal end couples applied in both the xy and the xz planes. If, in addition, we rotate the coordinate axes so that they become principal axes of the section, I_{yz} is zero, $C_y = H_y/I_y - 2e_z$, and $C_z = H_z/I_z - 2e_y$. In this case, Eqs. (7.68) reduce to

$$EI_z \frac{d^4v}{dx^4} - N_x \frac{d^2v}{dx^2} + (M_{ly} - N_x e_z) \frac{d^2\phi}{dx^2} = 0 \qquad (7.69a)$$

$$EI_y \frac{d^4w}{dx^4} - N_x \frac{d^2w}{dx^2} - (M_{lz} - N_x e_y) \frac{d^2\phi}{dx^2} = 0 \qquad (7.69b)$$

$$E\Gamma \frac{d^4\phi}{dx^4} - \left(GJ + \frac{I_E}{A} N_x + C_y M_{ly} + C_z M_{lz}\right) \frac{d^2\phi}{dx^2} - \frac{H_\omega}{\Gamma} \frac{d}{dx}\left(W_\omega \frac{d\phi}{dx}\right)$$
$$+ (M_{ly} - e_z N_x) \frac{d^2v}{dx^2} - (M_{lz} - e_y N_x) \frac{d^2w}{dx^2} = m_t \qquad (7.69c)$$

The bimoment W_ω in Eq. (7.69*c*) is, in general, a function of x. To evaluate this function, consider the thin-walled beam under eccentric axial loads shown in Fig. 7.16. From the definition of W_ω in Eq. (7.31), the external bimoments acting at the ends of this structure are

$$W_\omega(0) = N_x 2(\omega_0 - \omega_A) \quad \text{and} \quad W_\omega(L) = N_x 2(\omega_0 - \omega_B) \qquad (7.70)$$

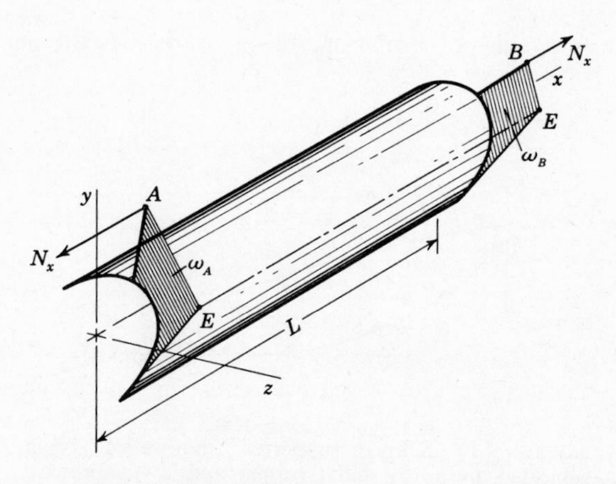

FIGURE 7.16 Thin-walled beam with end bimoments.

where ω_A and ω_B are the sectorial areas between $s = 0$ and the points at which N_x is applied. Referring to Eqs. (7.32), (7.45*b*), and (7.54), we see that

$$\frac{d^2 W_\omega}{dx^2} - c^2 W_\omega = 0 \tag{7.71}$$

where c is defined in Eq. (7.55). Solving this equation and introducing Eqs. (7.70) as boundary conditions, we find

$$W_\omega = \frac{2N_x}{\sinh cL}\left[(\omega_0 - \omega_B)\sinh cx + (\omega_0 - \omega_A)\sinh c(L - x)\right] \tag{7.72a}$$

In the present case, $\omega_A = \omega_B$, so that

$$W_\omega = \frac{W_{\omega_A}}{\sinh cL}\left[\sinh cx + \sinh c(L - x)\right] \tag{7.72b}$$

where W_{ω_A} is the applied bimoment. If point A is a sectorial centroid, $\omega_0 - \omega_A$ is zero and the bimoment in (7.69*c*) vanishes, but V_ω may exist.

Let us now examine the application of Eqs. (7.69) to a simple problem. Consider the simply supported beam in Fig. 7.17 which is subjected to a compressive axial force through its centroid and a sinusoidally distributed torque $m_t = m_0 \sin(\pi x/L)$. We assume that N_x is applied in such a way that W_ω is zero and that y and z are principal axes. In this case M_{ly} and M_{lz} are zero and the twelve boundary conditions are

$$v = w = \phi = 0$$

and

$$\frac{d^2 v}{dx^2} = \frac{d^2 w}{dx^2} = \frac{d^2 \phi}{dx^2} = 0$$

at $x = 0$ and at $x = L$. We assume the solution to Eqs. (7.69) to be of the form

$$v = v_0 \sin\frac{\pi x}{L} \qquad w = w_0 \sin\frac{\pi x}{L} \qquad \phi = \phi_0 \sin\frac{\pi x}{L} \tag{a}$$

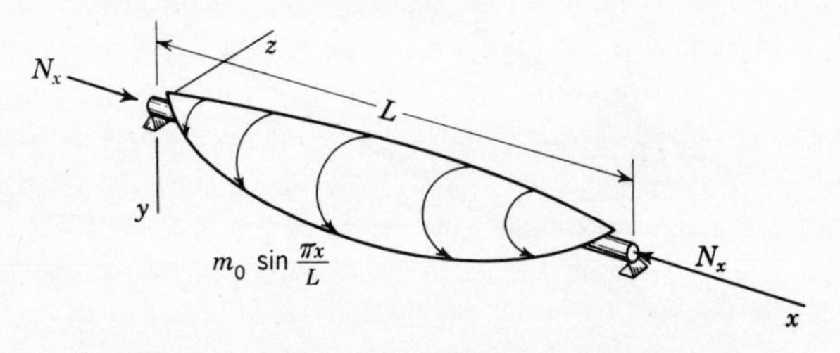

FIGURE 7.17 Simply supported thin-walled beam subjected to a sinusoidal torque and a compressive axial force.

where v_0, w_0, and ϕ_0 are constants yet to be determined. Equations (*a*) automatically satisfy all of the boundary conditions. Hence, if v_0, w_0, and ϕ_0 can be chosen so that Eqs. (*a*) also satisfy Eqs. (7.69), we shall have the solution to the problem.

Substituting Eqs. (*a*) into Eqs. (7.69) and recalling that N_x is now negative, we obtain

$$\left(EI_z \frac{\pi^2}{L^2} - N_x\right) v_0 - e_z N_x \phi_0 = 0$$

$$\left(EI_y \frac{\pi^2}{L^2} - N_x\right) w_0 + e_y N_x \phi_0 = 0 \tag{b}$$

$$\left(E\Gamma \frac{\pi^2}{L^2} + GJ - \frac{I_E}{A} N_x\right) \phi_0 - e_z N_x v_0 + e_y N_x w_0 = m_0$$

Solving these equations, we find

$$v_0 = -e_z N_x (N_x - N_2) \frac{m_0}{D}$$

$$w_0 = e_y N_x (N_x - N_1) \frac{m_0}{D} \tag{c}$$

$$\phi_0 = (N_x - N_1)(N_x - N_2) \frac{m_0}{D}$$

where N_1 and N_2 are the Euler buckling loads for instability of a pinned-end column in the *y* and *z* directions:

$$N_1 = \frac{\pi^2 EI_z}{L^2} \qquad N_2 = \frac{\pi^2 EI_y}{L^2} \tag{d}$$

and

$$D = (N_x - N_1)(N_x - N_2)\left(\frac{I_E}{A} N_x - GJ - E\Gamma \frac{\pi^2}{L^2}\right) - e_z^2 N_x^2 (N_x - N_2) - e_y^2 N_x^2 (N_x - N_1) \tag{e}$$

Final displacements are now obtained by substituting Eqs. (*c*) into Eqs. (*a*).

If N_x is steadily increased, the structure will eventually fail by buckling. If the beam bends without twisting, we found in Art. 6.9 that flexural instability will occur at a load of $\pi^2 EI_z/L^2$ or $\pi^2 EI_y/L^2$, whichever is smaller. These are the quantities N_1 and N_2 in the above equations. On the other hand, if the structure twists without bending, it may *twist-buckle* under a critical axial load N_ϕ. The magnitude of N_ϕ for a pinned-end column is easily obtained from Eqs. (*b*) by equating v_0, w_0, and m_0 to zero and noting that the last equation has a solution for ϕ_0 other than zero only if the coefficient of ϕ_0 vanishes:

$$E\Gamma \frac{\pi^2}{L^2} + GJ - \frac{I_E}{A} N_x = 0$$

Thus the critical load for twist buckling is

$$N_\phi = \frac{A}{I_E}\left(E\Gamma\frac{\pi^2}{L^2} + GJ\right) \tag{7.73}$$

The critical load in the case of combined bending and torsion, however, may be neither N_1, N_2, nor N_ϕ. In the classic sense, it is the smallest value of N_x which provides a nontrivial solution to Eqs. (*b*), m_0 being zero. This can occur only for values of N_x which cause the determinant of the coefficients in Eqs. (*b*) to vanish. In this case, this condition leads to the polynomial

$$D = 0 \tag{f}$$

where D is defined in Eq. (*e*). It is obvious that neither N_1, N_2, nor N_ϕ satisfies this polynomial; hence, the true critical load is a function of both the torsional and flexural stiffnesses of the structure.

For illustration purposes, let us assume that $N_1 < N_2$. Then if $N_x = N_1$ the left side of Eq. (*f*) is greater than zero. Thus, a smaller value of N_x is needed to satisfy this equation. This means that the true critical load is less than the Euler buckling load for the structure. Similarly, if $N_\phi < N_1 < N_2$ and $N_x = N_\phi$, D again is positive. Therefore the critical load is less than N_ϕ. This means that when we account for the effects of combined bending and torsion, we obtain a critical buckling load that is smaller than those which produce purely flexural or torsional buckling.

7.11 Thin-walled closed sections. We obtain an approximate analysis of thin-walled closed sections by assuming that the influence of shear deformation on normal stresses is negligible. The equations derived in Art. 5.9 for unrestrained closed tubes are also based on this assumption. This does not mean that we neglect shearing strains γ_{xs} as in the previous developments. On the contrary, the shearing-strain–displacement relations are essential in the extension of the theory to closed sections. The hypothesis is equivalent to assuming that the normal stress σ_x is given by Eq. (7.29) for both open and closed sections.

We recall from Chap. 3 that the torsional shearing stress in closed tubes is practically uniform over the thickness and that the linear stress $(\tau_{xs})_T$ is negligible. Thus, in the present case, the entire shearing-stress distribution can be described in terms of the shear flow q. Since we no longer regard γ_{xs} as being negligible, the strain-displacement relation for closed tubes is as follows:

$$\gamma_{xs} = \frac{q}{Gt} = \frac{\partial u}{\partial s} + \frac{\partial \eta}{\partial x} \tag{7.74}$$

where η is given in Eqs. (7.15). Solving this equation for u, we find

$$u = \int \frac{q}{Gt}\, ds - \int \frac{\partial \eta}{\partial x}\, ds + u_0$$

where, again, u_0 is the displacement at $s = 0$. If we carry out this integration around the entire closed contour of the section, we arrive at the compatibility condition

$$u - u_0 = 0 = \oint \frac{q}{Gt}\, ds - \oint \frac{\partial \eta}{\partial x}\, ds$$

Introducing η of Eqs. (7.15) into the above integral and recalling that $\sin \alpha = -dy/ds$ and $\cos \alpha = dz/ds$, we find that

$$\oint \frac{\partial \eta}{\partial x}\, ds = 2\Omega \frac{d\phi}{dx}$$

where Ω is the entire sectorial area enclosed by the tube wall. Therefore, the equation of consistent deformation is

$$\oint \frac{q}{Gt}\, ds - 2\Omega \frac{d\phi}{dx} = 0 \tag{7.75}$$

Note that the form of this equation is identical to that obtained by equating $\bar{u}$ to zero in Eqs. (3.53) and (5.87). In this instance, however, $d\phi/dx$ is not constant and q includes the effects of restrained warping.

To evaluate the shear flow q, we introduce Eq. (7.29) into Eq. (7.40) and perform the indicated integration. This gives

$$q = \bar{q} + q_0 - E \frac{d^3\phi}{dx^3} Q_\omega \tag{7.76}$$

where $\bar{q}$ is the shear flow at $s = 0$, q_0 is the shear flow due to bending:

$$q_0 = -\frac{Q_z I_y - Q_y I_{yz}}{I_y I_z - I_{yz}^2} V_y - \frac{Q_y I_z - Q_z I_{yz}}{I_y I_z - I_{yz}^2} V_z \tag{7.77}$$

and Q_ω is the sectorial moment defined by Eq. (7.42*a*). Note that q_0 can be calculated directly from Eq. (7.77) by assuming that the section is open at $s = 0$.

Substituting Eq. (7.76) into Eq. (7.75) and integrating, we get

$$\bar{q} \frac{1}{G} \oint \frac{ds}{t} + \frac{1}{G} \oint q_0 \frac{ds}{t} - \frac{E}{G} \frac{d^3\phi}{dx^3} \oint Q_\omega \frac{ds}{t} - 2\Omega \frac{d\phi}{dx} = 0 \tag{7.78}$$

We recognize the integral coefficient of $\bar{q}$ in this equation as the warping flexibility δ_1 defined in Eqs. (5.74). Thus, the first term in Eq. (7.78) represents the relative warping of opposite faces of a hypothetical cut in the tube wall at $s = 0$. Similarly, the term containing q_0 is the quantity

δ_0 in Eqs. (5.74), the relative warping at $s = 0$ due to bending. The remaining terms represent the relative warping displacements at $s = 0$ due to nonuniform twisting of the structure. Clearly, Eq. (7.78) is a condition of consistent deformation which states that $\bar{q}$ must be such that the total relative warping at $s = 0$ in the closed section is zero.

Solving Eq. (7.78) for $\bar{q}$, we get

$$\bar{q} = -\frac{\delta_0}{\delta_1} + \frac{2\Omega}{\delta_1}\frac{d\phi}{dx} + \frac{E}{G\delta_1}\frac{d^3\phi}{dx^3}\oint Q_\omega \frac{ds}{t} \tag{7.79}$$

On comparing this result with Eq. (5.75), we see that the first term on the right side is the shear flow due to bending at $s = 0$ in an unrestrained tube. The second term is $\bar{q}$ due to pure twisting of the tube, and the last term is the shear flow at $s = 0$ due to the nonuniform twisting produced by the restraints against warping.

Clearly, to evaluate $\bar{q}$ we must determine ϕ in terms of the applied loading. To accomplish this, we note that the twisting moment on a given section is again given by the formula

$$M_t = \oint qr\, ds$$

Substituting Eq. (7.76) into this relation gives

$$M_t = 2\Omega\bar{q} - E\Gamma\frac{d^3\phi}{dx^3} \tag{7.80}$$

Note that the moment due to q_0 is zero since we have selected the shear center as the pole of the arm r. Substituting Eq. (7.79) into (7.80) we find

$$M_t = -\frac{2\Omega\delta_0}{\delta_1} + \frac{4\Omega^2}{\delta_1}\frac{d\phi}{dx} - E\bar{\Gamma}\frac{d^3\phi}{dx^3} \tag{7.81}$$

where $\bar{\Gamma}$ is the *reduced warping constant* defined by

$$\bar{\Gamma} = \Gamma - \frac{2\Omega}{G\delta_1}\oint Q_\omega \frac{ds}{t} \tag{7.82}$$

Referring to Eq. (3.47) and noting that

$$\frac{4\Omega^2}{\delta_1} = \frac{G4\Omega^2}{\oint \frac{ds}{t}}$$

we see that the coefficient of $d\phi/dx$ in Eq. (7.81) is the torsional stiffness GJ for the closed section. Thus, in the case of general loading,

$$E\bar{\Gamma}\frac{d^4\phi}{dx^4} - GJ\frac{d^2\phi}{dx^2} = m_t - \frac{2\Omega}{\delta_1}\frac{d\delta_0}{dx} \tag{7.83}$$

where, again, m_t is the applied torque per unit length. Note that the parameter $c = \sqrt{GJ/E\Gamma}$ [see Eq. (7.55)] is many times larger than that of the open section. This means that the rate of decay of the restrained warping stresses is much more rapid for closed than for open sections.

We continue the analysis by solving Eq. (7.83) and applying the appropriate boundary conditions. Once ϕ is determined, we calculate $\bar{q}$ by means of Eq. (7.79) and then obtain the final shear-flow distribution from Eq. (7.76).

A similar procedure is followed in the case of multicell thin-walled beams. For example, let q_j be the shear flow in cell j of an n-celled beam under general loading and temporarily assume that cell j is the only closed cell in the beam. Then, from (7.76) we have

$$q_j = \bar{q}_j + q_0 - E\frac{d^3\phi}{dx^3}Q_\omega \tag{7.84}$$

where, in this case, $\bar{q}_j$ is the shear flow at an arbitrarily selected point m_j on the wall of cell j. Here q_0 and Q_ω are computed for the open-celled structure; that is, we assume that cuts are made in each cell rendering an open section. The shear center is then located as explained in Art. 5.9, and q_0 and Q_ω are evaluated using Eqs. (7.77) and (7.42*a*). If adjacent cells i and k are then closed by introducing corrective shear flows $\bar{q}_i$ and $\bar{q}_k$, the net corrective shear flows in webs s_{ij} and s_{jk} common to cells i and j and j and k are $\bar{q}_j - \bar{q}_i$ and $\bar{q}_j - \bar{q}_k$, respectively. Since the relative warping at m_j in the closed cell is zero, we have, from Eq. (7.75),

$$\delta_{ji}\bar{q}_i + \delta_{jj}\bar{q}_j + \delta_{jk}\bar{q}_k - 2\Omega_j\frac{d\phi}{dx} + \delta_{j0} - \frac{E}{G}\frac{d^3\phi}{dx^3}\oint_{s_j} Q_\omega \frac{ds}{t} = 0 \tag{7.85}$$

where δ_{ji}, δ_{jj}, and δ_{jk} are the warping flexibilities defined in Eqs. (5.81), δ_{j0} is defined in Eq. (5.80), and Ω_j is the sectorial area of cell j. The pole of the arm r in Q_ω is taken to be the shear center of the section. The last term in Eq. (7.84) represents the relative warping at m_j due to non-uniform twisting of the section. Except for this term and the fact that $d\phi/dx$ is not constant, Eq. (7.85) is identical to Eq. (5.88).

Equation (7.85) is an equation of consistent deformation simply stating that $\bar{q}_i$, $\bar{q}_j$, and $\bar{q}_k$ must be such that the total relative warping of opposite sides of a cut introduced in cell j at m_j is zero, if the cell is closed. One equation of this form is written for each cell and the resulting set is solved for the n flows $\bar{q}_1, \bar{q}_2, \ldots, \bar{q}_n$ in terms of $d\phi/dx$ and $d^3\phi/dx^3$. The corrective flow in cell j, for example, will be of the form

$$\bar{q}_j = C_{0j} + C_{1j}\frac{d\phi}{dx} + C_{2j}\frac{d^3\phi}{dx^3} \tag{7.86}$$

where C_{0j}, C_{1j}, and C_{2j} are constants. Then, to evaluate the twist of the section, we sum moments about the shear center and obtain the additional equation

$$M_t = 2\sum_{j=1}^{n} \bar{q}_j\Omega_j - E\Gamma \frac{d^3\phi}{dx^3} \tag{7.87}$$

where Γ is the warping constant of the entire section. Note that

$$2\sum_{j=1}^{n} \bar{q}_j\Omega_j = 2\sum_{j=1}^{n} C_{0j}\Omega_j + 2\frac{d\phi}{dx}\sum_{j=1}^{n} C_{1j}\Omega_j + 2\frac{d^3\phi}{dx^3}\sum_{j=1}^{n} C_{2j}\Omega_j \tag{7.88}$$

If M_t varies from section to section, we use the fourth-order equation

$$m_t = -2\sum_{j=1}^{n} \frac{d\bar{q}_j}{dx}\Omega_j + E\Gamma\frac{d^4\phi}{dx^4} \tag{7.89}$$

After solving Eq. (7.89) [or Eq. (7.87)] for ϕ and applying the appropriate boundary conditions, the corrective shear flows are determined from the equations of the form of Eq. (7.86) which were obtained earlier. We then obtain the final shear flow in each cell by introducing the quantities $\bar{q}_j$ into Eq. (7.84). The procedure is similar to that used in the analysis of unrestrained sections except that it is necessary to solve the differential equation (7.87) or (7.89) for the twist of the section.

It is interesting to examine briefly the consequences of abandoning the assumption that shear deformations do not affect σ_x. Then σ_x is no longer given by Eq. (7.29) and we must return to the general kinematic and static relations established earlier to formulate the problem.

Let us begin by putting Eq. (7.74) in the form

$$q = Gt\left(\frac{\partial u}{\partial s} + \frac{\partial \eta}{\partial x}\right) \tag{7.90}$$

and differentiating both sides with respect to s. We get

$$\frac{\partial q}{\partial s} = Gt\left(\frac{\partial^2 u}{\partial s^2} + \frac{\partial^2 \eta}{\partial x\, \partial s}\right)$$

Substituting this result into Eq. (7.39) and making use of Eq. (7.9) leads us to the partial differential equation

$$\frac{\partial^2 u}{\partial s^2} + \frac{E}{G}\frac{\partial^2 u}{\partial x^2} + \frac{\partial^2 \eta}{\partial x\, \partial s} = 0 \tag{7.91}$$

To eliminate the displacement η, we retain the assumption that the geometry of the cross section is unaltered during deformation so that Eqs. (7.15) are still applicable. We now locate the pole of r and the origin of s so that the following integrals vanish:

$$\oint t \cos\alpha\, r\, ds = \oint t \sin\alpha\, r\, ds = \oint t \sin\alpha \cos\alpha\, r\, ds = 0 \tag{7.92}$$

Coordinate axes possessing these properties are called the *principal shear axes*[9] of the section. The in-plane stress resultants are related to the shear flow as follows:

$$M_t = \oint qr\,ds$$
$$V_y = -\oint q \sin\alpha\,ds \tag{7.93}$$
$$V_z = \oint q \cos\alpha\,ds$$

We now substitute η of Eqs. (7.15) into Eq. (7.90) and introduce the result into Eqs. (7.93). This leads to three equations which are further simplified by introducing Eqs. (7.92). Solving these, we obtain the relations

$$\frac{d\phi}{dx} = \frac{M_t}{GI_c} - \frac{1}{I_c}\oint \frac{\partial u}{\partial s} r\,dA$$
$$\frac{dv}{dx} = \frac{V_y}{GA_y} + \frac{1}{A_y}\oint \frac{\partial u}{\partial s} \sin\alpha\,dA \tag{7.94}$$
$$\frac{dw}{dx} = \frac{V_z}{GA_z} - \frac{1}{A_z}\oint \frac{\partial u}{\partial s} \cos\alpha\,dA$$

where

$$I_c = \oint r^2\,dA$$
$$A_y = \oint \sin^2\alpha\,dA \tag{7.95}$$
$$A_z = \oint \cos^2\alpha\,dA$$

Thus

$$\frac{\partial\eta}{\partial x} = \frac{M_t r}{GI_c} - \frac{V_y \sin\alpha}{GA_y} + \frac{V_z \cos\alpha}{GA_z} - \frac{r}{I_c}\oint \frac{\partial u}{\partial s}\,dA - \frac{\sin\alpha}{A_y}\oint \frac{\partial u}{\partial s}\sin\alpha\,dA + \frac{\cos\alpha}{A_z}\oint \frac{\partial u}{\partial s}\cos\alpha\,dA \tag{7.96}$$

Finally, differentiating Eq. (7.96) with respect to s and substituting the result into Eq. (7.91) leads to the governing integrodifferential equation[10] for the displacement u:

$$\frac{\partial^2 u}{\partial s^2} + \frac{E}{G}\frac{\partial^2 u}{\partial x^2} - \frac{\partial}{\partial s}\left(\frac{r}{I_c}\oint \frac{\partial u}{\partial s}\,dA - \frac{\sin\alpha}{A_y}\oint \frac{\partial u}{\partial s}\sin\alpha\,dA + \frac{\cos\alpha}{A_z}\oint \frac{\partial u}{\partial s}\cos\alpha\,dA\right) = \frac{\partial}{\partial s}\left(-\frac{M_t r}{GI_c} + \frac{V_y \sin\alpha}{GA_y} - \frac{V_z \cos\alpha}{GA_z}\right) \tag{7.97}$$

[9] See Ref. 5, p. 13.

[10] Solutions to this equation can be found in Ref. 5. For a different approach to the problem, see Ref. 1. See also Refs. 68 and 70.

Once this equation is solved for u, η is determined from Eq. (7.96), σ_x is determined from Eq. (7.9), and q is evaluated from Eq. (7.90). Displacements are then evaluated using Eqs. (7.94) and the behavior of the structure is completely defined.

PROBLEMS

7.1–7.4. The wall thickness of each of the sections shown is constant and point A is selected as the sectorial pole. Find the following:

(*a*) the location of the shear center

(*b*) the location of the sectorial centroids

(*c*) the sectorial products of inertia

(*d*) the warping constant with respect to the shear center

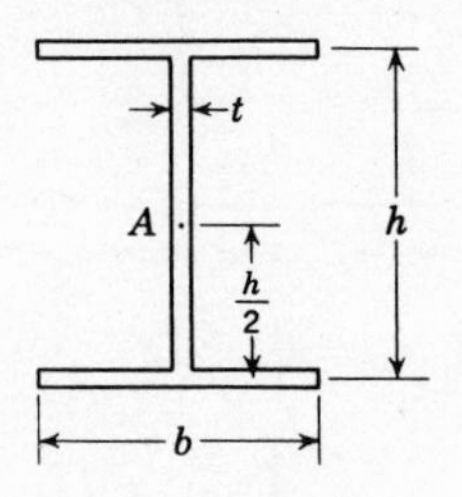

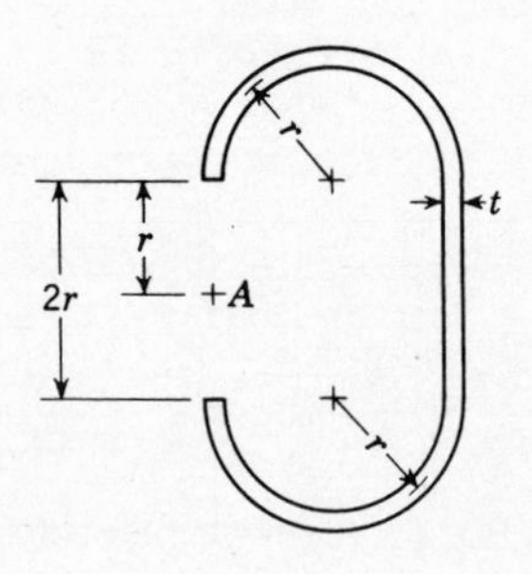

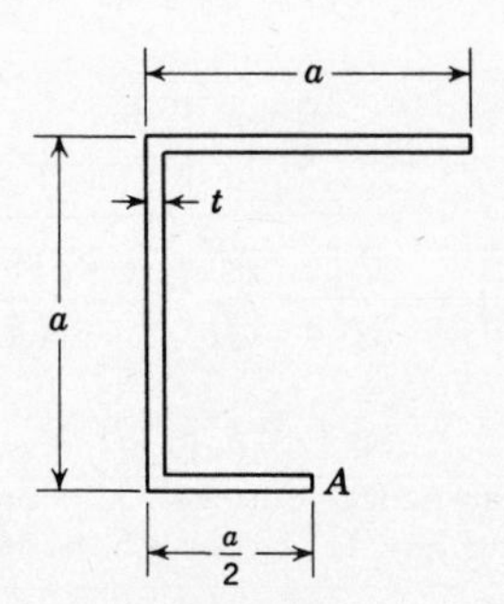

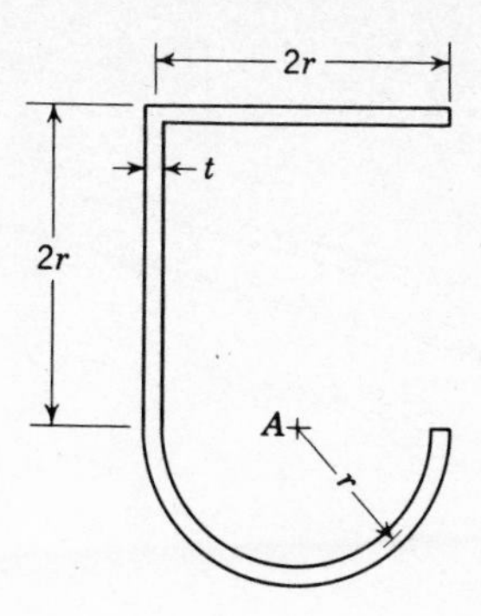

7.5. Derive Eq. (7.34*b*).

7.6. Derive the equation for Γ for the wide-flange section (section no. 1) in Table 7.1.

7.7. Derive the equation for Γ for the circular section (section no. 5) in Table 7.1.

7.8. Assume that the section shown in Fig. 7.1 is square and that the walls are of constant thickness. Locate the shear center and calculate the warping constant. Assume that the normal stresses are carried by the stringers and that the shear flow is carried by the webs. Express the result in terms of a, the stringer area; t, the wall thickness; and h, the width and depth of the section.

7.9. Calculate the bimoment applied to the free end of the section in Prob. 7.8 (and in Fig. 7.1) in terms of the applied load P.

7.10. The thin-walled section in Prob. 7.2 is that of a cantilevered beam. The beam is loaded by a prescribed normal-stress distribution σ_x at its free end. Calculate the applied end bimoment if $\sigma_x = Ks^2$, where K is a constant and s is measured counterclockwise from the lower free edge of the section.

7.11. The beam in Prob. 5.8 is subjected to an end torque of magnitude 20,000 in.-lb. Take $E = 5G/2 = 27 \times 10^6$ psi. Find the following:

(*a*) the warping constant for the section

(*b*) the normal-stress distribution at the fixed end

(*c*) the shearing-stress distribution at the fixed end and at the free end

(*d*) the angle of twist of the free end

7.12. The simply supported z section is acted upon by an eccentric transverse load, $P = 25{,}000$ lb. What is the maximum normal stress developed in the structure?

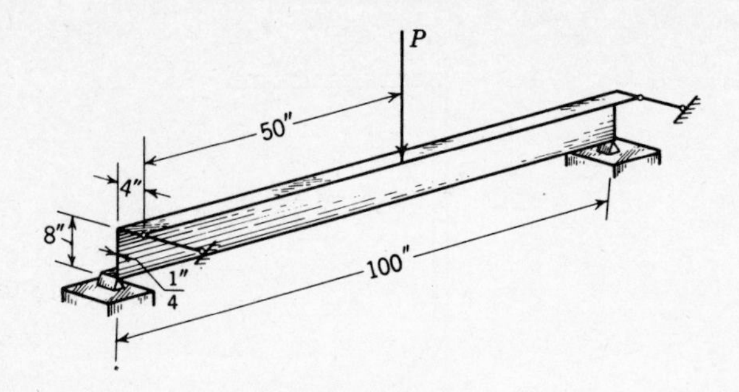

7.13. Evaluate the normal-stress and the shearing-stress distributions at the fixed end of a cantilevered wide-flange section subjected to an end torque T. The dimensions of the section are indicated in the diagram of section no. 1 in Table 7.1 except that $b_1 = b_2 = h = b$ and $t = b/8$. The length of the beam is $12b$. Obtain the stresses from purely physical arguments and elementary beam theory and then compare the results with those given by the general formulas.

7.14. The channel shown is subjected to a uniformly distributed torque of 1,000 in.-lb/in. Find the maximum normal stress and the maximum shearing stress developed in the structure. What is the angle of twist at the center of the beam?

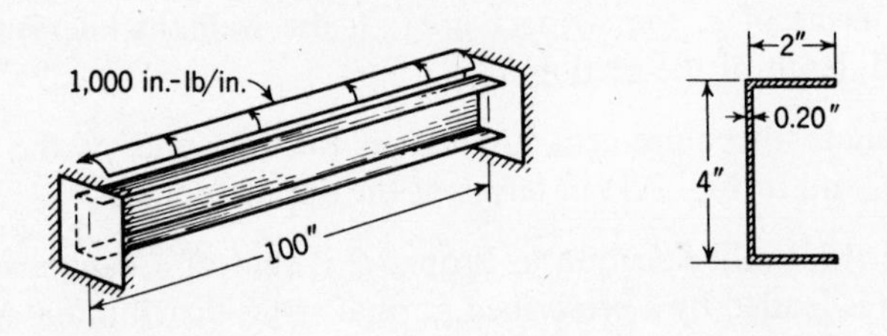

7.15. Verify that the secondary warping constant Γ^* for a rectangular section is $h^3t^3/144$.

7.16. Derive the formula for the secondary warping constant for the z section shown in Fig. 7.14.

7.17. Evaluate the geometrical quantities H_y, H_z, and H_ω [Eqs. (7.66)] for a channel section of constant thickness.

7.18. The circular section shown is subjected to eccentric axial loads. Calculate the end bimoments produced by these loads and W_ω as a function of x. The structure is assumed to be supported in such a way that it is in equilibrium.

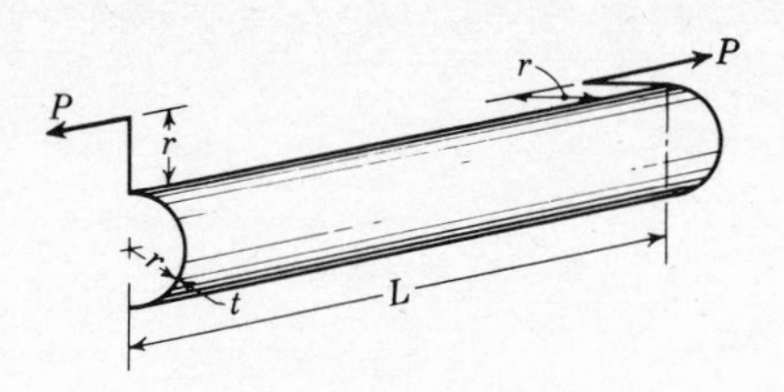

7.19. The thin-walled column shown is subjected to a load P at its centroid. Neglecting the W_ω term in the differential equations for the elastic curve, calculate the critical load for the structure. Take $E = 5G/2 = 27 \times 10^6$ psi. *Hint:* Assume the displacements are of the form

$$v = v_0\left(1 - \cos\frac{\pi x}{2L}\right) \qquad \text{etc.}$$

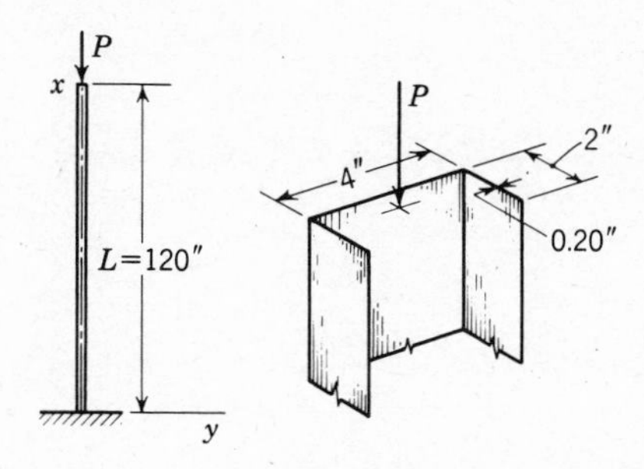

7.20. Find the equations for the elastic curve of the beam in Fig. 7.17 for the case in which it is free at $x = 0$ and fixed at $x = L$. The axial load is applied in such a way that W_ω and V_ω are zero or negligible in the general differential equations.

7.21. Compute the shear flow at the fixed end of the cantilevered beam in Prob. 5.18 for the case in which the 20,000-lb vertical force is applied at a corner rather than at the shear center. Account for the effects of restrained warping but neglect the influence of shear deformations on the normal stresses.

7.22. Compute and plot the normal-stress profile and the shear flow at the fixed end of the beam in Prob. 5.19. Also evaluate the angle of twist of the free end. Account for the effects of restrained warping but neglect the influence of shear deformations on the normal-stress distribution. Compare the results with those of Prob. 5.19.

eight

THE PRINCIPLES OF VIRTUAL WORK

8.1 Introduction. Thus far, we have established a number of relationships between the stresses and displacements developed in various structural elements and the stress resultants developed on a given cross section of a member. To arrive at these relationships, we have continually relied upon two basic requirements of static structural behavior, the requirement that the structure must be in equilibrium and the requirement that deformations must be compatible and consistent with natural boundary conditions. It is upon these two conditions that our entire theory is based. We have satisfied the first of these requirements, equilibrium, by simply applying the equations of statics. We have satisfied the second requirement, compatibility, by studying, in detail, the geometry of the elastic curve or the geometry of a deformed element. The resulting equations are sufficiently general to be applied to a wide range of both statically determinate and statically indeterminate structural problems.

There exist, however, alternative methods for establishing and satisfying these requirements of structural behavior which have as their basis some of the most fundamental principles of structural mechanics. It is to these principles that we now direct our attention.

The two characteristics of structural behavior, mentioned above, respectively, involve forces and stresses which must satisfy the equilibrium conditions and displacements and strains which must satisfy the compatibility conditions. It is logical that a general process through which both conditions are satisfied involves a quantity dependent on both forces and displacements, namely, *work*. Work, we recall, is the product of a force and the displacement of its point of application in the direction of the force. More generally, we define as the differential work dW done by a force F in moving through a differential displacement ds as the product of ds and F_s, the component of F in the direction ds. That is, F_s is the component of F tangent to the path traveled by the particle on which it acts. Thus, $dW = F_s\, ds$, and the total work done in displacing F from some point A to some point B is, by definition,

$$W = \int_A^B F_s\, ds \tag{8.1}$$

If F_s is constant and Δ is the total straight-line displacement,

$$W = F_s\Delta \tag{8.2}$$

Note, again, that work involves in a single expression both types of quantities that characterize structural behavior.

We now establish conditions involving the work done in mechanical and structural systems which encompass those that must be satisfied independently by forces and displacements.

8.2 Virtual work. The use of the concept of work as a basis for studying physical problems rather than directly applying Newton's laws of motion can be traced to the early philosophers of southern Europe. Leonardo da Vinci, almost 250 years before Newton was born, employed the rudiments of these principles in his studies of the mechanics of pulleys and levers; some say that the principle was used by Aristotle as early as 350 B.C. It is Jean Bernoulli,[1] however, who is credited with the first general formulation of the principle of "virtual displacements."

Although there appears to be little consistency in the literature in regard to terminology concerning the mechanical principles evolving from the concept of work, we shall refer to them as simply the *principles of virtual work*. The word virtual, as defined by Webster, means "being in

[1] Jean Bernoulli (1667–1748) was a prominent member of the famous Bernoulli family of mathematicians and physicists. He is often referred to as Jean Bernoulli I, to distinguish him from his son, Jean Bernoulli II (1710–1790), and his grandson, Jean Bernoulli III (1744–1807), who also contributed to mathematics and physics. The work of his older brother, Jaques (1654–1705), and another son, Daniel (1700–1782), led Euler to the discovery of the differential equation of the elastic curve; hence, the term "Bernoulli-Euler" theory of bending. See, for example, Ref. 62, pp. 25–36, or Ref. 57, pp. 426–433.

essence or effect, but not in fact," and we have interjected it to distinguish *virtual work*—that done by true forces moving through imaginary displacements (or vice versa)—from *real work* in which true forces move through true displacements.

The principle of virtual work may be divided into two parts, the *principle of virtual displacements* and the *principle of virtual forces.* In the first of these we deal with true forces and virtual, or imaginary, displacements; in the latter we use the work done by a system of virtual or fictitious forces in moving through the true displacements of a system. We shall find that conditions of equilibrium are established through the principle of virtual displacements and that conditions of compatibility are established through the principle of virtual forces. As before, our investigations are confined to structural systems at rest.

THE PRINCIPLE OF VIRTUAL DISPLACEMENTS

In the discussions to immediately follow, we first concern ourselves with the principle of virtual displacements. It is the oldest part of the principle of virtual work; and, since it is closely related to statics, the underlying concepts tie conveniently to our previous studies. We first demonstrate the principle for a single particle and then extend the ideas to more general systems.

8.3 Virtual displacements of a particle. Consider the particle in Fig. 8.1 which is under the action of a system of n concurrent forces $F_1, F_2, \ldots, F_j, \ldots F_n$. Now let us imagine that the particle undergoes a completely arbitrary virtual displacement δu during which all forces remain acting in their original directions. We use the symbol δ in front of u merely to remind us that δu is a virtual displacement and, for the present, has nothing to do with any true displacement u which defines the *true* motion of the particle. We give the symbol δ a more meaningful definition later. Suppose that the direction that we have given this virtual displacement and the line of action of the force F_i differ by an angle α_i. Then, F_{iu}, the component of F_i in the direction of δu ($F_{iu} = F_i \cos \alpha_i$), performs an amount of virtual work due to δu of magnitude $\delta u\, F_{iu}$. For simplicity, we assume that the displacement occurs adiabatically and

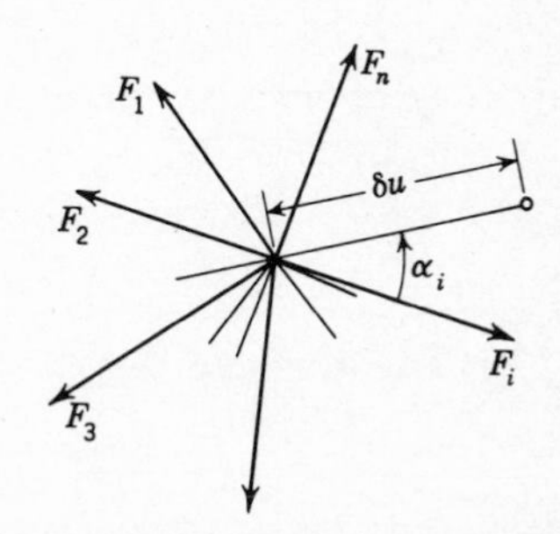

FIGURE 8.1 A particle acted upon by n concurrent forces.

infinitesimally slowly so that thermal and dynamic effects can be ignored. It follows that when all n forces are considered, the *total virtual work* done by the force system during the displacement δu is

$$\delta W = F_{1u}\,\delta u + F_{2u}\,\delta u + \cdots + F_{nu}\,\delta u$$

or, since δu is common to all terms on the right side of this expression,

$$\delta W = \delta u\left(\sum_{i=1}^{n} F_{iu}\right) \tag{8.3}$$

Again, the symbol δ is used to distinguish virtual work from the real work W.

We recognize the term in parentheses in Eq. (8.3) as the sum of components of all forces acting on the particle in the direction of the arbitrary virtual displacement δu. According to the familiar laws of statics, this sum is zero if the particle is in equilibrium. Conversely, if the particle is in equilibrium, then, for any choice of δu,

$$\delta W = 0 \tag{8.4}$$

This simple and seemingly trivial result is the mathematical statement of the principle of virtual displacements for a particle. It may be set forth as a theorem, as follows:

> If a particle is in equilibrium, the total virtual work done during any arbitrary virtual displacement of the particle is zero.
> (THEOREM I)

The converse of this theorem is not true; we cannot, in other words, conclude that a system is in equilibrium just because the virtual work is zero due to an arbitrary virtual displacement. This fact is amply demonstrated by the particle shown in Fig. 8.2, which, even though a displacement parallel to the x axis results in zero work, is obviously not in equilibrium.

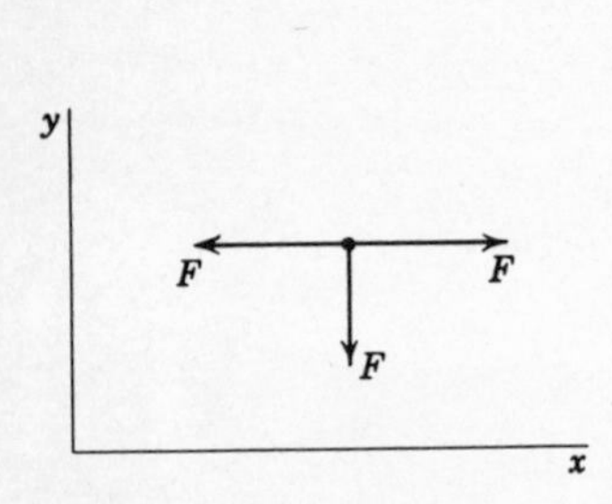

FIGURE 8.2 Particle not in equilibrium.

It may appear that all possible virtual displacements must be investigated in order to establish the equilibrium of a particle. If this were true, to use the principle to study equilibrium would be an extremely laborious and impractical task, since, even for a single particle, the number of possible virtual displacements is infinite. Fortunately, this is not the case. A quick calculation will show that *every possible displacement* of the particle in Fig. 8.2 can be expressed in terms of two *independent* virtual displacements—say one in the x direction and one in the y direction—and that it is only when the virtual work done during these displacements is zero that we can conclude that the particle is in equilibrium. The use of two independent virtual displacements corresponds to the two degrees of freedom of a single particle in the plane. A particle in space, for example, has three degrees of freedom and, consequently, the virtual work done during three independent virtual displacements must vanish for equilibrium to exist. Similarly, two particles in a plane have four degrees of freedom and, for equilibrium, δW must vanish for any four independent virtual displacements. In view of this, we now write a useful corollary to Theorem I:

> A system of particles is in equilibrium if the total virtual work done is zero for every independent virtual displacement.
>
> (THEOREM II)

This condition is both necessary and sufficient for equilibrium. It should come as no surprise when we recall that the principle of virtual displacements represents, in effect, an alternative to the equations of statics, though it encompasses far broader ideas than simple statics. To use statics to investigate the equilibrium of the particle shown in Fig. 8.2, for instance, we would write two independent equations, one involving the sum of forces in, say, the x direction and another the sum of forces in the y direction. In fact, we shall find it informative from time to time to look upon the equations of virtual work as being merely the equations of statics multiplied by some arbitrary function which, in our initial studies, we have called a "virtual displacement." If δW in Eq. (8.3) is zero, for example, then that equation simply represents an equation of statics $\Sigma F_{iu} = 0$ multiplied by some irrelevant function δu. The quantity δu need not even be a displacement—we could have referred to it as a velocity, as did Galileo, or a pressure, a temperature, an acceleration, or just a purely mathematical function. Our choice of calling δu a displacement was solely to assign physical significance to the process that was demonstrated.

8.4 Virtual displacements of a rigid body. Since a rigid body is no more than a collection of particles constrained to remain a constant distance from one another, the extension of the principles discussed in the previous paragraphs to apply to rigid bodies is a trivial matter. We need only replace the word *particle* (or *system of particles*) with *rigid body* (or *system of rigid bodies*) in the theorems that we established. Nevertheless, certain details of the theory can be easily demonstrated by applying it to some simple examples involving rigid bodies.

We have established that, although it is perfectly permissible to introduce every conceivable virtual displacement in a system to arrive at the virtual work, only a number of independent virtual displacements equal to the degrees of freedom of the system is necessary to establish equilibrium. Now two particles in the plane which are constrained to remain a constant distance apart form, in essence, a rigid body; their original four degrees of freedom have been reduced to only three because of the introduction of a *condition of constraint*. Only the virtual work due to three independent virtual displacements needs to be considered in order to establish the equilibrium of the system. This same reasoning applies to the most general mechanical system. Thus, if we find that δW is zero due to equal virtual displacements of the two particles in the x and y directions and due to a virtual rotation, in the plane, of a straight line connecting them we can conclude, without doubt, that the system is in equilibrium. We need not evaluate the virtual work done in displacing one particle relative to another (and thereby violating the condition of constraint), although such calculations may be perfectly acceptable and valid. It follows, therefore, that in order to establish equilibrium we need consider only those virtual displacements that are consistent with the constraints. With this in mind, we rewrite the principle of virtual displacements in the following form:[2]

> A mechanical system is in equilibrium if the total virtual work done is zero for every virtual displacement consistent with the constraints.
> (THEOREM III)

By "mechanical" system, we mean a collection of particles and rigid bodies as distinguished from a system whose members may deform. Again, this condition is both necessary and sufficient for equilibrium.

According to this form of the theorem, to establish the equilibrium of the block of weight Q, shown constrained to move in a horizontal plane in Fig. 8.3*a*, it is sufficient to consider only the virtual work done during the displacement δu. A vertical displacement δv would violate the constraint, and, therefore, we need not consider it. Note, however, that the virtual work done by δv in Fig. 8.3*b* is $(N - Q)\,\delta v$ and is clearly zero *if the block is in equilibrium*. Thus, if we know in advance that the system is in equilibrium, we can obtain additional information by considering

[2] We confine our attention to conservative systems; that is, we neglect effects of dissipative forces such as those due to friction. If friction were present, δW could be less than zero.

virtual displacements which *appear* to violate the constraints. Equilibrium conditions obtained in this manner involve forces at the boundary and, hence, provide us with static instead of kinematic boundary conditions at the constrained points. The force N, for example, can be regarded as a force on the boundary of the block which is in contact with the horizontal plane. The relation $N - Q = 0$, then, may be thought of as a static boundary condition.

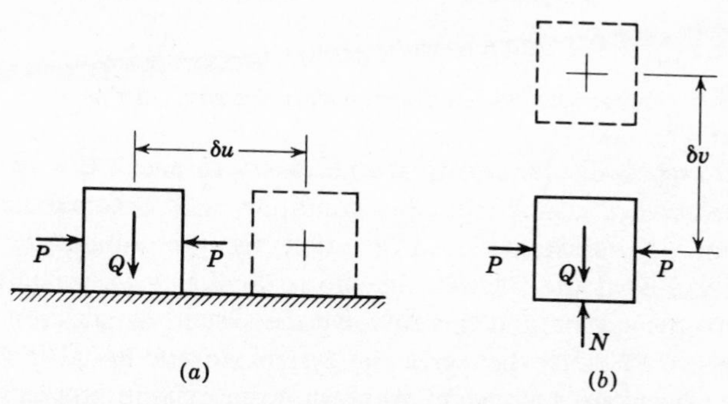

FIGURE 8.3 Rigid block of weight Q on a rigid horizontal plane. (*a*) The vertical displacement is prescribed; (*b*) the vertical force is prescribed.

Technically speaking, such virtual displacements do not actually violate the constraints; for "constraint," in the present sense, means that the displacements of certain points on the body are *prescribed*. The displacement of the block normal to the horizontal plane, for example, is prescribed as zero in Fig. 8.3*a*. However, by bringing the normal force N into consideration as in Fig. 8.3*b*, we have, in effect, prescribed a force rather than a displacement at the boundary. In so doing, we have given the system an additional degree of freedom, and it becomes necessary to consider the additional condition $(N - Q)\,\delta v = 0$ in order to ensure equilibrium.

This process of removing a constraint by prescribing a force rather than a displacement is called *releasing* the system, and by prescribing N in Fig. 8.3*b* we are said to have *introduced a release* into the system. Clearly, the introduction of each release increases by one the number of degrees of freedom of the system.

A similar example is provided by the rigid lever in Fig. 8.4*a*. Since the displacement at the fulcrum is prescribed ($u = v = 0$), the system has only one degree of freedom. A single virtual rotation $\delta\theta$ of the lever about its fulcrum is consistent with the constraints, and the condition for equilibrium is that the virtual work $(P_2 b - P_1 a) \sin \delta\theta$ be zero. This leads us to the age-old "law of the lever" $P_2 b - P_1 a = 0$. No

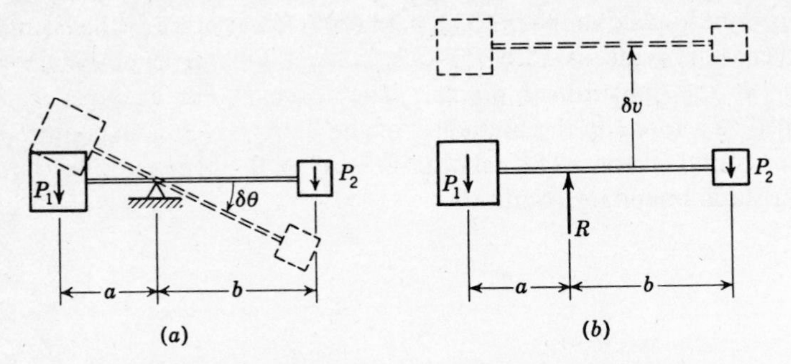

FIGURE 8.4 Virtual displacements of a rigid fulcrum.

additional virtual displacements are necessary to check this result—any set of forces P_1, P_2 satisfying this condition will automatically be in equilibrium. Note again, however, that by prescribing the vertical reactive force R at the fulcrum, as shown in Fig. 8.4*b*, we introduce a release into the system and thereby give the system an additional degree of freedom. We must then give the system an additional independent virtual displacement, such as δv, to be sure that equilibrium exists. The vertical forces perform virtual work of magnitude $(R - P_1 - P_2)\,\delta v$, which, if zero, merely proves that the reaction R is $P_1 + P_2$.

It follows from these observations that a system which is completely constrained, such as the planar rigid bar in Fig. 8.5*a*, is in equilibrium under *any* set of forces. In other words, complete constraint, by

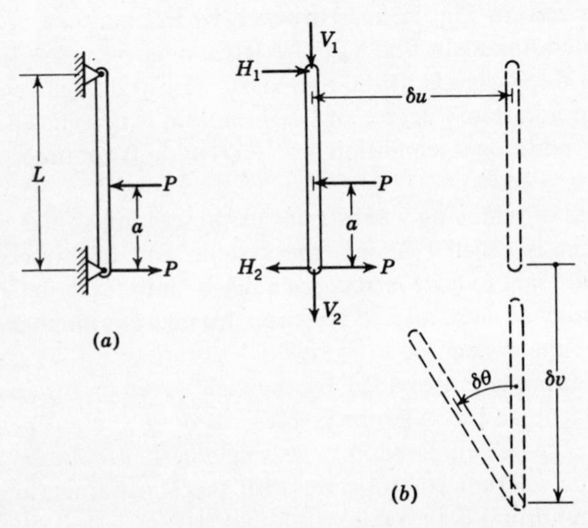

FIGURE 8.5 (*a*) Completely constrained bar; (*b*) virtual displacements after forces have been prescribed.

definition, means no motion. Again, if we release the constraints by prescribing the forces shown in Fig. 8.5*b*, the virtual work done during the three independent virtual displacements δu, δv, and $\delta\theta$ must vanish for equilibrium to exist. Assuming, only for the sake of simplicity, that $\delta\theta$ is small, the virtual work done during these displacements is

$$\delta W = (H_1 - H_2)\,\delta u + (V_1 + V_2)\,\delta v + (Pa - H_1 L)\,\delta\theta$$

If the bar is in equilibrium, δW must be zero for any choice of δu, δv, and $\delta\theta$. Therefore, the three expressions in parentheses must vanish independently, and we have

$$H_1 - H_2 = 0 \qquad V_1 + V_2 = 0 \qquad Pa - H_1 L = 0$$

We recognize these relations as the equations of statics for the bar.

As before, we may also interpret the above expression for the total virtual work as being merely equations of statics multiplied by arbitrary functions δu, δv, and $\delta\theta$ and added together.

8.5 Virtual displacements of a deformable body. We are finally ready to generalize the concepts developed previously so that they apply to general, deformable bodies. The reasoning is quite similar except that now points within a body may move relative to one another without violating conditions of constraint. Thus, in the case of a deformable body, both the external and the internal forces may do virtual work. Because of this, we shall see that it is necessary to consider *internal* as well as *external* virtual work.

First, let us examine more closely what is meant by prescribed and unprescribed forces and displacements for relatively general systems. Consider, for example, the two-dimensional deformable body in Fig. 8.6*a*. We assume that the body is fixed along portions $\overline{AB}$ and $\overline{FE}$ of the boundary curve and that no body forces are present. We now deform the body in two ways: first, by applying a system of known surface forces $P(s)$ along portion $\overline{CD}$ of the boundary and, second, by giving the supported portion of the boundary $\overline{EF}$ a known displacement pattern $\Delta(s)$ (Fig. 8.6*b*). Regardless of the loading, the displacements of portions $\overline{AB}$ and $\overline{FE}$ of the boundary are *prescribed;* in fact, they are always zero along $\overline{AB}$ and are given by $\Delta(s)$ along $\overline{EF}$. The *reactive* forces on these parts of the boundary are not prescribed; they are initially zero in the unstressed state of the body and build up to a certain distribution which depends upon the final configuration of the system. On the other hand, surface forces rather than displacements are prescribed along

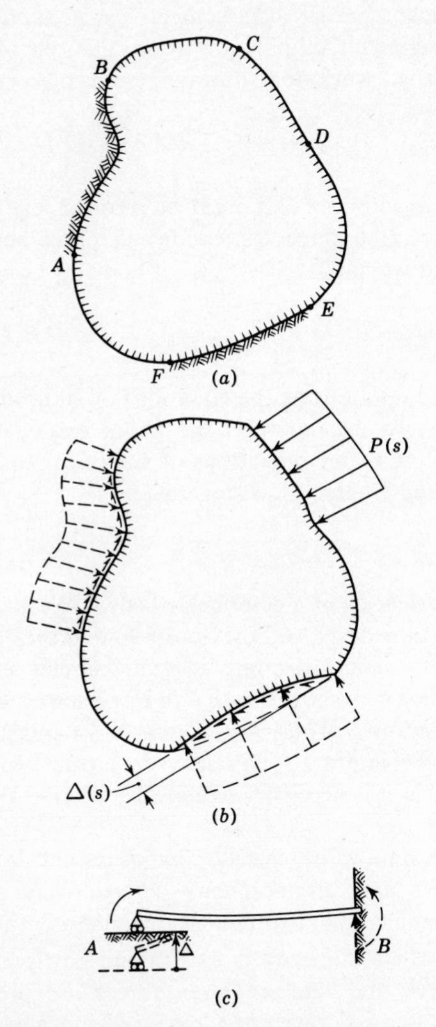

FIGURE 8.6 Deformable systems. (*a*) A two-dimensional deformable body; (*b*) displacements are prescribed along portions $\overline{AB}$ and $\overline{EF}$ of the boundary and forces are prescribed along portions $\overline{BCDE}$ and $\overline{FA}$, as indicated; (*c*) beam with prescribed end displacements and prescribed end moment.

$\overline{BCDE}$ and $\overline{FA}$; in fact, they are always zero along $\overline{BC}$, $\overline{DE}$, and $\overline{FA}$ and are given by $P(s)$ along $\overline{CD}$. The displacements of these portions of the body are not prescribed; they are initially zero in the undeformed state of the body and build up to a certain distribution which depends upon the final deformed state of the body.

An analogous situation exists in the case of the beam in Fig. 8.6*c*. The slope and deflection at B and the vertical deflection at A are prescribed, but the slope at A is unprescribed. The reactive forces at A and B and the reactive moment at B are unprescribed, but the applied moment at A is prescribed.

From these observations, we conclude that the total boundary surface of a deformable body can be divided into portions on which either the displacements or the forces are prescribed. In general, on parts of the boundary where the displacements are prescribed the forces are not prescribed, and vice versa. Thus, if S denotes the total surface area and S_1 and S_2 denote the portions of S on which forces and displacements are prescribed, respectively, then $S = S_1 + S_2$. Further, since prescribing the displacement of a point is the same as constraining the point, it follows that in applying the principle of virtual displacements we must consider virtual displacements which satisfy the kinematic boundary conditions on S_2; otherwise the displacements would violate the constraints. As before, if we then wish to obtain static conditions involving the forces on S_2, it is necessary to prescribe forces rather than displacements on this part of the boundary. This gives the system greater freedom and requires that we use nonzero virtual displacements here as well as on S_1.

It is important to note that in the case of deformable systems an infinite number of material points are free to displace. Hence, continuous systems such as those in Fig. 8.6 possess an infinite number of degrees of freedom. To handle such infinite numbers of quantities, we assume that the material is continuously distributed within the system and that the components of displacement u, v, and w are continuous functions of the coordinates x, y, and z. We also assume that the virtual displacements are continuous functions of these coordinates and that they satisfy the kinematic boundary conditions on S_2 (that is, they are consistent with the constraints). Functions which have these properties are referred to as *kinematically admissible functions*. Thus, *any kinematically admissible function can be used as a virtual displacement.*

To continue our investigation of deformable systems, let us consider the two-dimensional body of general shape shown in Fig. 8.7. The results can easily be extended to the three-dimensional case. We assume, for simplicity, that the derivatives of the displacements are sufficiently small that the linear strain-displacement relations given by Eqs. (2.18) are valid. In other words, we assume that for the two-dimensional case

$$\epsilon_x = \frac{\partial u}{\partial x} \qquad \epsilon_y = \frac{\partial v}{\partial y} \qquad \gamma_{xy} = \frac{\partial u}{\partial y} + \frac{\partial v}{\partial x} \tag{8.5}$$

where ϵ_x, ϵ_y, and γ_{xy} are the components of strain and u and v are the true continuous, single-valued components of displacement. Furthermore, we again assume that the body deforms so slowly that dynamic effects are negligible; we neglect thermal effects and we assume that the body is in equilibrium under the action of a general system of external forces. The body need not be elastic and may follow any nonlinear (but monotonically increasing) stress-strain law.

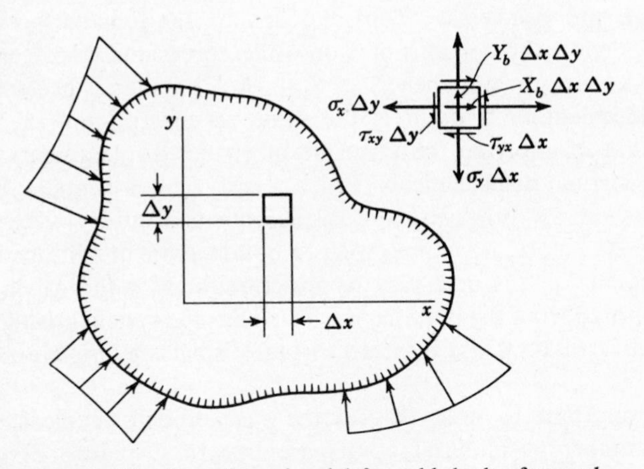

FIGURE 8.7 Two-dimensional deformable body of general shape.

We pointed out in Chap. 2 that the forces which act on a body may, in general, be classified as external forces and internal forces. External forces, we recall, are generally independent of the material properties and the deformation of the body, though they may depend upon the distribution of mass of the body. They are represented by body forces which act within the body and surface forces which act on the boundary of the body (see Art. 2.2). Body forces are expressed in units of force per unit volume and surface forces are expressed in units of force per unit surface area. The intensities of the x and y components of the body forces and the surface forces are denoted X_b, Y_b, and X_s, Y_s, respectively. Internal forces, on the other hand, are defined in terms of stresses that are developed within the body. Within the body, the internal forces (stresses) are related to the external (body) forces through the conditions of static equilibrium given by Eqs. (2.4). For two-dimensional problems, these equations become

$$\begin{aligned}\frac{\partial \sigma_x}{\partial x} + \frac{\partial \tau_{xy}}{\partial y} + X_b = 0 \\ \frac{\partial \tau_{xy}}{\partial x} + \frac{\partial \sigma_y}{\partial y} + Y_b = 0\end{aligned} \tag{8.6}$$

Similarly, on the boundary of the body the internal forces are related to the external (surface) forces according to Eqs. (2.6). In two dimensions, these equations become

$$\begin{aligned}X_s = \sigma_x l + \tau_{xy} m \\ Y_s = \tau_{xy} l + \sigma_y m\end{aligned} \tag{8.7}$$

where l and m are the direction cosines of a normal to the boundary curve. Equations (8.7), we recall, are referred to as the static boundary conditions.

We also obtain both of these sets of equilibrium conditions, Eqs. (8.6) and (8.7), by using the principle of virtual displacements. For example, Eqs. (8.6) follow directly from equating to zero the virtual work done in giving the element in Fig. 8.7 virtual displacements δu and δv in the x and y directions. We refer to the portion of this work performed by the stresses as the internal virtual work.

Physically, we may interpret internal virtual work as follows: a given deformable body, originally unstressed, is subjected to a system of self-equilibrating external forces. The body slowly deforms so that strains and consequently stresses are produced. The displacements, strains, and stresses finally reach a unique set of values which simultaneously correspond to a stress distribution in equilibrium with the external forces, and a pattern of compatible displacements which are consistent with given kinematic boundary conditions. We now superimpose on the true displacements a system of arbitrary virtual displacements δu and δv, which are continuous functions of the coordinates x and y and which satisfy the kinematic boundary conditions. We may also think of the situation as one in which the true displacements u and v take on arbitrary increments δu and δv, or we may simply regard δu and δv as arbitrary kinematically admissible functions with no special physical significance. In either case, it is clear that the virtual work done by moving a single differential element of the body through these virtual displacements is

$$\left[\left(\frac{\partial \sigma_x}{\partial x} + \frac{\partial \tau_{xy}}{\partial y} + X_b\right)\delta u + \left(\frac{\partial \tau_{xy}}{\partial x} + \frac{\partial \sigma_y}{\partial y} + Y_b\right)\delta v\right]\Delta x\,\Delta y$$

which, according to Eqs. (8.6), is zero if the body is in equilibrium. The *total* virtual work is now obtained by allowing Δx and Δy to approach zero and simultaneously summing the work done on each element as their number becomes infinitely large. In the limit,

$$\delta W = \iint_A \left[\left(\frac{\partial \sigma_x}{\partial x} + \frac{\partial \tau_{xy}}{\partial y} + X_b\right)\delta u + \left(\frac{\partial \tau_{xy}}{\partial x} + \frac{\partial \sigma_y}{\partial y} + Y_b\right)\delta v\right] dx\, dy \quad (8.8)$$

In order to adequately interpret this result, we make use of Green's theorem[3] and write the integral in an alternative form. According to Green's theorem in two dimensions, if $\phi(x,y)$ and $\psi(x,y)$ and their first and second partial derivatives are continuous single-valued functions over some region bounded by a closed curve C, then

$$\iint_A \left(\frac{\partial \phi}{\partial x}\frac{\partial \psi}{\partial x} + \frac{\partial \phi}{\partial y}\frac{\partial \psi}{\partial y}\right) dx\, dy = -\iint_A \phi\left(\frac{\partial^2 \psi}{\partial x^2} + \frac{\partial^2 \psi}{\partial x^2}\right) dx\, dy + \oint_C \phi\left(\frac{\partial \psi}{\partial x} l + \frac{\partial \psi}{\partial y} m\right) ds \quad (8.9)$$

where A is the area of the region, ds is an element of arc of the curve C, and l and m are the direction cosines of normals to C. To use Eq. (8.9) to integrate Eq. (8.8), we first let $\sigma_x = \phi$ and assume that ψ is such that $\partial\psi/\partial x = \delta u$ and $\partial\psi/\partial y = 0$. Then,

$$\iint_A \frac{\partial \sigma_x}{\partial x}\delta u\, dx\, dy = -\iint_A \sigma_x \frac{\partial \delta u}{\partial x} dx\, dy + \oint_C \sigma_x l\, \delta u\, ds$$

Similarly

$$\iint_A \left(\frac{\partial \tau_{xy}}{\partial x}\delta v + \frac{\partial \tau_{xy}}{\partial y}\delta u\right) dx\, dy = -\iint_A \tau_{xy}\left(\frac{\partial \delta v}{\partial x} + \frac{\partial \delta u}{\partial y}\right) dx\, dy + \oint_C (\tau_{xy} m\, \delta u + \tau_{xy} l\, \delta v)\, ds$$

By proceeding in this manner, we transform Eq. (8.8) into the following form:

$$\delta W = -\iint_A \left[\sigma_x \frac{\partial \delta u}{\partial x} + \sigma_y \frac{\partial \delta v}{\partial y} + \tau_{xy}\left(\frac{\partial \delta v}{\partial x} + \frac{\partial \delta u}{\partial y}\right)\right] dx\, dy + \iint_A (X_b\, \delta u + Y_b\, \delta v)\, dx\, dy + \oint_C [(\sigma_x l + \tau_{xy} m)\, \delta u + (\tau_{xy} l + \sigma_y m)\, \delta v]\, ds \quad (8.10)$$

Each integral in this equation has a very important physical significance. The second double integral on the right side of Eq. (8.10) obviously represents the virtual work done by the (external) body forces due to δu

[3] See, for example, Ref. 59, p. 493.

and δv. Furthermore, on referring to Eqs. (8.7), we recognize the terms in parentheses in the line integral as being the surface forces which act along the boundary of the body. Thus, the line integral represents the virtual work done by the surface forces due to the virtual displacements at the boundary:

$$\oint_C [(\sigma_x l + \tau_{xy} m)\,\delta u + (\tau_{xy} l + \sigma_y m)\,\delta v]\,ds = \oint_C (X_s\,\delta u + Y_s\,\delta v)\,ds \quad (8.11)$$

If we choose those virtual displacements which are zero on the portions of the boundary on which displacements are prescribed,

$$\oint_C (X_s\,\delta u + Y_s\,\delta v)\,ds = \int_{C_1} (X_s\,\delta u + Y_s\,\delta v)\,ds + \int_{C_2} (X_s\,\delta u + Y_s\,\delta v)\,ds$$
$$= \int_{C_1} (X_s\,\delta u + Y_s\,\delta v)\,ds + 0 \quad (8.12)$$

where C_1 and C_2 are, respectively, the parts of C on which forces and displacements are prescribed. It is clear that the last two integrals in Eqs. (8.10) represent the virtual work done by the external forces. Thus, denoting by δW_e the *total external virtual work* and taking into account Eqs. (8.11) and (8.12), we have

$$\delta W_e = \iint_A (X_b\,\delta u + Y_b\,\delta v)\,dA + \int_{C_1} (X_s\,\delta u + Y_s\,\delta v)\,ds \quad (8.13)$$

Now the first integral on the right side of Eq. (8.10) contains stresses and therefore represents the *internal virtual work* due to δu and δv. We shall refer to this quantity as δU. By comparing the virtual displacement terms in δU with Eqs. (8.5), we note an amazing similarity between the expressions for strains in terms of true displacements and those involving derivatives of the virtual displacements. In fact, it is this similarity that encourages us to interpret δu and δv as displacements—or even better, *variations* in the true displacements—rather than arbitrary functions. It is this interpretation that brings physical significance to our argument.

Consider the first term in this integral, for example. If we give an element subjected only to a normal stress σ_x an arbitrary virtual displacement δu, then the virtual work done at some point x is $-\sigma_x\,\Delta y\,\delta u$, while that done at a point $x + \Delta x$ is $\sigma_x\,\Delta y[\delta u + (\partial \delta u/\partial x)\,\Delta x]$. The result is a quantity of internal virtual work of magnitude $\sigma_x(\partial \delta u/\partial x)\,\Delta x\,\Delta y$. Furthermore, if $\epsilon_x = \partial u/\partial x$ is the *true* strain undergone by this element, we may also regard internal virtual work as being produced by true stresses "moving through" *virtual strains* (or arbitrary variations in true strains) provided $\partial \delta u/\partial x = \delta(\partial u/\partial x)$. This, indeed, is true; for if

$\bar{u} = u + \delta u$ is a continuous single-valued function obtained by giving u a small variation, and $\partial\bar{u}/\partial x$ is the result of varying $\partial u/\partial x$, clearly[4]

$$\frac{\partial \delta u}{\partial x} = \frac{\partial \bar{u}}{\partial x} - \frac{\partial u}{\partial x} = \delta\left(\frac{\partial u}{\partial x}\right) \tag{8.14}$$

Thus, we may write the *total internal virtual work* in the form

$$\delta U = \iint_A (\sigma_x\,\delta\epsilon_x + \sigma_y\,\delta\epsilon_y + \tau_{xy}\,\delta\gamma_{xy})\,dA \tag{8.15}$$

By replacing dA in Eqs. (8.13) and (8.15) by the differential volume dV, and ds in Eq. (8.13) by dS, a differential element of surface area, and by adding the appropriate stresses and forces, we easily extend these results to the three-dimensional case. The external and internal virtual work, in three dimensions, are, respectively,

$$\delta W_e = \iiint_V (X_b\,\delta u + Y_b\,\delta v + Z_b\,\delta w)\,dV + \iint_{S_1} (X_s\,\delta u + Y_s\,\delta v + Z_s\,\delta w)\,dS \tag{8.16}$$

and

$$\delta U = \iiint_V (\sigma_x\,\delta\epsilon_x + \sigma_y\,\delta\epsilon_y + \sigma_z\,\delta\epsilon_z + \tau_{xy}\,\delta\gamma_{xy} + \tau_{xz}\,\delta\gamma_{xz} + \tau_{yz}\,\delta\gamma_{yz})\,dV \tag{8.17}$$

With the above notation, Eq. (8.10) becomes

$$\delta W = -\delta U + \delta W_e \tag{8.18}$$

so that for equilibrium to exist,

$$\delta W_e = \delta U \tag{8.19}$$

Equation (8.19) is the mathematical statement of the principle of virtual displacements. As it now stands, we have only proved that Eq. (8.19) is a necessary condition for equilibrium. Briefly, to show that it is also sufficient for equilibrium, we regard Eqs. (8.15) and (8.13) as *definitions* of the internal and external virtual work. Then, using Green's theorem, we work backward to obtain δW of Eq. (8.8), which we conclude is zero only if the system is in equilibrium. It follows that the condition $\delta U = \delta W_e$ is both necessary and sufficient for equilibrium. Thus, the

[4] This relationship may be set forth as a basic theorem occurring in that branch of mathematics called the calculus of variations. See, for example, Ref. 35.

general principle of virtual displacements may be stated as follows:

> A deformable system is in equilibrium if the total external virtual work is equal to the total internal virtual work for every virtual displacement consistent with the constraints.
>
> (THEOREM IV)

Since, for rigid bodies, a virtual strain violates the constraints (and, in fact, is meaningless), δU in this case is automatically zero. Thus, the theorems stated earlier for particles and rigid bodies may be regarded as special cases of the above theorem. Note also that all of the results thus far obtained are independent of the material properties of the body.

8.6 Virtual displacements of simple structural systems. We acquire a clearer understanding of Eq. (8.19) by considering its application to some simple structural systems.

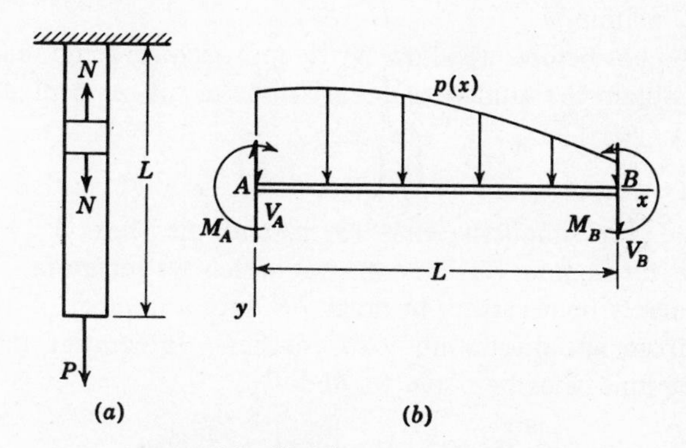

FIGURE 8.8 Simple structural systems.

The result obtained for the straight bar in Fig. 8.8*a* is almost trivial. If N is the internal force in the bar, P the external force, and A the cross-sectional area, $\sigma_x = N/A$ and $X_s = P/A$. Thus, a uniform virtual elongation $\delta\Delta$ results in a virtual strain $\delta\epsilon_x = \delta\Delta/L$. According to Eqs. (8.16) and (8.17)

$$\delta U = \iiint_V \sigma_x \, \delta\epsilon_x \, dV = \frac{N}{A}\frac{\delta\Delta}{L} AL = N\,\delta\Delta$$

and, neglecting the weight of the bar,

$$\delta W_e = \iint_A X_s \, \delta\Delta \, dS = P\,\delta\Delta$$

Thus, $N = P$ for equilibrium.

Similarly, we could give the elastic curve of the beam in Fig. 8.8*b* a virtual displacement, calculate $\delta\epsilon_x$, introduce $M_z y/I_z$ for σ_x in Eq. (8.17), and arrive at certain conditions of equilibrium. A much more rewarding approach, however, is to start, as before, by considering the work done by forces acting on an "internal" element of the bar. If a typical bar element is given a transverse virtual displacement δv, the vertical forces perform virtual work of an amount

$$\left(\frac{dV}{dx} + p\right) \Delta x \, \delta v$$

where p is the external load per unit length and V is the *internal* shear.[5] This may also be written

$$\left(\frac{d^2M}{dx^2} + p\right) \Delta x \, \delta v$$

where M is the *internal* bending moment in the beam. The terms within the parentheses are recognized as those appearing in the governing equation of equilibrium of a straight bar [Eqs. (5.17)]. Hence, virtual work may again be interpreted as merely an equation of statics multiplied by an arbitrary function δv.

Proceeding as before, we allow Δx to approach zero and integrate from A to B to obtain the total virtual work done in the segment AB:[6]

$$\delta W = \int_A^B \left(\frac{d^2M}{dx^2} + p\right) \delta v \, dx \tag{8.20}$$

We assume, for simplicity, that M, p, and the shear $V = dM/dx$ are continuous throughout the interval over which we integrate. Otherwise, it would merely be necessary to break δW into a number of integrals and integrate from one discontinuity to another. Integrating the first term in the integrand twice by parts, we find

$$\begin{aligned}\int_A^B \frac{d^2M}{dx^2} \delta v \, dx &= \int_A^B \delta v \, dV = V \, \delta v \Big|_A^B - \int_A^B \frac{d\delta v}{dx} \, dM \\ &= (V_B \bar{v}_B - V_A \bar{v}_A - M_B \bar{\theta}_B + M_A \bar{\theta}_A) + \int_A^B M \frac{d^2\bar{v}}{dx^2} \, dx\end{aligned}$$

where, for simplicity in writing, we have denoted

$$\delta v = \bar{v}$$
$$\frac{d\delta v}{dx} = \bar{\theta}$$
$$\frac{d^2\delta v}{dx^2} = \frac{d^2\bar{v}}{dx^2}$$

[5] Since we are dealing with a coplanar bar subjected to in-plane loading, we omit the subscripts on V_y, M_z, and p_y, for simplicity.

[6] We neglect the virtual work done by true shearing forces in moving through virtual shear displacements.

Thus $\bar{v}_A$ and $\bar{v}_B$ denote the virtual displacements at A and B and $\bar{\theta}_A$ and $\bar{\theta}_B$ denote the virtual rotations at A and B, respectively. V_A, V_B, M_A, and M_B are the *true* shears and moments at points A and B. Equation (8.20) becomes

$$\delta W = \int_A^B M \frac{d^2\bar{v}}{dx^2}\, dx + \int_A^B p\bar{v}\, dx + (V_B\bar{v}_B - V_A\bar{v}_A - M_B\bar{\theta}_B + M_A\bar{\theta}_A) = 0 \tag{8.21}$$

This result may be referred to as the principle of virtual displacements for a straight beam. Comparing Eq. (8.21) with Eq. (8.10), we see that p is equivalent to a body force in this one-dimensional case; clearly, the second integral in (8.21) is the virtual work done by the external forces p. Similarly, the term within the parentheses is the external virtual work done by forces and moments at the boundaries A and B. V_A, for example, performs work by moving through the virtual displacement $\bar{v}_A$, and M_A performs work by moving through the virtual rotation $\bar{\theta}_A$. Thus, the term in parentheses corresponds to the external virtual work done by the surface forces in Eq. (8.10). Again, if the beam were simply supported at A and B, we could legitimately set $\bar{v}_A$ and $\bar{v}_B$ equal to zero because these displacements would then violate the constraints.

Finally, since bending moments are developed internally in the structure, the first integral on the right side of Eq. (8.21) is (minus) the internal virtual work. Hence, it is the one-dimensional equivalent of δU in Eqs. (8.15) and (8.17). Physically, this integral arises from the fact that a moment M at some point x does virtual work $M(d\bar{v}/dx)$, while that done at $x + \Delta x$ is $-(M + \Delta M)[d\bar{v}/dx + (d^2\bar{v}/dx^2)\,\Delta x]$, or, in the limit, $-M[d\bar{v}/dx + (d^2\bar{v}/dx^2)\, dx]$. Thus, $-M(d^2\bar{v}/dx^2)\, dx$ is the resulting internal work done in displacing a single element. Integrating this over the entire bar gives the total internal virtual work. On comparing the developments of the principle for the general body and for the straight bar, it is also interesting to note that ordinary integration by parts may be considered as a one-dimensional form of Green's theorem.

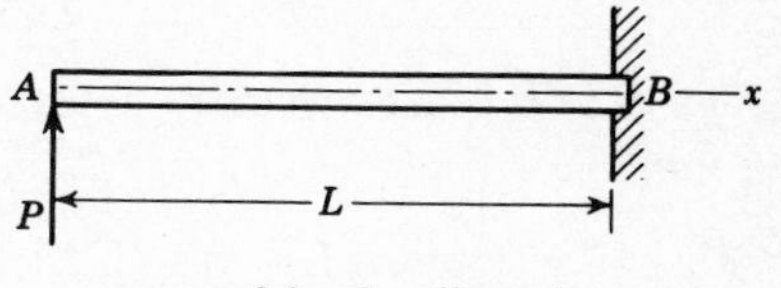

FIGURE 8.9 Cantilever beam.

Let us apply Eq. (8.21) to a simple problem. Consider the cantilever beam shown in Fig. 8.9 subjected to a concentrated force P at its free end. In this case $p = M_A = 0$ and $M = Px$; but V_A, V_B, and M_B are not zero. We begin by choosing a pattern of virtual displacements of the form

$$\delta v = \bar{v} = a\left(1 - \sin \frac{\pi x}{2L}\right) \tag{a}$$

where a is an arbitrary constant. Since, for this choice, both $\bar{v}$ and $d\bar{v}/dx$ are zero at $x = L$, this virtual displacement pattern does not violate the constraints. The external virtual work is [Eq. (8.21)]

$$\delta W_e = V_B \times 0 - V_A \times a - M_B \times 0 + 0 \times \left(-\frac{\pi a}{2L}\right) = -V_A a \quad (b)$$

and the internal virtual work is

$$\delta U = -\int_0^L Px\left(\frac{a\pi^2}{4L^2}\sin\frac{\pi x}{2L}\right)dx = -Pa \quad (c)$$

Substituting these results into Eq. (8.19), we get the trivial condition for equilibrium, $V_A = P$.

Note that the magnitude of the virtual displacement at any point on the elastic curve depends upon the parameter a in Eq. (a). Since there are an infinite number of points, Eq. (a) defines an infinity of conditions of constraint relating the deflections of every point to a single parameter a. The least number of independent variables required to specify the configuration of a system corresponds to the number of degrees of freedom of the system. These independent variables are called *generalized coordinates*, and a change in a generalized coordinate is called a *generalized displacement*. It follows that the parameter a behaves as a generalized displacement and, by using it in Eq. (a), we have, in effect, reduced the number of degrees of freedom to only one. Equation (a), of course, defines an arbitrary virtual displacement which may have little in common with the true displacement of the beam. However, a similar line of reasoning is the basis of some important approximate methods of structural analysis which enable us to reduce continuous systems to systems with a finite number of degrees of freedom. We discuss one of these methods in the following chapter.[7]

Returning to the beam in Fig. 8.9, let us now select a virtual displacement pattern which bears no resemblance to the true shape of the elastic curve. Take, for example, the function

$$\delta v = \bar{v} = a\cosh x + b\sinh x \quad (d)$$

where a and b are arbitrary constants. Introducing δv into Eq. (8.21), we find

$$\delta W_e = V_B(a\cosh L + b\sinh L) - V_A a - M_B(a\sinh L + b\cosh L)$$

and

$$\delta U = -\int_0^L Px\frac{d^2\bar{v}}{dx^2}dx = -Pa(L\sinh L - \cosh L + 1) - Pb(L\cosh L - \sinh L)$$

[7] See Art. 9.8.

Substituting these expressions into Eq. (8.19) and rearranging terms, we get

$$\delta W = 0 = (V_B - P)(a \cosh L + b \sinh L) + (PL - M_B)(a \sinh L + b \cosh L) + (P - V_A)a$$

Since this relationship must hold for any choice of a and b, we conclude that for equilibrium to exist,

$$V_B - P = 0 \qquad PL - M_B = 0 \qquad P - V_A = 0 \tag{e}$$

which we recognize, of course, as the equations of statics for the bar.

It is important to note that, in effect, we prescribed the end forces and moments V_A, V_B, and M_B to obtain Eqs. (*e*). The virtual displacement in Eq. (*a*) is consistent with the constraints indicated in Fig. 8.9, but that given in Eq. (*d*) violates these constraints. However, we managed to obtain useful equilibrium conditions involving boundary forces and moments by prescribing forces rather than displacements at points A and B, as is indicated in Fig. 8.8*b*.

8.7 The unit-dummy-displacement method. By now it is clear that when we apply the principle of virtual displacements we obtain conditions of equilibrium. On examining the example problems outlined in the previous section, it also becomes clear that specifically which portion of the total external force system appears in the final equilibrium conditions depends entirely on the nature of the virtual displacement pattern given the structure and the forces we prescribe on the boundaries. It is this property of the virtual work expressions that we take full advantage of when confronted with the following problem: At a given point and direction on a deformable body under a stress system, $\sigma_x, \sigma_y, \sigma_z, \ldots, \tau_{yz}$, what force P must exist in order that the system be in equilibrium? The answer follows directly from Eq. (8.16). We simply let P be the only external force acting on the structure to undergo a virtual displacement. We then give its point of application a virtual displacement $\delta\Delta$ in the direction in which P acts. Interpreted another way, we give the point under consideration a virtual displacement while keeping the position of all other loads fixed (that is, we give all other external loads zero virtual displacements). The external virtual work, then, is simply $P\,\delta\Delta$. This virtual displacement, in turn, develops a system of virtual strains, $\delta\epsilon_x{}^\Delta, \delta\epsilon_y{}^\Delta, \delta\epsilon_z{}^\Delta, \ldots, \delta\gamma_{yz}{}^\Delta$. It follows from the principle of virtual displacements that

$$P\,\delta\Delta = \iiint_V (\sigma_x\,\delta\epsilon_x{}^\Delta + \sigma_y\,\delta\epsilon_y{}^\Delta + \cdots + \tau_{yz}\,\delta\gamma_{yz}{}^\Delta)\,dV$$

where $\sigma_x, \ldots, \tau_{yz}$ are the true stresses. Now $\delta\Delta$ is completely arbitrary, and, for simplicity, we can assume that it is unity. We then have

$$P = \int_V \sigma \, \delta\epsilon^{\Delta} \, dV \tag{8.22}$$

where, for convenience in writing, we express the volume integral in the concise symbolic form. This process is called the *unit-dummy-displacement method.* Again, σ denotes the true stress system and $\delta\epsilon^{\Delta}$ the virtual strain system due to a unit virtual displacement of the point of application of P and in the direction of P. An example of the method is provided by simply letting the parameter a be unity in Eq. (a) of the preceding article.

Physically, P may also be interpreted as a measure of the *stiffness* of the structure; i.e., P is the force developed at some point i due to a unit (virtual) displacement of i in the direction of P. The significance of this interpretation and of Eq. (8.22) is important in the analysis of complex structures.[8]

The unit-dummy-displacement method is extremely useful for analyzing indeterminate structures. As a simple example, consider the pin-connected coplanar truss shown in Fig. 8.10, which consists of n linearly

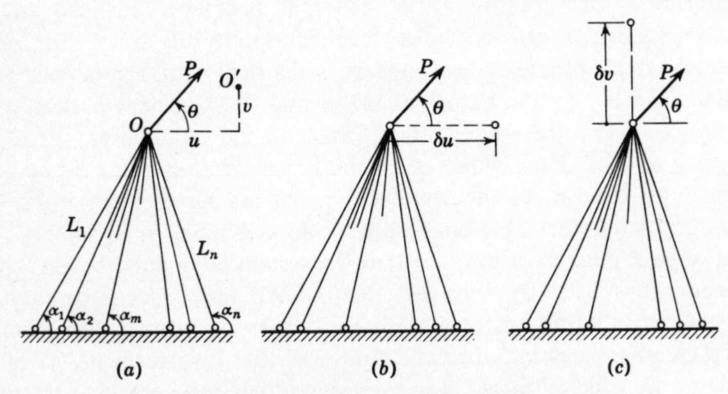

FIGURE 8.10 True and virtual displacements of a statically indeterminate coplanar truss.

elastic members. The degree of statical indeterminacy of the structure is $n - 2$. A typical member m of the truss is of length L_m, of area A_m, and is inclined an angle α_m with respect to the horizontal.

When the external force P is applied to the structure, joint O undergoes *true* displacements u and v before equilibrium is established. As a result,

[8] This interpretation is discussed in more detail in the following chapter. See Art. 9.10.

member m of the truss is elongated an amount $u \cos \alpha_m + v \sin \alpha_m$. Consequently, the *true* strain in m due to P is

$$\epsilon_m = \frac{u \cos \alpha_m + v \sin \alpha_m}{L_m} \tag{a}$$

and the stress is

$$\sigma_m = \frac{E}{L_m}(u \cos \alpha_m + v \sin \alpha_m) \tag{b}$$

We now give joint O a unit *virtual* displacement in the horizontal direction $\delta u = 1$, as shown in Fig. 8.10*b*. There results in member m a virtual strain given by

$$\delta \epsilon_m{}^\Delta = \frac{1 \cos \alpha_m}{L_m}$$

so that the internal virtual work done in member m is

$$\int_V \sigma_m \, \delta \epsilon_m{}^\Delta \, dV = \frac{EA_m}{L_m}(u \cos^2 \alpha_m + v \sin \alpha_m \cos \alpha_m)$$

Thus, the total internal virtual work is

$$\delta U = u \sum_{m=1}^{n} \frac{EA_m}{L_m} \cos^2 \alpha_m + v \sum_{m=1}^{n} \frac{EA_m}{L_m} \sin \alpha_m \cos \alpha_m \tag{c}$$

The total external virtual work is

$$\delta W_e = P \cdot 1 \cos \theta \tag{d}$$

where θ is the angle indicated in the figure.

From the principle of virtual displacements, it follows that

$$P \cos \theta = u \sum_{m=1}^{n} \frac{EA_m}{L_m} \cos^2 \alpha_m + v \sum_{m=1}^{n} \frac{EA_m}{L_m} \sin \alpha_m \cos \alpha_m \tag{e}$$

Joint O has two degrees of freedom. Therefore, we can obtain a second independent equilibrium condition by also computing the virtual work done in giving O a unit virtual displacement vertically, as indicated in Fig. 8.10*c*. We arrive at the condition

$$P \sin \theta = u \sum_{m=1}^{n} \frac{EA_m}{L_m} \cos \alpha_m \sin \alpha_m + v \sum_{m=1}^{n} \frac{EA_m}{L_m} \sin^2 \alpha_m \tag{f}$$

We now evaluate the true displacements u and v by solving Eqs. (*e*) and (*f*) simultaneously. True stresses are then evaluated using Eq. (*b*).

It is important that we observe the following characteristics of the above procedure:

1. Displacements were taken as unknowns.

2. Compatibility requirements were satisfied at the onset [Eq. (*a*)] for any choice of u and v.

3. A number of independent conditions equal to the number of degrees of freedom[9] of the structure were established [Eqs. (*e*) and (*f*)].
4. Final displacements u and v were chosen so as to provide equilibrium.

These characteristics are typical of procedures called *stiffness methods*[10] of structural analysis.

THE PRINCIPLE OF VIRTUAL FORCES

Both forces and displacements appear in the definition of work. As we noted earlier, it is natural to expect that some useful results may be obtained by calculating virtual work done by an imaginary virtual force system in moving through true displacements. This, of course, is true. In fact, we shall find that it is from this complementary concept that some of the most important principles of structural analysis evolve.

8.8 Virtual forces on particles and rigid bodies. The application of the principle of virtual forces to systems of particles and rigid bodies is comparatively trivial; the true significance of the principle is best appreciated when deformed bodies are considered. This is because the material properties and the nature of the stresses and strains play an important role in the results. The concepts of stress and strain are usually meaningless in the study of rigid undeformable bodies. Nevertheless, for the sake of completeness and in order to parallel the development of the principle of virtual displacements, we shall demonstrate the application of virtual forces to rigid systems by a simple example.

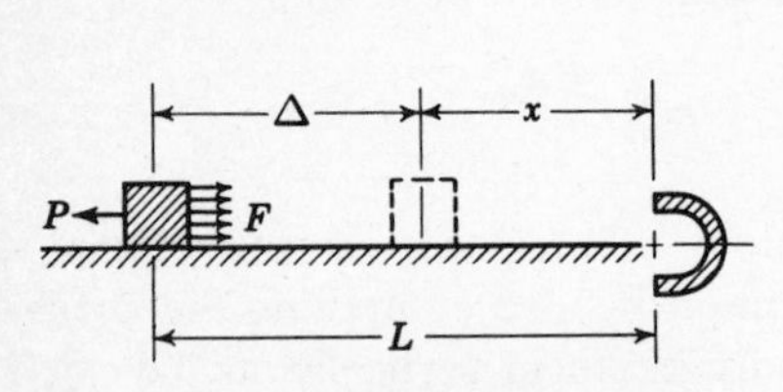

FIGURE 8.11 Rigid body subjected to a magnetic field.

Consider the rigid block resting on a frictionless plane in Fig. 8.11. The block is subjected to an external force P and an external force F created by a force field due to a magnet placed a distance L from the block. Let F be inversely proportional to the square of the distance x

[9] The degrees of freedom are often referred to as degrees of kinematical indeterminacy and the unknown displacements are called kinematic redundants. See Art. 2.6.

[10] See Art. 9.10.

from the block to the magnet. Then, everywhere on the plane $F = k/x^2$, or $x = \sqrt{k/F}$, where k is a known constant. Suppose that the block is put in equilibrium after undergoing a *true* displacement Δ, in which case $F = P$. We now introduce a pair of fictitious forces δP, in place of P, and δF, in place of F, *which satisfy the equations of equilibrium of the system* (that is, $\delta P = \delta F$) and compute the virtual work done by these *virtual forces* in moving through the true displacement Δ. We refer to this work as the *complementary virtual work*, denoted δW^* to distinguish it from virtual work due to virtual displacements. Thus,

$$\delta W^* = (-\delta P + \delta F)\Delta \tag{a}$$

or

$$\delta W^* = -\delta P\Delta + \left(L - \sqrt{\frac{k}{P}}\right)\delta F \tag{b}$$

δW^* in Eq. (a) is clearly zero, since we require δP and δF to be in equilibrium. Replacing δF in Eq. (b) with δP, we find that

$$\left(-\Delta + L - \sqrt{\frac{k}{P}}\right)\delta P = 0 \tag{c}$$

for any choice of δP. Therefore, if the block is in equilibrium, the displacement must satisfy the condition

$$-\Delta + L - \sqrt{\frac{k}{P}} = 0 \tag{d}$$

Note that, in contrast with the principle of virtual displacements, use of virtual forces leads to conditions involving true displacements. Note also that the directions of δP and δF are parallel to the rigid horizontal plane. If we apply a self-equilibrating virtual force system normal to the plane, we arrive at the trivial condition that the vertical displacement of the block is zero.

We postpone a more detailed discussion of these ideas until the following article. Without further comment, we record the principle of virtual forces for particles and rigid bodies:

> The true displacements of a system of particles and rigid bodies in reaching an equilibrium configuration are such that the total complementary virtual work is zero for every system of virtual forces satisfying the equilibrium conditions.
>
> (THEOREM V)

8.9 Virtual forces on a deformable body. Let us now return to the two-dimensional deformable body of general shape shown in Fig. 8.7. We recall that each point in the body undergoes true displacements u and v until a system of stresses is developed which satisfy Eqs. (8.6) at every point. It is only then that the body comes to rest.

We now introduce a number of continuous single-valued functions $\delta\sigma_x$, $\delta\sigma_y$, $\delta\tau_{xy}$, δX_b, δY_b, δX_s, and δY_s, which satisfy the equilibrium equations both within the body and on its boundaries:

$$\begin{aligned} \frac{\partial \delta\sigma_x}{\partial x} + \frac{\partial \delta\tau_{xy}}{\partial y} + \delta X_b = 0 \\ \frac{\partial \delta\tau_{xy}}{\partial x} + \frac{\partial \delta\sigma_y}{\partial y} + \delta Y_b = 0 \end{aligned} \tag{8.23}$$

and

$$\begin{aligned} \delta\sigma_x l + \delta\tau_{xy} m = \delta X_s \\ \delta\tau_{xy} l + \delta\sigma_y m = \delta Y_s \end{aligned} \tag{8.24}$$

We refer to functions with these properties as *statically admissible functions*, in analogy with kinematically admissible virtual displacements. Except for requiring that they be continuous and self-equilibrating, the functions $\delta\sigma_x$, $\delta\sigma_y$, . . . , and δY_s are completely arbitrary. We identify them with symbols resembling those used for stresses only to bring physical meaning into the development. Physically, we can regard these functions as a system of self-equilibrating *virtual forces* and *stresses* which are independent of the true forces and stresses existing in the structure. Or, we can think of them as *variations* in the true forces and stresses of the system.

Multiplying Eqs. (8.23) by $u\,\Delta x\,\Delta y$ and $v\,\Delta x\,\Delta y$, where u and v are components of displacement, and then adding the results, we get

$$\left[\left(\frac{\partial \delta\sigma_x}{\partial x} + \frac{\partial \delta\tau_{xy}}{\partial y} + \delta X_b\right)u + \left(\frac{\partial \delta\tau_{xy}}{\partial x} + \frac{\partial \delta\sigma_y}{\partial y} + \delta Y_b\right)v\right]\Delta x\,\Delta y$$

We interpret this quantity as the *complementary virtual work* done by the virtual forces and stresses acting on a single element in moving through the displacements u and v. We obtain the total complementary virtual work by allowing Δx and Δy to approach zero as we sum the work done on all elements of the body:

$$\delta W^* = \iint_A \left[\left(\frac{\partial \delta\sigma_x}{\partial x} + \frac{\partial \delta\tau_{xy}}{\partial y} + \delta X_b\right)u + \left(\frac{\partial \delta\tau_{xy}}{\partial x} + \frac{\partial \delta\sigma_y}{\partial y} + \delta Y_b\right)v\right]dx\,dy \tag{8.25}$$

We now return to Green's theorem [Eq. (8.9)] and proceed in the manner similar to that used in the case of virtual displacements. As a result, Eq. (8.25) is transformed into the alternative form

$$\delta W^* = -\iint_A \left[\delta\sigma_x \frac{\partial u}{\partial x} + \delta\sigma_y \frac{\partial v}{\partial y} + \delta\tau_{xy}\left(\frac{\partial v}{\partial x} + \frac{\partial u}{\partial y}\right)\right] dA + \iint_A (\delta X_b u + Y_b\, \delta v)$$
$$\times\, dA + \oint_C [(\delta\sigma_x l + \delta\tau_{xy} m)u + (\delta\tau_{xy} l + \delta\sigma_y m)v]\, ds \qquad (8.26)$$

where A, C, l, m, and s have the same meaning as in Eq. (8.10).

Comparing the terms in parentheses in the line integral with Eqs. (8.24), we find that they can be replaced by the virtual surface forces δX_s and δY_s. Hence, this integral represents the external complementary virtual work done along the boundaries of the body by δX_s and δY_s. Similarly, the second integral is the external complementary virtual work done by the virtual body forces. Therefore, the *total external complementary virtual work* is given by the equation

$$\delta W_e^* = \iint_A (\delta X_b u + \delta Y_b v)\, dA + \oint_C (\delta X_s u + \delta Y_s v)\, ds \qquad (8.27)$$

Now if u and v are the actual displacements of points of the body, we know that along portions C_2 of the boundary they must take on prescribed values. Since we are to use the principle of virtual forces to obtain conditions involving such displacements, we need consider only those virtual surface forces which vanish on the portions of the boundary where the forces are prescribed. With such choices of δX_s and δY_s, Eq. (8.27) becomes

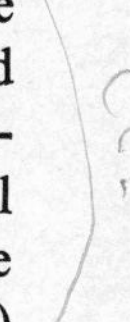

$$\delta W_e^* = \iint_A (\delta X_b u + \delta Y_b v)\, dA + \int_{C_2} (\delta X_s u + \delta Y_s v)\, ds \qquad (8.28)$$

In analogy with virtual displacements, we shall find that kinematic boundary conditions are obtained if we decide to prescribe displacements rather than forces on certain parts of the boundary. We must then use nonzero virtual forces on these parts.

Returning now to Eq. (8.26), we observe that the first integral on the right side of this equation is the virtual work done by the internal virtual stresses. Since these virtual stresses satisfy Eqs. (8.6), δW^* in Eq. (8.26) is automatically zero, regardless of the interpretation that we give the quantities u and v. Indeed, as the equation now stands u and v need not even correspond to compatible strains. However, if, out of all the choices of u and v at our disposal, we choose u and v to be those single-valued continuous functions which satisfy Eqs. (8.5), then u and v are the true compatible displacements of points in the body. We now define as the *internal complementary virtual work*, denoted δU^*, the work done by the *virtual stresses* in acting through the *true strains* which satisfy the conditions of compatibility within the body. Thus,

$$\delta U^* = \iint_A (\epsilon_x \, \delta\sigma_x + \epsilon_y \, \delta\sigma_y + \gamma_{xy} \, \delta\tau_{xy}) \, dA \tag{8.29}$$

For the general three-dimensional body, Eqs. (8.28) and (8.29) become

$$\delta W_e^* = \iiint_V (\delta X_b u + \delta Y_b v + \delta Z_b w) \, dV + \iint_{S_2} (\delta X_s u + \delta Y_s v + \delta Z_s w) \, dS \tag{8.30}$$

and

$$\delta U^* = \iiint_V (\epsilon_x \, \delta\sigma_x + \epsilon_y \, \delta\sigma_y + \epsilon_z \, \delta\sigma_z + \gamma_{xy} \, \delta\tau_{xy} + \gamma_{xz} \, \delta\tau_{xz} + \gamma_{yz} \, \delta\tau_{yz}) \, dV \tag{8.31}$$

the integration now being over the volume and the surface area of the body. Again note that the virtual surface forces are defined only on those portions of the boundary on which the displacements are prescribed (i.e., that portion on which the surface forces are not prescribed).

With the above notation, Eq. (8.26) becomes

$$\delta W^* = -\delta U^* + \delta W_e^* = 0 \tag{8.32}$$

or, in analogy with Eq. (8.19),

$$\delta W_e^* = \delta U^* \tag{8.33}$$

Equation (8.33) is a mathematical statement of the principle of virtual forces. As it now stands, we have only shown that this is a necessary condition for compatible deformations. To show that it is also sufficient, consider the two-dimensional case in which no body forces are present. Let us introduce a continuously differentiable function Ψ defined so that

$$\frac{\partial^2 \Psi}{\partial x^2} = \delta\sigma_y \qquad \frac{\partial^2 \Psi}{\partial y^2} = \delta\sigma_x \qquad \frac{\partial^2 \Psi}{\partial x \, \partial y} = -\delta\tau_{xy} \tag{8.34}$$

This function is called the *Airy stress function.*[11] Any function satisfying Eqs. (8.34) and continuous through at least its third partial derivatives is automatically statically admissible within the body; since $\delta X_b = \delta Y_b = 0$, it automatically satisfies Eqs. (8.23). Now from Eq. (2.23) we recall that the strains are compatible provided

$$\frac{\partial^2 \epsilon_y}{\partial x^2} + \frac{\partial^2 \epsilon_x}{\partial y^2} - \frac{\partial^2 \gamma_{xy}}{\partial x \, \partial y} = 0 \tag{8.35}$$

Thus

$$\iint_A \left(\frac{\partial^2 \epsilon_y}{\partial x^2} + \frac{\partial^2 \epsilon_x}{\partial y^2} - \frac{\partial^2 \gamma_{xy}}{\partial x \, \partial y} \right) \Psi \, dx \, dy = 0 \tag{8.36}$$

[11] See Prob. 2.13.

only if ϵ_x, ϵ_y, and γ_{xy} satisfy the compatibility equation. To transform this integral into a more meaningful form, we successively apply Green's theorem and obtain the identities

$$\begin{aligned}
\iint_A \Psi \frac{\partial^2 \epsilon_y}{\partial x^2}\, dx\, dy &= \iint_A \frac{\partial^2 \Psi}{\partial x^2} \epsilon_y\, dx\, dy + \oint_C \left(\Psi \frac{\partial \epsilon_y}{\partial x} - \epsilon_y \frac{\partial \Psi}{\partial y} \right) dy \\
\iint_A \Psi \frac{\partial^2 \epsilon_x}{\partial y^2}\, dx\, dy &= \iint_A \frac{\partial^2 \Psi}{\partial y^2} \epsilon_x\, dx\, dy + \oint_C \left(\Psi \frac{\partial \epsilon_x}{\partial y} + \epsilon_x \frac{\partial \Psi}{\partial x} \right) dx \qquad (8.37) \\
\iint_A \Psi \frac{\partial^2 \gamma_{xy}}{\partial x\, \partial y}\, dx\, dy &= \iint_A \frac{\partial^2 \Psi}{\partial x\, \partial y} \gamma_{xy}\, dx\, dy + \oint_C \left(\Psi \frac{\partial \gamma_{xy}}{\partial y}\, dy + \Psi \frac{\partial \gamma_{xy}}{\partial x}\, dx \right)
\end{aligned}$$

Since Ψ is arbitrary, we set it equal to zero on the boundary after integrating the line integrals in Eqs. (8.37) once again by parts. Then noting that $dy = l\, ds$ and $dx = -m\, ds$, we introduce Eqs. (8.5) and (8.34) into these results and find that

$$\begin{aligned}
\iint_A \left(\frac{\partial^2 \epsilon_y}{\partial x^2} + \frac{\partial^2 \epsilon_x}{\partial y^2} - \frac{\partial^2 \gamma_{xy}}{\partial x\, \partial y} \right) \Psi\, dx\, dy &= \iint_A \left[\delta\sigma_x \frac{\partial u}{\partial x} + \delta\sigma_y \frac{\partial v}{\partial y} \right. \\
&\quad \left. + \delta\tau_{xy} \left(\frac{\partial v}{\partial x} + \frac{\partial u}{\partial y} \right) \right] dx\, dy - \oint_C [(\delta\sigma_x l + \delta\tau_{xy} m) u \\
&\quad + (\delta\tau_{xy} l + \delta\sigma_y m) v]\, ds \qquad (8.38)
\end{aligned}$$

or, in view of Eqs. (8.28) and (8.29),

$$\iint_A \left(\frac{\partial^2 \epsilon_y}{\partial x^2} + \frac{\partial^2 \epsilon_x}{\partial y^2} - \frac{\partial^2 \gamma_{xy}}{\partial x\, \partial y} \right) \Psi\, dx\, dy = \delta U^* - \delta W_e^* \qquad (8.39)$$

It follows that Eq. (8.33) holds only if the integral on the left side of this equation vanishes; and, as we noted previously, this happens only if the strain components satisfy the compatibility equation. Therefore, we conclude that Eq. (8.33) is both a necessary and a sufficient condition for compatible deformations.

In words, the principle of virtual forces is stated as follows:

> The strains and displacements in a deformable system are compatible and consistent with the constraints if the total external complementary virtual work is equal to the total internal complementary virtual work for every system of virtual forces and stresses that satisfy the equations of equilibrium.
>
> (THEOREM VI)

Thus, virtual forces provide us with conditions of compatibility, whereas virtual displacements provide conditions of equilibrium. Note that in rigid systems δU^* is zero, so that the above theorem reduces to Theorem V of the preceding article.

In applications of this principle, true strains depend upon the material properties of the body. Therefore, unlike the principle of virtual displacements, in applications of the principle of virtual forces the material properties of the structure must be known in advance.

It is important to note the interesting analogy between the virtual work principles. The principle of virtual forces is obtained, in essence, by replacing the words stress and force with strain and displacement and interchanging the words equilibrium and compatibility in our previous discussion of virtual displacements.

8.10 Virtual forces on simple structural systems. In the field of structural analysis, the principle of virtual forces provides the basis for many useful theorems and methods. To illustrate its application, we begin, as before, with the simple example of the straight bar in tension in Fig. 8.12*a*. This time, however, we assume that the bar is composed of a nonlinearly elastic material having a stress-strain law of the form[12] $\sigma_x = E\epsilon_x^{\frac{1}{2}}$ (see Fig. 8.12*b*).

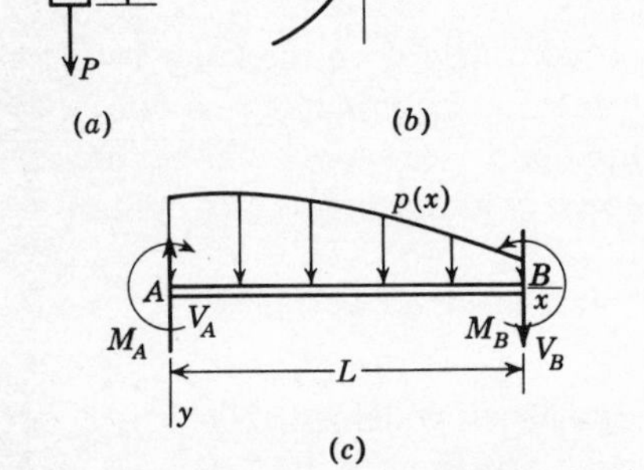

When the bar reaches a state of equilibrium, the true stress is P/A and the true strain at every point is $(P/AE)^2$. The external load is displaced an amount Δ before the internal force finally balances it and the bar is in equilibrium. We now introduce a constant virtual stress $\delta\sigma_x = \delta N/A$. Since the δN on the cross section of a typical longitudinal element is equal and opposite to that on the other cross section of the element, the virtual forces and stresses satisfy the equilibrium conditions. Therefore, if we assume that the weight of the bar is negligible, we have,

FIGURE 8.12 (*a*) Nonlinearly elastic bar; (*b*) nonlinear stress-strain diagram; (*c*) straight beam subjected to a general system of loads.

[12] We assume, of course, that a negative strain corresponds to a compressive stress. The law, therefore, is more correctly stated as $\sigma_x = E(\text{sgn}\ \epsilon_x)\epsilon_x^{\frac{1}{2}}$, where sgn (pronounced "signum") is defined by sgn $x = 1$, 0, or -1 when x is >0, 0, <0, respectively.

according to Eqs. (8.30) and (8.31),

$$\delta W_e^* = \iint_{S_2} \frac{\delta N}{A} \Delta \, dS = \delta N \Delta$$

$$\delta U^* = \iiint_V \frac{\delta N}{A} \left(\frac{P}{AE}\right)^2 dx \, dA = \frac{P^2 L}{(AE)^2} \delta N$$

Substituting these results into Eq. (8.33) gives

$$\Delta \, \delta N = \frac{P^2 L}{(AE)^2} \delta N$$

for any δN in equilibrium. We conclude that the strains and displacements are compatible only if

$$\Delta = \frac{P^2 L}{(AE)^2}$$

Let us now use the principle of virtual forces to arrive at compatibility conditions for the straight beam shown in Fig. 8.12*c*. For the present, we assume that the influence of shearing strains on the true transverse displacements v is negligible and that all cross sections are symmetrical with respect to the plane of bending. As before, it is rewarding to start the development by considering a typical element of the beam rather than directly applying the general equations. The results, of course, are the same.

We begin by applying a virtual external load of intensity $\bar{p}$ per unit length of the beam as well as a virtual stress distribution $\delta\sigma_x$ on every cross section which results in a virtual bending moment $\bar{m}$. Or, for an equivalent interpretation, we can imagine that the true loading, stress, and bending moment p, σ_x, and M acquire arbitrary variations $\delta p = \bar{p}$, $\delta\sigma_x$, and $\delta M = \bar{m}$, respectively. According to the principle of virtual forces, these variations must satisfy the equations of equilibrium. Therefore, for every element of the beam $\bar{m}$ and $\bar{p}$ must be such that

$$\frac{d^2\bar{m}}{dx^2} + \bar{p} = 0 \tag{8.40}$$

Following the same procedure that we used in the general case, we multiply Eq. (8.40) by $v\,dx$, v being the true transverse displacement of the element, and obtain the complementary virtual work done on one element. Assuming, only for simplicity, that v, $dv/dx, \ldots, d^n v/dx^n$ are continuous and denoting by A and B the end points of the segment under consideration, we find for the total complementary virtual work

$$\delta W^* = \int_A^B \left(\frac{d^2\bar{m}}{dx^2} + \bar{p}\right) v \, dx \tag{8.41}$$

Integrating twice by parts (which we are again reminded is equivalent to a one-dimensional version of Green's theorem), we obtain

$$\delta W^* = \int_A^B \bar{m}\frac{d^2v}{dx^2}\,dx + \int_A^B \bar{p}v\,dx + (\bar{V}_B v_B - \bar{V}_A v_A - \bar{m}_B\theta_B + \bar{m}_A\theta_A) \tag{8.42}$$

where $\bar{V}_A$ and $\bar{V}_B$ are the virtual shearing forces at A and B (that is, $\bar{V} = d\bar{m}/dx$); $\bar{m}_A$ and $\bar{m}_B$ are the virtual moments at A and B; and v_A, θ_A and v_B, θ_B are the true displacements and slopes ($\theta = dv/dx$) at A and B, respectively.

The second integral in Eq. (8.42) plus the term in parentheses obviously represents the external complementary virtual work. The load $\bar{p}$, for example, corresponds to a virtual body force, and $\bar{V}_A$, $\bar{V}_B$, $\bar{m}_A$, and $\bar{m}_B$ correspond to the virtual surface forces since they act at the boundaries of the beam. The first integral on the right side of Eq. (8.42), however, is *not* the internal complementary virtual work unless we choose that particular v (or d^2v/dx^2) which satisfies the compatibility conditions of the beam. The form of these conditions depends entirely upon the material properties of the beam. Hence, it is necessary that the material properties be known before the specific form of δU^* can be determined. Thus,

$$\delta U^* = -\int_A^B \bar{m}\frac{d^2v}{dx^2}\,dx \tag{8.43}$$

only if v is the true displacement function satisfying the compatibility conditions.

Take, as a simple example, the case of a linearly elastic beam whose cross sections remain plane during deformation. Then the condition of compatibility is [see Eqs. (6.15)]

$$\frac{d^2v}{dx^2} = -\frac{M}{EI}$$

so that in this case

$$\delta U^* = \int_A^B \frac{M\bar{m}}{EI}\,dx \tag{8.44}$$

Equation (8.44) may be interpreted physically as the virtual work done by the virtual moments $\bar{m}$ in acting through the true changes in slope of an element $d\theta = M\,dx/EI$ integrated over the length of the beam.

To illustrate the application of Eq. (8.44), let us consider the uniformly loaded cantilever beam of constant cross section shown in Fig. 8.13*a*. The beam is made of a linearly elastic material, and the tip deflection is prescribed as Δ_0. The applied load p_0 and the bending moment $M = -p_0x^2/2$ together with the end reactions constitute a system of true forces and stress resultants satisfying the conditions of equilibrium.

We must introduce into this system a set of virtual forces and stress resultants which also satisfy the equilibrium conditions. If we treat the displacements through which the virtual forces act as prescribed displacements, any self-equilibrating force system will qualify as a virtual force system. Therefore, we have an infinite set of virtual systems to choose from. Indeed, the virtual system need not resemble the true force system in any way. Any of the virtual systems shown in Fig. 8.13*b*, *c*, and *d*, for example, can be used if we regard the appropriate displacements as prescribed displacements, simply because each system shown is in equilibrium.

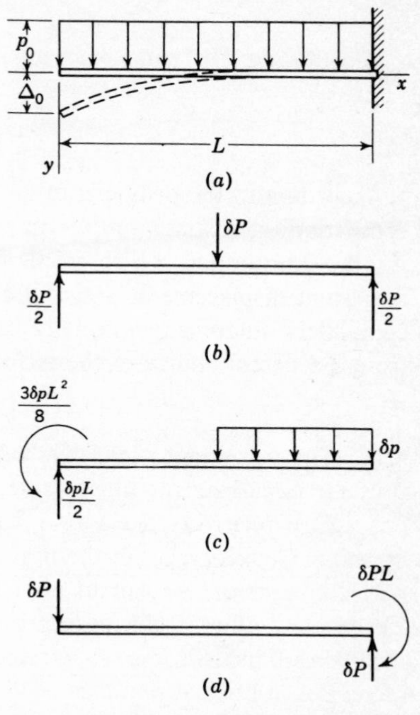

FIGURE 8.13 (*a*) Cantilever beam with real forces; (*b–d*) some possible virtual force systems.

For illustration, let us use the virtual force system shown in Fig. 8.13*b*. For this system $\bar{p}$ is zero at every point except $x = L/2$, where it equals δP (in the limit). Also, $\bar{V}(0) = -\bar{V}(L) = \delta P/2$; $m(0) = m(L) = 0$; and

$$\bar{m} = \begin{cases} \dfrac{\delta P x}{2} & 0 \leq x \leq \dfrac{L}{2} \\[2ex] \dfrac{\delta P(L - x)}{2} & \dfrac{L}{2} \leq x \leq L \end{cases}$$

Referring to Eq. (8.42), we find for the total external complementary virtual work

$$\delta W_e^* = \int_0^L \bar{p} v \, dx$$

$$+ \left(-\frac{\delta P}{2} \times 0 - \frac{\delta P}{2} \times \Delta_0 - 0 \times \theta_0 + 0 \times 0 \right) = \delta P \, \Delta_1 - \frac{\delta P}{2} \Delta_0 \quad (a)$$

where $\Delta_1 = v(L/2)$ is the center-line deflection. Note that this result can be obtained by inspection.

Referring to Eq. (8.44), we have

$$\delta U^* = \int_0^L \frac{-p_0 x^2/2}{EI} \bar{m} \, dx$$

$$= \int_0^{L/2} \frac{-p_0 x^2/2}{EI} \frac{\delta P x}{2} \, dx + \int_{L/2}^L \frac{-p_0 x^2/2}{EI} \frac{\delta P}{2} (L - x) \, dx$$

or, after performing the integration,

$$\delta U^* = -\frac{7p_0L^4}{384EI}\,\delta P \tag{b}$$

Substituting δW_e^* and δU^* into Eq. (8.33) and canceling δP, we get

$$\Delta_1 - \frac{\Delta_0}{2} = -\frac{7p_0L^4}{384EI} \tag{c}$$

According to the principle of virtual forces, this result is a condition of compatibility; it states simply that of all the conceivable values of Δ_1 and Δ_0, the unique pair which result from the true, compatible strains and consistent displacements satisfy Eq. (*c*). It is important to notice that a completely different (yet equally valid) compatibility condition will result from a different choice of the virtual force system.

8.11 The unit-dummy-load method. Although the results of our labors thus far have been at times quite interesting, they have been in almost every case relatively useless so far as calculating displacements is concerned. Condition (*c*) in the previous article is a good example—though perfectly correct, we can in no way use it to evaluate either Δ_0 or Δ_1. Fortunately, the virtual force principle can be used to evaluate displacements; and, indeed, it is an extremely efficient tool for this purpose. The secret lies in the last sentence of the preceding article—namely, the choice of the virtual force system.

We would not have to work many problems before we noticed the advantage in choosing as the virtual system one which is identical to or geometrically similar to the real system—in other words, it is often advantageous to introduce virtual surface forces (i.e., static boundary conditions in the virtual system) only at points where the true displacements are prescribed (i.e., where kinematic boundary conditions are imposed in the real system).

The virtual force system in Fig. 8.13*c*, for example, does not satisfy this condition; virtual surface forces (and moments) exist at the end where true forces and moments are prescribed as zero rather than at the other end where true displacements are prescribed [that is, $v(L) = \theta(L) = (0)$]. The virtual force system in Fig. 8.13*d*, however, does satisfy this condition—virtual surface "forces" δP and δPL act at a boundary point where the true displacements are prescribed.

In view of the above discussion, in the developments to follow we limit our choice of virtual force systems to those with virtual boundary forces existing only at points where the true boundary displacements are prescribed. We shall keep in mind, however, that virtual systems not possessing this property may still lead to correct compatibility equations.

It is not enough, however, to merely choose this type of virtual system in order to evaluate displacements. An examination of Eq. (*a*) in the preceding article will reveal that the compatibility condition involves the true displacements corresponding to every point on which an external virtual force is applied. Therefore, to evaluate a specific displacement we must select a virtual force system which leads to an expression for δW_e^* containing only the desired displacement. Clearly, to accomplish this we apply a single virtual force corresponding to the displacement which is to be found. Since, by definition, the magnitude of this force is arbitrary, it is customary to assume that it is unity; and it is for this reason that the procedure discussed above is called the *unit-dummy-load method.* Obviously, the method is merely a special case of the general principle of virtual forces.

The basic idea can be generalized in analogy with the unit-dummy-displacement method. Consider a general deformable body in equilibrium and subjected to strains $\epsilon_x, \epsilon_y, \ldots, \gamma_{yz}$ that satisfy the compatibility equations. At some point i on this body we apply a virtual external force δP which develops a system of virtual stresses $\delta\sigma_x{}^p, \delta\sigma_y{}^p, \ldots, \delta\tau_{yz}{}^p$ that satisfy the equilibrium equations. If Δ_i is the true displacement of i in the direction of δP, the total internal complementary virtual work is obtained by introducing $\delta\sigma_x{}^p, \delta\sigma_y{}^p, \ldots, \delta\tau_{xz}{}^p$ into Eq. (8.31). Finally, from the principle of virtual forces [Eq. (8.33)], we have

$$\Delta_i\, \delta P = \iiint_V (\epsilon_x\, \delta\sigma_x{}^p + \epsilon_y\, \delta\sigma_y{}^p + \epsilon_z\, \delta\sigma_z{}^p + \gamma_{xy}\, \delta\tau_{xy}{}^p + \gamma_{xy}\, \delta\tau_{xz}{}^p + \gamma_{yz}\, \delta\tau_{yz}{}^p)\, dV$$

Since the magnitude of δP is arbitrary, we assume that it is unity and obtain

$$\Delta_i = \int_V \epsilon\, \delta\sigma^p\, dV \tag{8.45}$$

where, as before, we have expressed the volume integral in a concise symbolic form for convenience in writing.

Equation (8.45) is a general statement of the unit-dummy-load method. Accordingly, the true displacement of a point i in a deformable body in a given direction is equal to the integral over the volume of the body of the product of true strains and the virtual stresses produced by a unit virtual force at i in the direction of Δ_i. Physically, Δ_i is a measure of the *flexibility*[13] of the structure.

Suppose, for example, that we wish to evaluate the tip displacement Δ_0 of the beam shown in Fig. 8.13*a*. We must choose the virtual force

[13] This interpretation is discussed in more detail in the following chapter. See Art. 9.18.

system shown in Fig. 8.13*d* for this purpose, this time setting $\delta P = 1$ for simplicity. Then

$$\delta W_e^* = \Delta_0$$

and from Eq. (8.44),

$$\delta U^* = \int_0^L \frac{-p_0 x^2}{2EI}(x \cdot 1)\,dx = \frac{p_0 L^4}{8EI}$$

Thus

$$\Delta_0 = \frac{p_0 L^4}{8EI}$$

If we follow this procedure in beam analysis, the external work will always be simply the desired displacement times unity. Thus, for the linearly elastic beam with negligible shear deformation, we write, in general,

$$\Delta_i = \int_A^B \frac{M\bar{m}\,dx}{EI} \tag{8.46}$$

The virtual force system must correspond to the type of displacement sought. For example, to evaluate the slope θ_i at some point i on the beam, we apply a unit external virtual moment at i. The external complementary virtual work is, then, simply $\theta_i \cdot 1$. Thus,

$$\theta_i = \int_A^B \frac{M\bar{m}}{EI}\,dx \tag{8.47}$$

where, in this case, $\bar{m}$ is the virtual bending moment due to a unit moment at i.

Equations (8.46) and (8.47) represent the one-dimensional form of Eq. (8.45) applied to the special case of a linearly elastic beam.

The same procedure can be used for structures composed of nonlinear materials except that δU^* is no longer given by Eq. (8.44). For example, suppose that the material used to construct a straight beam of constant depth obeys the parabolic stress-strain law[14] $\sigma_x = E\epsilon_x^{\frac{1}{2}}$. Neglecting shear deformations and assuming that planes remain plane, we recall that purely geometric considerations lead to the relation

$$\epsilon_x = -y\frac{d^2v}{dx^2}$$

Thus, for this nonlinear material,

$$M = \int_A \sigma_x y\,dA = -EI^*\left(\frac{d^2v}{dx^2}\right)^{\frac{1}{2}}$$

where I^* is a cross-sectional constant defined by

$$I^* = \int_A y^{\frac{3}{2}}\,dA$$

[14] Page 258.

Therefore, we obtain the nonlinear[15] compatibility equation

$$\frac{d^2v}{dx^2} = \left(\frac{M}{EI^*}\right)^2 \tag{a}$$

Substituting this result into Eq. (8.43) gives

$$\delta U^* = -\int_A^B \left(\frac{M}{EI^*}\right)^2 \bar{m}\, dx \tag{b}$$

Let us assume that the beam in Fig. 8.13*a* is made of such a parabolic material and that we wish to determine its tip deflection. Again, the virtual system in Fig. 8.13*d* is used with $\delta P = 1$ so that $\delta W_e^* = \Delta_0$ and $\bar{m} = -x$, as before. From Eq. (*b*),

$$\delta U^* = -\int_0^L \left(\frac{-p_0 x^2}{2EI^*}\right)^2 (-x)\, dx = \frac{p_0{}^2 L^6}{24E^2 I^{*2}}$$

Therefore $$\Delta_0 = \frac{p_0{}^2 L^6}{24E^2 I^{*2}} \tag{c}$$

If the beam is of rectangular cross section with width b and depth h, we find $I^* = (b/5)(h^5/2)^{\frac{1}{2}}$, so that

$$\Delta_0 = \frac{25 p_0{}^2 L^6}{12E^2 b^2 h^5} \tag{d}$$

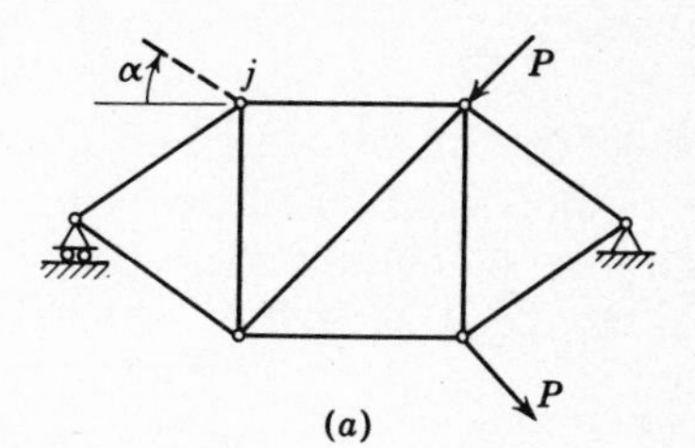

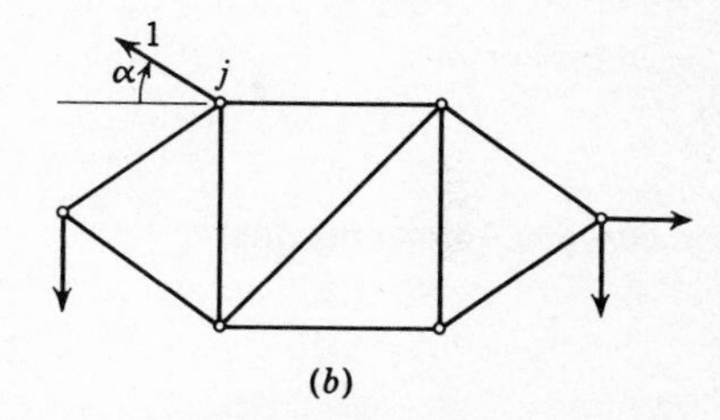

FIGURE 8.14 (*a*) A coplanar truss; (*b*) unit dummy load.

Similarly, to evaluate the displacement of joint j in the direction α of the truss in Fig. 8.14*a*, we apply a unit dummy (virtual) load at j, as shown in Fig. 8.14*b*. The total external complementary virtual work is then

$$W_e^* = 1 \cdot \Delta_j \tag{e}$$

where Δ_j is the desired displacement.

Let $\delta\sigma_m{}^p$ be the virtual stress in member m produced by the unit virtual force at j. If ϵ_m is the true strain in member m, then the internal complementary virtual work done in this member is

$$\int_V \epsilon_m\, \delta\sigma_m{}^p\, dV = \epsilon_m\, \delta\sigma_m{}^p A_m L_m \tag{f}$$

[15]In the strict sense of the word, this condition is linear since derivatives of the dependent variable v appear only to the first degree.

where A_m and L_m are the area and the length of member m. The true elongation of member m is

$$e_m = \epsilon_m L_m$$

and the *virtual axial force* in m due to $\delta\sigma_m{}^p$ is

$$n_m = A_m\,\delta\sigma_m{}^p$$

Thus, we may also express the internal work in Eq. (*f*) as simply $n_m e_m$. For a truss with n members, the total internal complementary virtual work is

$$\delta U^* = \sum_{m=1}^{n} e_m n_m$$

Therefore
$$\Delta_j = \sum_{m=1}^{n} e_m n_m \tag{8.48}$$

Note that we introduced no stress-strain relation in deriving Eq. (8.48); the result is valid for trusses composed of any nonlinearly elastic material with a monotonically increasing stress-strain law. For the special case of a Hookean material,

$$e_m = N_m \lambda_m$$

where N_m is the true axial force in member m and λ_m is the *axial flexibility*[16] of member m:

$$\lambda_m = \frac{L_m}{EA_m} \tag{g}$$

Thus, Eq. (8.48) becomes

$$\Delta_j = \sum_{m=1}^{n} N_m n_m \lambda_m \tag{8.49}$$

If, in addition to axial forces, members of the structure also develop bending moments M_m, the internal complementary virtual work due to virtual bending moments $\bar{m}_m$ produced by the unit load at point j would also have to be considered. In view of Eq. (8.43) and (8.48), it is clear that in such cases

$$\Delta_j = \sum_{m=1}^{n}\left(-\int_{L_m} \bar{m}_m \frac{d^2v}{dx^2}\,dx + e_m n_m\right) \tag{8.50}$$

where L_m is the length of member m and x is a local coordinate measured along the axis of member m. Referring to Eqs. (8.46) and (8.49), we see that in the case of linearly elastic materials Eq. (8.50) becomes

$$\Delta_j = \sum_{m=1}^{n}\left(\int_{L_m} \frac{M_m \bar{m}_m}{EI_m}\,dx + N_m n_m \lambda_m\right) \tag{8.51}$$

where I_m is the moment of inertia of member m.

[16] Physically, λ_m is the elongation of member m due to a unit axial force.

8.12 Statically indeterminate structures. It is important to recognize the great utility of the unit-dummy-load method in the analysis of statically indeterminate structures. As a simple example, let us consider the uniformly loaded continuous beam in Fig. 8.15*a*. The structure is statically indeterminate to the first degree owing to the presence of the intermediate support at *B*. From simple statics we find

$$R_A = R_C = p_0 L - \frac{R}{2} \qquad (h)$$

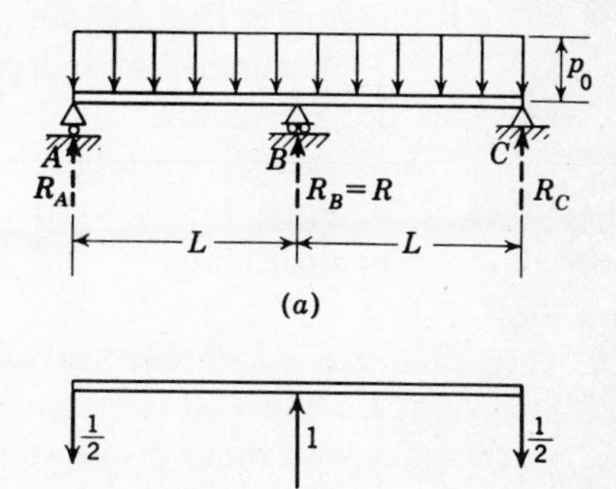

FIGURE 8.15 (*a*) Continuous beam; (*b*) virtual force system.

where R_A and R_C are the reactions at supports *A* and *C*, and *R* is the unknown (redundant) reaction at *B*. Thus, the true curvature of the beam between *A* and *B* is

$$\frac{M}{EI} = \frac{1}{EI}\left[\left(p_0 L - \frac{R}{2}\right)x - \frac{p_0 x^2}{2}\right]$$

We now introduce the virtual force system shown in Fig. 8.15*b*. The virtual bending moment is

$$\bar{m} = -\frac{x}{2}$$

so that from Eq. (8.44) we find that the total internal complementary virtual work is

$$\delta U^* = 2\int_0^L \frac{(p_0 L - R/2)x - p_0 x^2/2}{EI}\left(-\frac{x}{2}\right) dx$$

$$= \frac{L^3}{24EI}(5p_0 L - 4R) \qquad (i)$$

Here we have taken advantage of symmetry and doubled the integral over half the length of the beam.

Owing to our choice of a virtual force system, the true displacements corresponding to each external virtual load is seen to be zero. Therefore, δW_e^* is zero, which means that δU^* is also zero. Thus, from Eq. (*i*) we have

$$5p_0 L - 4R = 0 \qquad (j)$$

or

$$R = \frac{5p_0 L}{4} \qquad (k)$$

It is important that we observe the following characteristics of the above procedure:

1. Forces were taken as unknowns.
2. Equilibrium requirements were satisfied at the onset [Eq. (*h*)] for any choice of *R*.
3. A number of independent conditions equal to the number of degrees of statical indeterminacy were established [Eq. (*j*)].
4. Final forces were chosen so as to provide compatibility.

These characteristics are typical of procedures called *flexibility methods*[17] of structural analysis. It is of considerable interest to compare the above characteristics with those of the stiffness methods, listed in Art. 8.7.

The procedure used to analyze the statically indeterminate beam in Fig. 8.15 can be generalized so that it applies to structural systems with

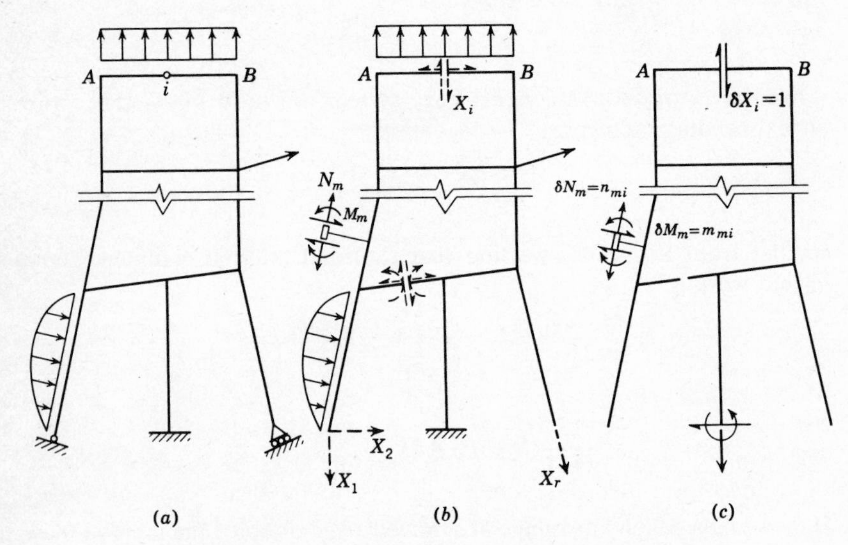

FIGURE 8.16 (*a*) Coplanar framework which is statically indeterminate to the *r*th degree; (*b*) the *r* redundants; (*c*) virtual force system corresponding to a unit virtual force at the point of application of X_i.

any number of redundants. Consider, for example, the coplanar system of straight beams and ties shown in Fig. 8.16*a* which is assumed to be statically indeterminate to the *r*th degree. If $X_1, X_2, \ldots, X_r$ denote the *r* independent forces and moments which we select for the redundants

[17] See also Art. 9.18.

(Fig. 8.16*b*), then we can express the bending moment and the axial force in any member m in the form

$$\begin{aligned} M_m &= M_{0m} + X_1 m_{m1} + X_2 m_{m2} + \cdots + X_r m_{mr} \\ N_m &= N_{0m} + X_1 n_{m1} + X_2 n_{m2} + \cdots + X_r n_{mr} \end{aligned} \tag{8.52}$$

Here M_{0m} and N_{0m} are the bending moment and the axial force that would be developed in member m if all of the redundants were zero (that is, if the structure were statically determinate); m_{mi} and n_{mi} are the bending moment and the axial force in member m due to a unit value of the redundant X_i. Equations (8.52) are obtained from statics and, in general, the functions M_{0m}, N_{0m}, $m_{m1}, \ldots, m_{mr}$, $n_{m1}, \ldots, n_{mr}$ vary from point to point along the member.

Let us now introduce a virtual force system corresponding to a unit virtual force $\delta X_i = 1$, applied at the point of action of the ith redundant. Even though the actual structure is statically indeterminate, the structure on which we apply this virtual force system need not be indeterminate, for the only condition that we are to impose on this system is that it be in equilibrium. For simplicity, and because it is the most convenient, we obtain a statically admissible virtual force system corresponding to $\delta X_i = 1$ by applying unit virtual forces at point i of the statically determinate structure obtained by setting all r redundants equal to zero. This structure, indicated in Fig. 8.16*c*, is called the *basic structure* corresponding to the redundants $X_1, X_2, \ldots, X_r$. By choosing this particular virtual force system, we obtain for the virtual moments and axial forces in any member m

$$\delta M_m = \bar{m}_m = m_{mi} \qquad \text{and} \qquad \delta N_m = \bar{n}_m = n_{mi} \tag{8.53}$$

where m_{mi} and n_{mi} are the functions defined in Eqs. (8.52). Although the virtual force system must satisfy only static conditions and is independent of the true deformation of the structure, it is often instructive to think of the basic determinate system as being obtained by *releasing* all of the redundants. In this case, the basic structure has r degrees of freedom more than the actual structure, since it is obtained by removing r constraints. At point i, for example, we imagine that a hypothetical cut is made so that side iA can displace relative to side iB, as indicated in Fig. 8.16*c*. The magnitudes of the redundants must be such that the relative displacement of opposite sides of the cut is zero.

Assuming that shear deformations are negligible and that material is linearly elastic, we find that the internal complementary virtual work done in member m due to the virtual force system corresponding to $\delta X_i = 1$ is

$$\delta U_m^* = \int_{L_m} \left(\frac{M_m m_{mi}}{EI_m} + \frac{N_m n_{mi}}{EA_m} \right) dx \tag{8.54}$$

where M_m and N_m are given by Eqs. (8.52). Thus, the total internal complementary virtual work due to $\delta X_i = 1$ is

$$\delta U^* = \sum_{m=1}^{n} \left[\int_{L_m} \left(\frac{M_m m_{mi}}{EI_m} + \frac{N_m n_{mi}}{EA_m} \right) dx \right] \tag{8.55}$$

where n is the total number of structural members. If N_m is constant throughout the length of each member,

$$\delta U^* = \sum_{m=1}^{n} \int_{L_m} \frac{M_m m_{mi}}{EI_m} dx + \sum_{m=1}^{n} N_m n_{mi} \lambda_m \tag{8.56}$$

where $\lambda_m = L_m/EA_m$.

We assume that the displacement of the point of application of the unit virtual force δX_i either is prescribed as zero or is not prescribed. Then, owing to our particular choice of virtual force systems, the virtual forces perform no external complementary virtual work. The redundant X_r in Fig. 8.16*b*, for example, is the reactive force at a support. Since the true displacement of the point is prescribed as zero, a unit virtual force $\delta X_r = 1$ is not displaced and, hence, performs no work. A similar situation exists in the case of an "internal" redundant, such as X_i in Fig. 8.16*b*. Since the virtual force system corresponding to X_i must be self-equilibrating, it is developed by the pair of equal and opposite unit forces on either side of cross section i of member AB. In the actual structure, the transverse displacement of member AB is continuous throughout the length of the member; no discontinuity occurs at point i, and sides iA and iB undergo identical displacements Δ_i at i. Thus, the opposing pair of unit forces must move through the same displacement and the net external virtual work must be zero. This is illustrated in Fig. 8.17.

It follows from the principle of virtual forces that the compatibility

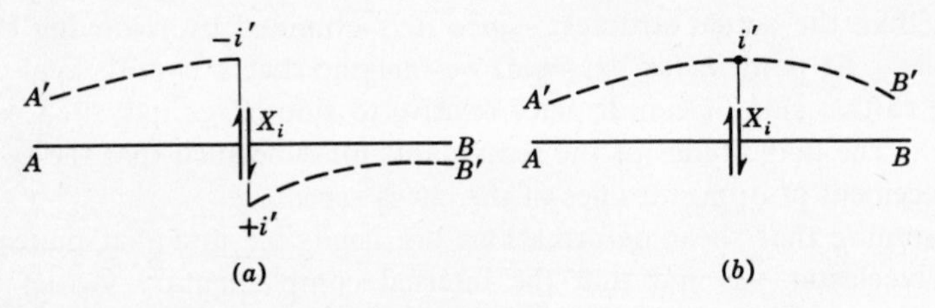

FIGURE 8.17 (*a*) Incompatible and (*b*) compatible displacements of opposite sides of a hypothetical cut at point i. In the first case, the net external complementary virtual work δW_e^* is not zero, whereas, in the case of compatible displacements, $\delta W_e^* = 0$.

condition for such structures is

$$\delta U^* = 0 \tag{8.57}$$

Thus, according to Eq. (8.56),

$$\sum_{m=1}^{n} \int_{L_m} \frac{M_m m_{mi}}{EI_m} dx + \sum_{m=1}^{n} N_m n_{mi} \lambda_m = 0 \tag{8.58}$$

Introducing Eqs. (8.52) into this condition and simplifying, we get[18]

$$f_{0i} + X_1 f_{i1} + X_2 f_{i2} + \cdots + X_i f_{ii} + \cdots + X_j f_{ij} + \cdots + X_r f_{ir} = 0 \tag{8.59}$$

where
$$f_{0i} = \sum_{m=1}^{n} \int_{L_m} \frac{M_{0m} m_{mi}}{EI_m} dx + \sum_{m=1}^{n} N_{0m} n_{mi} \lambda_m \tag{8.60a}$$

and
$$f_{ij} = \sum_{m=1}^{n} \int_{L_m} \frac{m_{mj} m_{mi}}{EI_m} dx + \sum_{m=1}^{n} n_{mj} n_{mi} \lambda_m \tag{8.60b}$$

The quantities $f_{0i}, f_{i1}, \ldots, f_{ir}$ are the *flexibilities* of the structure corresponding to the r redundants. Comparing Eqs. (8.60) with Eq. (8.51), we see that the flexibilities can be interpreted as displacements of point i produced by loads and by unit values of the redundants. The quantity f_{0i}, for example, is the displacement of point i of the statically determinate (basic) structure obtained by setting all of the redundants equal to zero. f_{ij} is the displacement of point i of the determinate structure in the direction of X_i due to a unit virtual force $\delta X_j = 1$. Thus, Eq. (8.59) is an *equation of consistent deformation;* it states that if compatibility is to exist, the accumulative effects of the applied loads and all of the redundants must be such that the relative displacement of either side of the cut at point i is zero.

We now repeat this procedure for each redundant, each time using the determinate structure as a basic system and applying a unit virtual force corresponding to the appropriate redundant. As a result, we generate a system of r linearly independent equations of consistent deformation of the form

$$\begin{aligned}
X_1 f_{11} + X_2 f_{12} + \cdots + X_i f_{1i} + \cdots + X_r f_{1r} + f_{01} &= 0 \\
X_1 f_{21} + X_2 f_{22} + \cdots + X_i f_{2i} + \cdots + X_r f_{2r} + f_{02} &= 0 \\
\cdots\cdots\cdots\cdots\cdots\cdots\cdots\cdots\cdots\cdots\cdots\cdots \\
X_1 f_{i1} + X_2 f_{i2} + \cdots + X_i f_{ii} + \cdots + X_r f_{ir} + f_{0i} &= 0 \\
\cdots\cdots\cdots\cdots\cdots\cdots\cdots\cdots\cdots\cdots\cdots\cdots \\
X_1 f_{r1} + X_2 f_{r2} + \cdots + X_i f_{ri} + \cdots + X_r f_{rr} + f_{0r} &= 0
\end{aligned} \tag{8.61}$$

[18] A slightly different approach is used to obtain this equation in Chap. 9. See Art. 9.18.

Once these equations are solved for r unknowns $X_1, X_2, \ldots, X_r$, the results are substituted into Eqs. (8.52) to obtain final bending moments and axial forces.

As a simple example of this procedure, consider the square framework shown in Fig. 8.18*a*. We assume that all of the members have the same

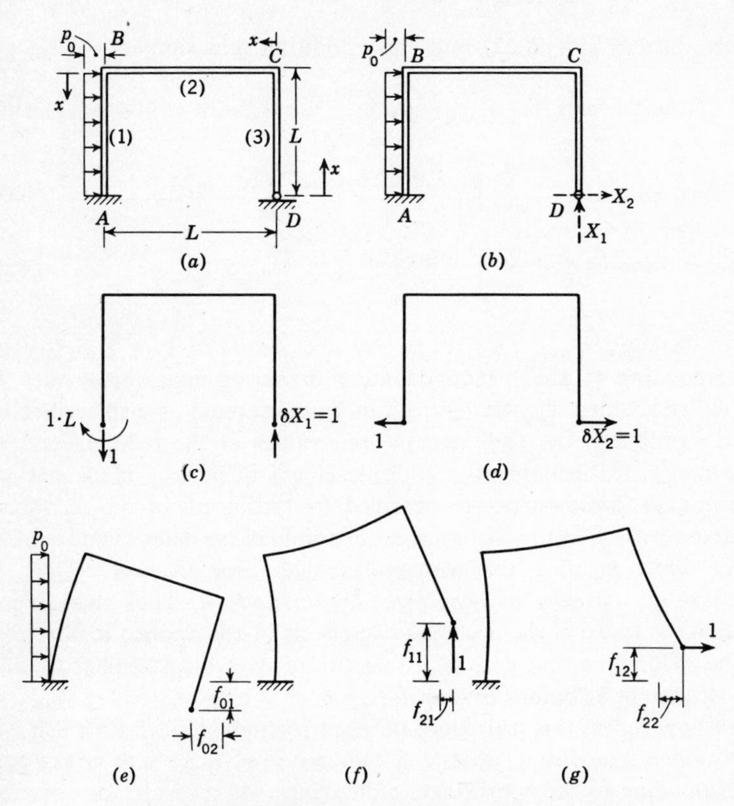

FIGURE 8.18 (*a*) Statically indeterminate frame; (*b*) redundant forces X_1 and X_2; (*c, d*) virtual force systems corresponding to unit virtual forces at the points of application of the redundants; (*e–g*) physical interpretations of the flexibilities corresponding to X_1 and X_2.

dimensions and are linearly elastic. For identification purposes, we label the members 1, 2, and 3, as indicated in the figure. The origin of the local coordinate x is also indicated. This structure is twice statically indeterminate, and we select as the redundants the horizontal and vertical reactive forces at D (Fig. 8.18*b*). Then following the procedure discussed previously, we use for this choice of redundants the virtual force systems indicated in Fig. 8.18*c* and *d*. Thus, in this case the basic system is a bent cantilevered bar fixed at point A and free at point D. Then, from the statics of the basic system, we find:

Member 1		Member 2	
$M_{01} = -p_0 \frac{x^2}{2}$	$N_{01} = 0$	$M_{02} = 0$	$N_{02} = 0$
$m_{11} = L$	$n_{11} = 1$	$m_{21} = x$	$n_{21} = 0$
$m_{12} = L - x$	$n_{12} = 0$	$m_{22} = L$	$n_{22} = 1$

(l)

Member 3

$$M_{03} = 0 \qquad N_{03} = 0$$
$$m_{31} = 0 \qquad n_{31} = -1 \tag{m}$$
$$m_{32} = x \qquad n_{32} = 0$$

Introducing these functions into Eqs. (8.60), we get

$$f_{01} = -\int_0^L \frac{p_0 x^2 L}{2EI}\,dx = -\frac{p_0 L^4}{6EI}$$
$$f_{02} = -\int_0^L \frac{p_0 x^2 (L - x)}{2EI}\,dx = -\frac{p_0 L^4}{24EI} \tag{n}$$

$$f_{11} = \left(\int_0^L \frac{L^2}{EI}\,dx + \frac{L}{AE}\right) + \int_0^L \frac{x^2}{EI}\,dx + \frac{L}{AE} = \frac{4L^3}{3EI} + \frac{2L}{AE}$$
$$f_{12} = f_{21} = \int_0^L \frac{L(L - x)}{EI}\,dx + \int_0^L \frac{Lx}{EI}\,dx = \frac{L^3}{EI} \tag{o}$$
$$f_{22} = \int_0^L \frac{(L - x)^2}{EI}\,dx + \left(\int_0^L \frac{L^2}{EI}\,dx + \frac{L}{AE}\right) + \int_0^L \frac{x^2}{EI}\,dx = \frac{5L^3}{3EI} + \frac{L}{AE}$$

Physical interpretations of these quantities are indicated in Fig. 8.18*e* to *g*.

We now obtain the equations of consistent deformation by introducing Eqs. (*n*) and (*o*) into Eqs. (8.61):

$$\left(\frac{4L^3}{3EI} + \frac{2L}{AE}\right)X_1 + \frac{L^3}{EI}X_2 - \frac{p_0 L^4}{6EI} = 0$$
$$\frac{L^3}{EI}X_1 + \left(\frac{5L^3}{3EI} + \frac{L}{AE}\right)X_2 - \frac{p_0 L^4}{24EI} = 0 \tag{p}$$

Taking, for illustration purposes, $L^2 A/I = 300$, we find

$$X_1 = 0.1912 p_0 L \qquad \text{and} \qquad X_2 = -0.0895 p_0 L \tag{q}$$

The final bending moments and axial forces in each member are

$$M_m = M_{0m} + 0.1912 p_0 L m_{m1} - 0.0895 p_0 L m_{m2}$$
$$N_m = N_{0m} + 0.1912 p_0 L n_{m1} - 0.0895 p_0 L n_{m2} \tag{r}$$

where M_{0m}, N_{0m}, m_{m1}, m_{m2}, n_{m1}, and n_{m2} are given in Eqs. (*l*) and (*m*) for the appropriate member.

PROBLEMS

8.1. The two rigid bars shown are connected by frictionless hinges. Each bar is of length a and weight Q. Using the principle of virtual displacements, determine the values of the angles α_1 and α_2 corresponding to equilibrium configurations.

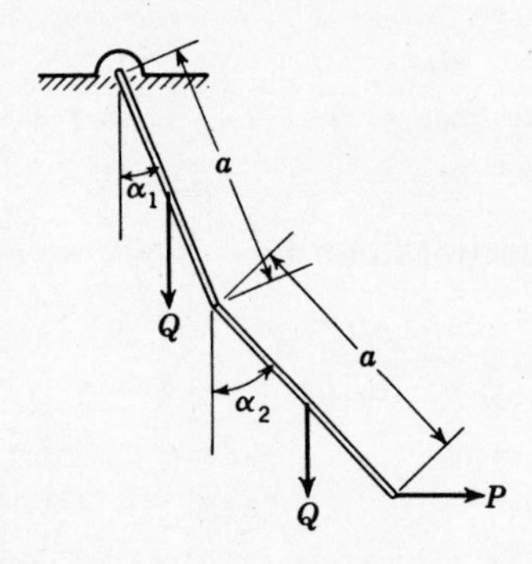

8.2. A rigid bar of length a and weight Q leans against frictionless walls, as shown. Use virtual displacements to compute the angle α for which the bar will be in equilibrium and the reactive forces at the ends of the bar.

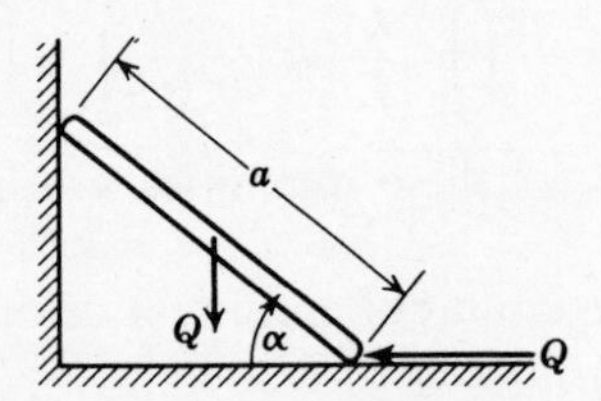

8.3–8.6. Determine the equilibrium configurations of the rigid bars shown below using the principle of virtual displacements. In each case, neglect friction and assume that each bar is of weight Q.

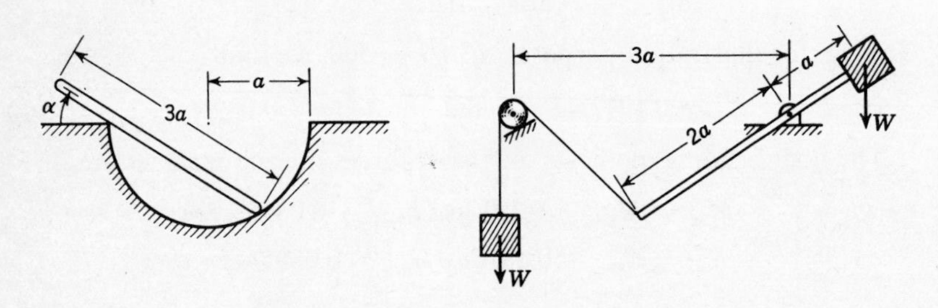

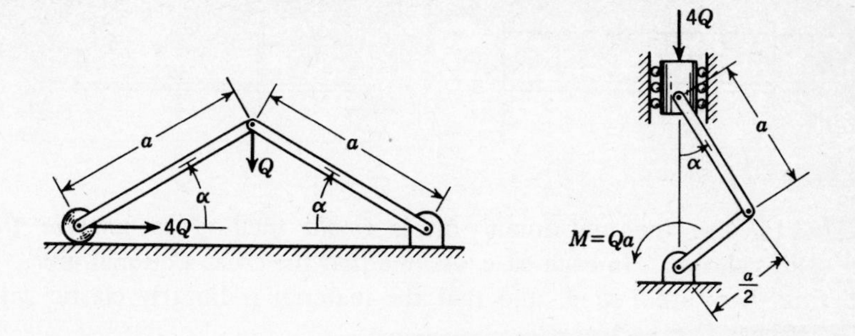

8.7. The rigid bar shown is suspended by three linearly elastic wires of equal cross-sectional area. If $E = 10^7$ psi and each wire is 0.1 sq in. in area, determine the force in each wire and the inclination of the bar at the equilibrium configuration.

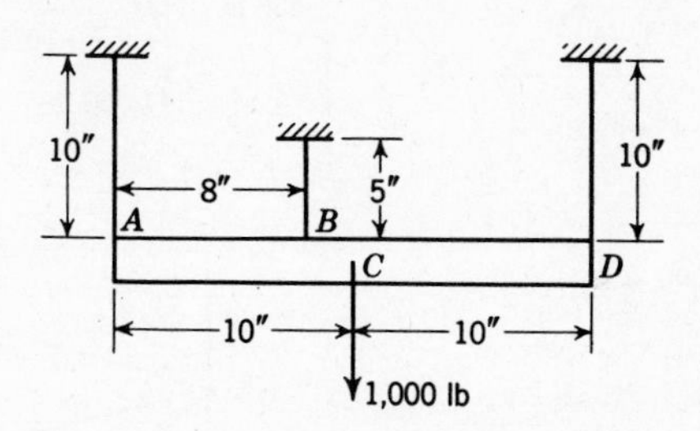

8.8. The rigid bar shown is supported by three elastic wires of cross-sectional area 0.05 sq in. For each wire, $E = 30 \times 10^6$ psi. Determine the stresses in each wire and the deflection of point A at the equilibrium configuration.

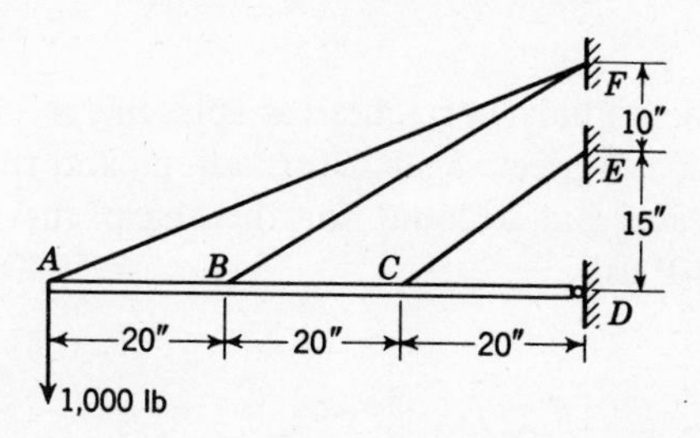

8.9–8.10. By assuming appropriate virtual displacement functions, use the principle of virtual displacements to obtain equilibrium conditions for the structures shown. Neglect shear deformation.

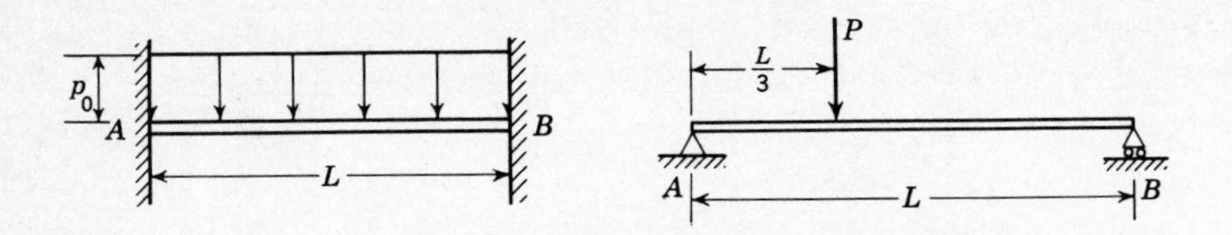

8.11–8.14. Use the unit-dummy-displacement method to analyze the structures shown. In each case, assume that the cross-sectional areas of all truss bars are 1 sq in. and that the material is linearly elastic with $E = 10^7$ psi.

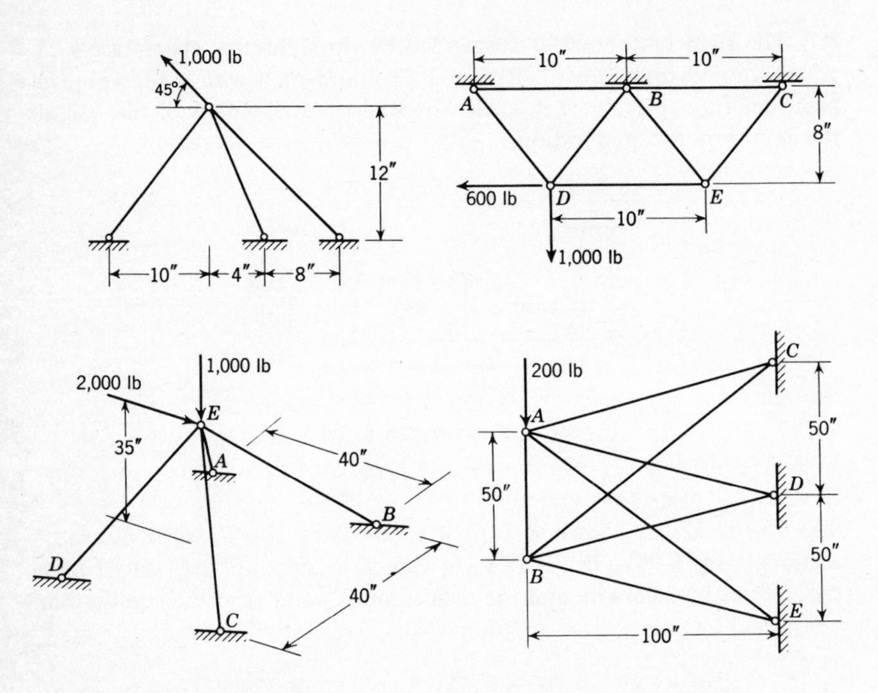

8.15. Compute the radial (horizontal) displacement of point A of the curved bar shown. Neglect shear deformations and the effects of transverse normal stresses but account for the sharp curvature of the bar. Take $E = 30 \times 10^6$ psi.

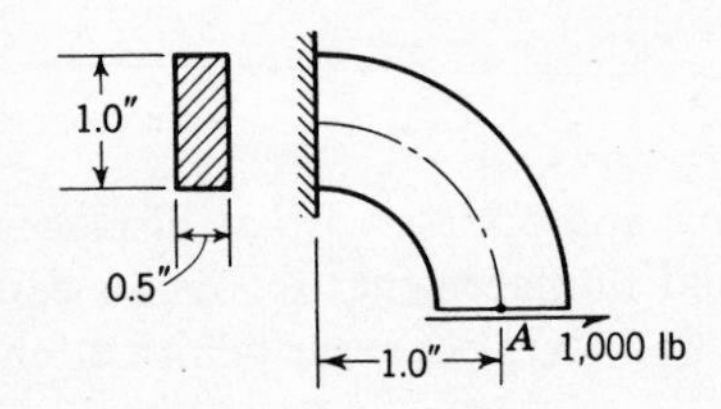

8.16. Use the principle of virtual forces to obtain compatibility conditions involving (*a*) Δ_0 and θ_0, (*b*) Δ_0 and Δ_1, for the beam shown.

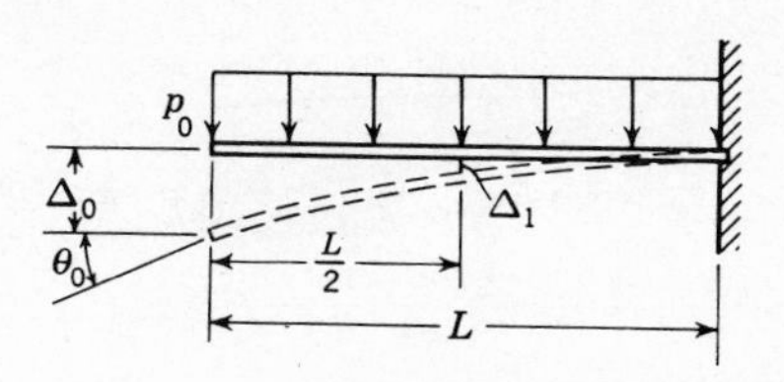

8.17. Compute the horizontal displacement of point *A* of the bar shown. Neglect shear deformation.

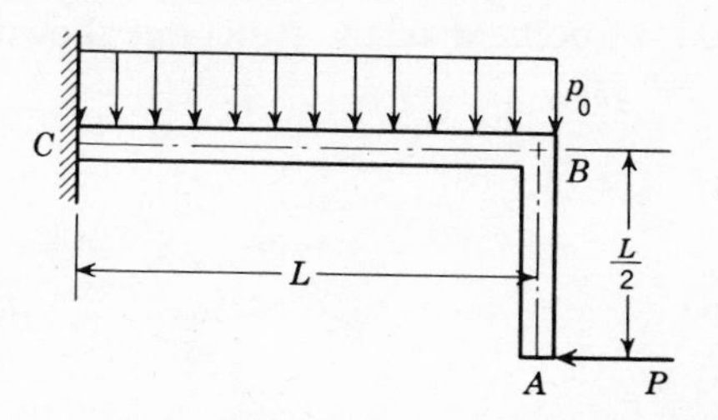

8.18–8.20. Use the unit-dummy-load method to compute the horizontal and vertical displacements of points *A* of the structures shown. All members are linearly elastic and the truss members have equal cross-sectional areas. Neglect shear deformations.

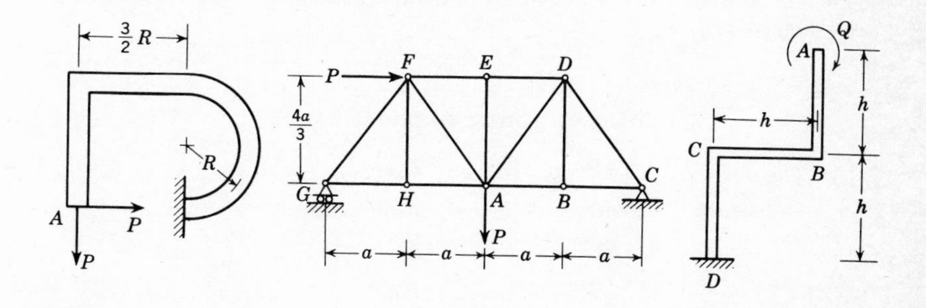

8.21. Show that if shear deformations and the effects of restrained warping are negligible, the displacement in the *n* direction of point *A* of the three-dimensional bar shown is

$$\Delta_{An} = \int_A^B \left(\frac{N_x n_x}{EA} + \frac{M_y m_y}{EI_y} + \frac{M_z m_z}{EI_z} + \frac{Tt}{GJ} \right) dx$$

where n_x, m_y, m_z, and t are the virtual stress resultants (axial force, bending moments, and twisting moments) produced by a unit virtual force at A in the direction n.

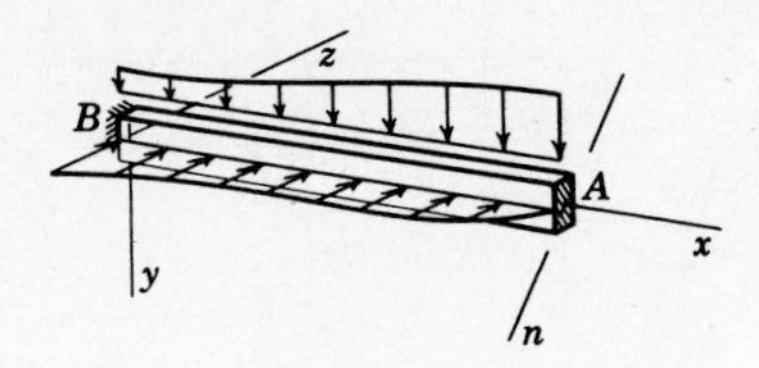

8.22–8.23. Use the results of Prob. 8.21 to determine (*a*) the displacement normal to the plane of the structure and (*b*) the angular rotation (slope) θ_A about the x axis at point A of the structures shown. Take $E = 10^7$ psi and $G = 4 \times 10^6$ psi.

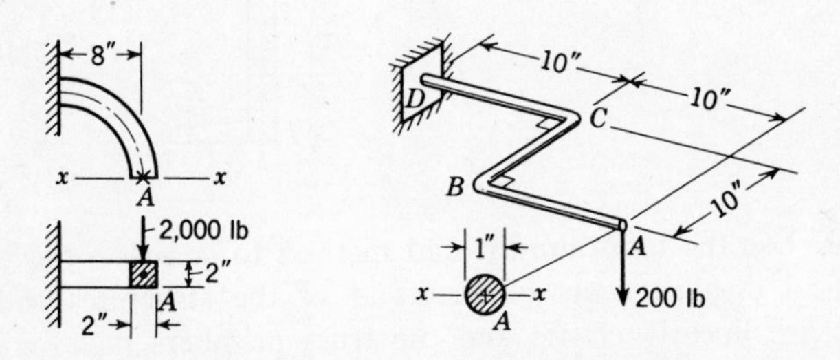

8.24. The stress-strain relations for various members of the truss shown are listed as follows:

Member	Stress-strain relation, σ
AB	$E_1\epsilon^{\frac{1}{2}}$
AC	$E\epsilon$
BC	$E\epsilon$
CD	$E_2\epsilon^{\frac{1}{3}}$
DA	$E\epsilon$

where E, E_1, and E_2 are material constants. All members of the truss have the same cross-sectional area A. Determine the vertical displacement of point C.

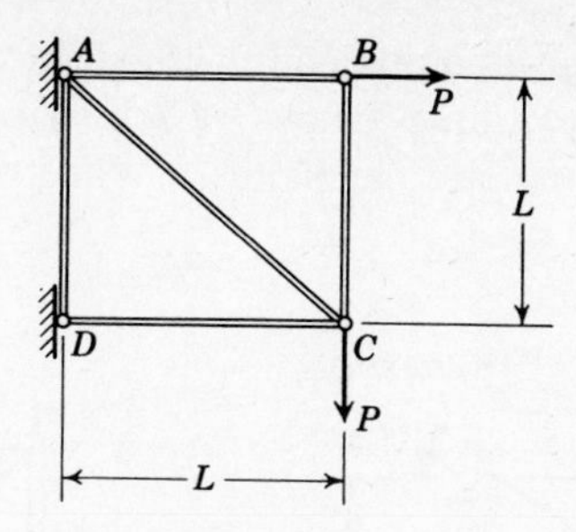

8.25. The stress-strain diagram given below corresponds to the material used to construct the truss shown. It is assumed that this diagram indicates the stress-strain relation for both tension and compression. All members of the truss have the same cross-sectional area $A = 1$ sq in. Use the unit-dummy-load method to compute the vertical displacement of point C.

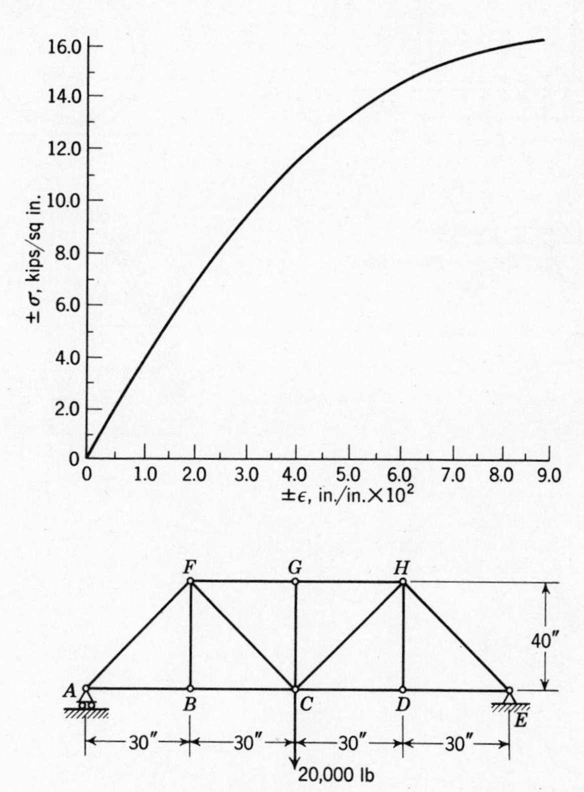

8.26–8.31. Use the unit-dummy-load method to analyze the statically indeterminate structures shown below. Each structure is linearly elastic and shear deformations are negligible. When numerical values are required, take $E = 10^7$ psi.

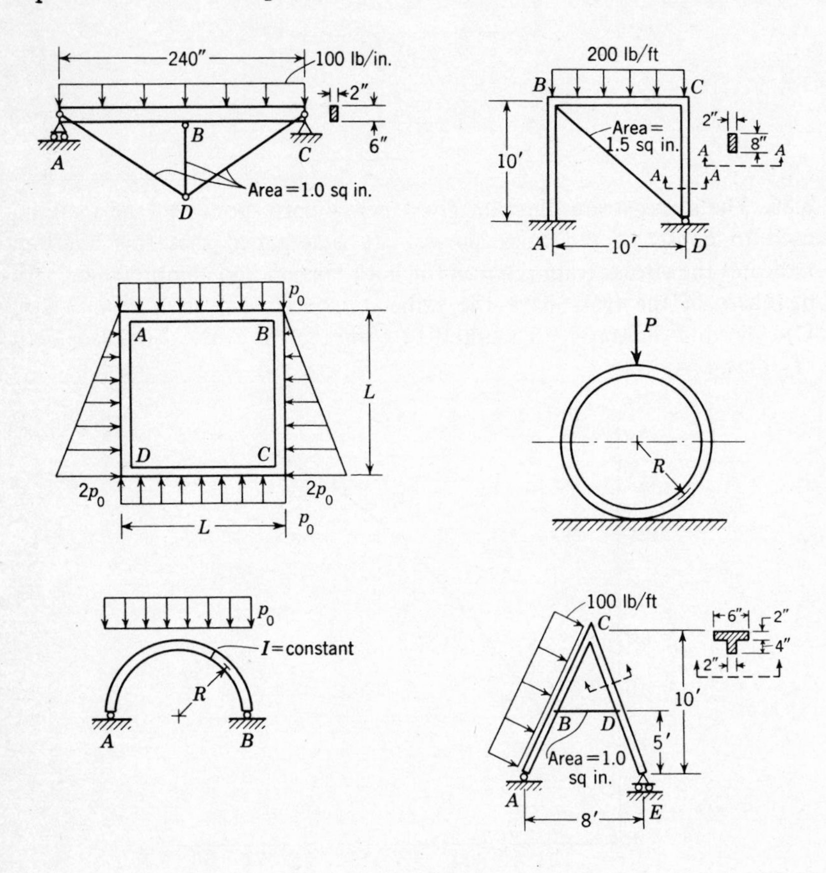

nine

THE ENERGY PRINCIPLES

9.1 Introduction. The frequent reference to work in the preceding chapter suggests the importance of a closely related concept, *energy*. In every type of system, even that of a single atom, forces are present to which may be associated a "capacity" to displace and, thereby, a capacity to perform work. It is this capacity that is indicative of the "energy possessed" by the system. Thus, energy, in simple terms, is the *capacity* to do work; it is manifested in all physical systems in virtually every way that we can conceive of such a capacity—in fact, it may be considered as a quantity inherent in matter itself.

Energy can display itself in a variety of forms depending upon the nature of the forces existing in the system. For example, a system may have *kinetic* energy (from the Greek *kinētikos*, meaning *to move*) by virtue of its motion—a volume of gas has kinetic energy because forces exist which impart to its molecules a continuous motion; a single particle of constant mass m moving through space at some continually changing velocity v has kinetic energy because, according to Newton, an inertia force, $F = m\, dv/dt$, is present which has a capacity to do work. An electric system possesses electric energy because of the work capacity of the attractive and repulsive forces which can propel electrons through the circuit and create current. Similarly, we may speak of chemical energy, thermal energy (heat), acoustical energy, etc., each term referring to a certain capacity to unleash forces and perform work.

Though related through definition, work and energy are distinctly different concepts. Forces in a system may "perform" work, but the system "possesses" energy. Work is done only when a form of energy is changed; this has been referred to as energy in transition.[1] A charge of dynamite, for example, "possesses" chemical energy; but this energy is used to perform work only by converting it to another form—say to heat and to kinetic energy when the charge is detonated. A body of weight Q located a distance h above some datum is said to have *gravitational potential energy* of magnitude Qh because of the capacity of Q, the gravity force, to do work by virtue of its location with reference to the plane. Again, work is done when this energy is changed, for example, by moving the body closer to the plane.

It follows that in order to evaluate the amount of work done in a physical process we need know only the *change* in energy; therefore, since only changes are involved, *the reference or datum with respect to which we measure energy is completely arbitrary.* Furthermore, we have also implied in the foregoing discussion the validity of the *law of the conservation of energy*, which states that *energy can be neither created nor destroyed but it can be transformed from one form to another.* This is a natural law based on physical observations; other than possibly the conversion of mass to energy and energy to mass in atomic processes, no exceptions to this law have been experienced.

Consider, for example, the work done by the force $F = m\,dv/dt$ acting on a particle of constant mass m and velocity v as it moves from point 1 to another point 2 along its path. The work done, by definition [see Eq. (8.1)], is

$$W = \int_1^2 m \frac{dv}{dt}\,dx = m\int_1^2 v\,dv = \tfrac{1}{2}mv_2^2 - \tfrac{1}{2}mv_1^2 \tag{9.1}$$

where v_1 and v_2 are the velocities of the particle at points 1 and 2. We may interpret this as a change from point 1 to 2 of the quantity $\frac{1}{2}mv^2$, which, accordingly, must represent the kinetic energy of the particle. Thus the kinetic energy T of the particle at point 1 was $\frac{1}{2}mv_1^2$, while that at point 2 was $\frac{1}{2}mv_2^2$. The magnitude of the change is obviously independent of the datum used to define the location of the particle. Since energy is conserved, this difference in kinetic energy must be transformed into some other form; some may be converted into heat, and, if the location of the particle with respect to an arbitrary datum has changed, some may be converted into a change in the gravitational potential energy of the particle.

[1] Ref. 20.

Now the work done in bringing a system from configuration A to a configuration B is, by definition,

$$W = \int_A^B F_s \, ds$$

where F_s is the net force tangent to the path traveled from A to B. If the system is conservative,[2] the work done is independent of the path from A to B. Further, we recall from Art. 2.6[3] that if the integral is independent of the path, the quantity $F_s \, ds$ is an *exact differential* of some function Π. Thus, for conservative systems,

$$W = \int_A^B F_s \, ds = \int_A^B d\Pi = \Pi_B - \Pi_A = -\Delta\Pi \qquad (9.2)$$

where $\Delta\Pi$ is the change in Π from A to B.

The function Π is called the *potential function*, or, in the present context, the *potential energy* of the system. Physically, the potential energy is the capacity of a conservative system to perform work by virtue of its configuration with respect to an arbitrary datum. Equation (9.2) shows that a change in potential energy is the negative of the work done. For example, the rigid body of weight Q mentioned previously has potential energy Qh with reference to a plane a distance h below it. If the body is dropped so that it comes in contact with the plane, an amount of work Qh is done; the potential energy is reduced to zero and there is a *change* in potential energy of $-Qh$, which is converted into a *change* in kinetic energy, ΔT. It follows from Eqs. (9.1) and (9.2) that

$$-\Delta\Pi = \Delta T \qquad \text{or} \qquad \Delta(T + \Pi) = 0$$

which leads us to the familiar "law of kinetic energy" from elementary dynamics:

$$T + \Pi = \text{constant} \qquad (9.3)$$

9.2 Potential energy in structural systems. In the remainder of this chapter, we are concerned with the use of certain forms of energy to study the static behavior of structures. To clarify the principles involved, we consider only those forms of energy which ordinarily are significant in structural behavior. Thus, unless noted otherwise, we assume that on the application of loads, elements of a structure are displaced infinitesimally slowly and adiabatically from their undeformed configuration to their

[2] Again, we disregard dissipative forces such as those due to friction. In general, a system is conservative if the work done in moving the system around any closed path is zero. This implies that the work done in moving the system from A to B is independent of the path taken from A to B.

[3] See Eqs. (2.19) to (2.22).

deformed equilibrium configuration. We neglect, in other words, the small change in kinetic energy and the small change in thermal energy which occur in the deformation process. We are naturally concerned with the forms of energy indicative of the nature of the forces which possess the capacity to do work. For structural systems, we classified these forces as being either external, such as loads, or internal, as represented by stresses and stress resultants. Thus, within the limits of our initial assumptions, the total energy of a deformable system is the sum of the potential energy of the external and internal forces.

Let us first examine the potential energy of external forces which act on a three-dimensional deformable body. To this end, we begin by establishing as a reference the undeformed configuration 0 of the body so that initially the external forces have zero potential energy. As the body slowly deforms, external forces move through true displacements and perform "external" work until the final configuration F is reached (see Fig. 9.1). Again, we assume that the work done by the external forces

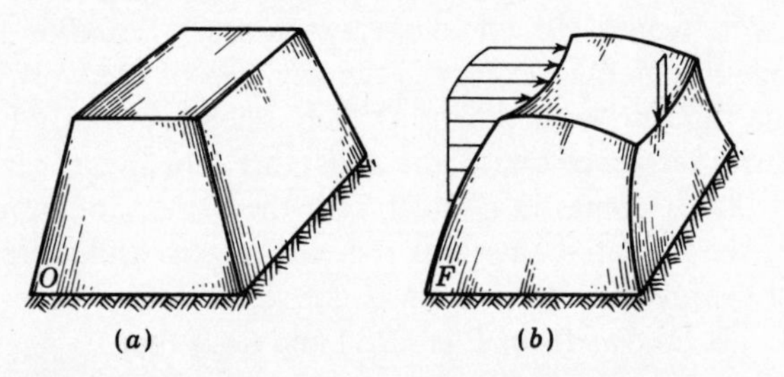

FIGURE 9.1 (*a*) Three-dimensional body in its initial (undeformed) configuration O; (*b*) the same body in its final configuration F after it has deformed under the action of a general system of external forces.

in deforming the body is independent of the "path" taken from configuration 0 to F. Thus, if W_e is the total work done by the external forces, dW_e is the exact differential of a potential function H, which we refer to as the *potential energy of the external forces*. The change in H from configuration 0 to F is $-W_e$. Thus, from the definition of work, it follows that

$$H = -\iiint_V \left[\int_0^F (X_b\,du + Y_b\,dv + Z_b\,dw)\right] dV - \iint_{S_1} \left[\int_0^F (X_s\,du + Y_s\,dv + Z_s\,dw)\right] dS \quad (9.4)$$

where the integrals inside the brackets are intended to signify that the integration is to be carried from the initial state to the final deformed state of the body and S_1 is the portion of the surface area on which the surface forces are prescribed. The external forces are usually independent of the displacements, so that

$$H = -\iiint_V (X_b u + Y_b v + Z_b w)\, dV - \iint_{S_1} (X_s u + Y_s v + Z_s w)\, dS \quad (9.5)$$

where u, v, and w represent the final displacements of points in the system.

Frequently, the body forces are negligible and the surface forces are represented by a system of concentrated forces (and moments) $P_1, P_2, \ldots, P_n$. If $\Delta_1, \Delta_2, \ldots, \Delta_n$ are the displacements (and rotations) corresponding to these forces, in this case Eq. (9.5) becomes

$$H = -(P_1\Delta_1 + P_2\Delta_2 + \cdots + P_n\Delta_n) \quad (9.6)$$

Here the forces P_i are regarded as being independent of the displacements Δ_i. Clearly, Eq. (9.6) is a special case of Eqs. (9.5) and (9.4).

The internal forces developed in a deformable body also possess a capacity to perform work. As a body deforms, stresses are developed which result in internal forces. These forces perform work in moving through internal displacements until the final configuration is reached. If the strained body were allowed to slowly return to its unstrained state, it would be capable of returning at least a portion of the work done by the external forces. This capacity of the internal forces to perform work by virtue of the strained state of the body is called *strain energy*.

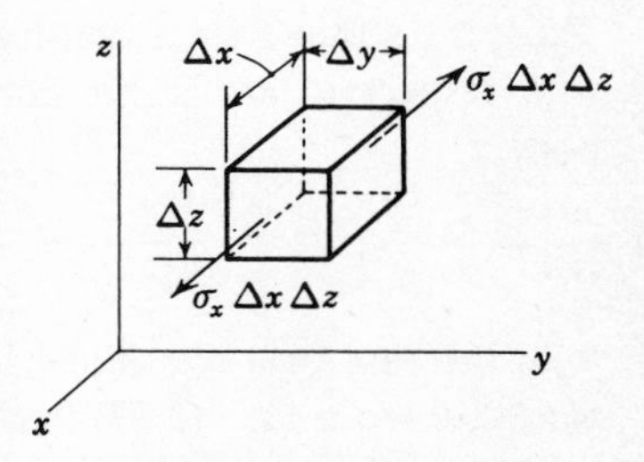

FIGURE 9.2 Internal forces in three-dimensional solid.

To evaluate strain energy, we need only compute the work done by the internal forces during the deformation process. Consider, for example, the work done by the internal force $\sigma_x\, \Delta x\, \Delta y$ resulting from the normal stress component[4] σ_x on an element in a three-dimensional solid (Fig. 9.2). The displacement of the x face of the element is, by definition, $\Delta x\, \epsilon_x$, while that of the face $x + \Delta x$ is $\Delta x\,(\epsilon_x + d\epsilon_x)$. If the volume of the element approaches zero, the differential work done due to the increment in strain is $\sigma_x\, d\epsilon_x\, dV$, where dV is the differential volume. Similarly, the internal force corresponding to the component of shearing stress τ_{xy} performs an amount of work $\tau_{xy}\, d\gamma_{xy}\, dV$ due to an increment in shearing strain, and so forth for the remaining components. If follows that the

[4] It is not necessary to compute the work done by the variation σ_x from x to $x + \Delta x$. The total work done by these variations is zero. See Ref. 72, p. 37.

total strain energy developed in the element as the strains increase from zero to their final values is

$$\left(\int_0^{\epsilon_x} \sigma_x \, d\epsilon_x + \int_0^{\epsilon_y} \sigma_y \, d\epsilon_y + \int_0^{\epsilon_z} \sigma_z \, d\epsilon_z \right.$$
$$\left. + \int_0^{\gamma_{xy}} \tau_{xy} \, d\gamma_{xy} + \int_0^{\gamma_{yz}} \tau_{yz} \, d\gamma_{xz} + \int_0^{\gamma_{xz}} \tau_{xz} \, d\gamma_{xz} \right) dV$$

or symbolically,

$$\left[\int_0^F (\sigma_x \, d\epsilon_x + \sigma_y \, d\epsilon_y + \sigma_z \, d\epsilon_z + \tau_{xy} \, d\gamma_{xy} + \tau_{yz} \, d\gamma_{yz} + \tau_{xz} \, d\gamma_{xz})\right] dV \tag{9.7}$$

in which it is understood that the integration is to be carried out from the initial to the final state of strain of the body. Since the integral in Eq. (9.7) is independent of the path from 0 to F, we conclude that the differential work dW_i done by the internal forces acting on a volume element is the exact differential of a potential function U_0. U_0 is called the *strain energy density*, and in view of Eq. (9.7),

$$dU_0 = \sigma_x \, d\epsilon_x + \sigma_y \, d\epsilon_y + \sigma_z \, d\epsilon_z + \tau_{xy} \, d\gamma_{xy} + \tau_{yz} \, d\gamma_{yz} + \tau_{xz} \, d\gamma_{xz} \tag{9.8}$$

Since dU_0 is an exact differential,

$$dU_0 = \frac{\partial U_0}{\partial \epsilon_x} d\epsilon_x + \frac{\partial U_0}{\partial \epsilon_y} d\epsilon_y + \cdots + \frac{\partial U_0}{\partial \gamma_{xz}} d\gamma_{xz} \tag{9.9}$$

Thus, the strain energy density has the property that its partial derivative with respect to any strain component is the corresponding stress component:

$$\sigma_x = \frac{\partial U_0}{\partial \epsilon_x} \qquad \sigma_y = \frac{\partial U_0}{\partial \epsilon_y} \qquad \cdots \qquad \tau_{xz} = \frac{\partial U_0}{\partial \gamma_{xz}} \tag{9.10}$$

In the case of a two-dimensional isotropic body of Hookean material, we recall from Eqs. (2.27) that

$$\begin{aligned} \epsilon_x &= \frac{\sigma_x}{E} - \nu \frac{\sigma_y}{E} \\ \epsilon_y &= \frac{\sigma_y}{E} - \nu \frac{\sigma_x}{E} \\ \gamma_{xy} &= \frac{2(1+\nu)}{E} \tau_{xy} = \frac{\tau_{xy}}{G} \end{aligned} \tag{9.11}$$

According to Eq. (9.8), for such materials

$$dU_0 = \sigma_x\left(\frac{d\sigma_x}{E} - \nu \frac{d\sigma_y}{E}\right) + \sigma_y\left(\frac{d\sigma_y}{E} - \nu \frac{d\sigma_x}{E}\right) + \tau_{xy} \frac{d\tau_{xy}}{G} \tag{9.12}$$

Thus $$U_0 = \frac{1}{2E}(\sigma_x^2 + \sigma_y^2) - \frac{\nu}{E}\sigma_x\sigma_y + \frac{1}{2G}\tau_{xy}^2 \tag{9.13}$$

or, in view of Eqs. (9.11),

$$U_0 = \tfrac{1}{2}(\sigma_x\epsilon_x + \sigma_y\epsilon_y + \tau_{xy}\gamma_{xy}) \tag{9.14}$$

In the case of a three-dimensional linearly elastic isotropic body,

$$U_0 = \frac{1}{2E}(\sigma_x^2 + \sigma_y^2 + \sigma_y^2) - \frac{2\nu}{E}(\sigma_x\sigma_y + \sigma_y\sigma_z + \sigma_x\sigma_z) + \frac{1+\nu}{E}(\tau_{xy}^2 + \tau_{yz}^2 + \tau_{xz}^2) \tag{9.15a}$$

or $$U_0 = \tfrac{1}{2}(\sigma_x\epsilon_x + \sigma_y\epsilon_y + \sigma_z\epsilon_z + \tau_{xy}\gamma_{xy} + \tau_{yz}\gamma_{yz} + \tau_{xz}\gamma_{xz}) \tag{9.15b}$$

A strain energy function U_0 does not exist for all types of materials because the deformation is not, in general, independent of the path. It does exist in the case of Hookean materials since we were able to integrate Eq. (9.12) to obtain Eq. (9.13); but even this result is valid only for small strains. Often the reverse procedure is used in solid mechanics; that is, the strain energy density function for a material is given (or determined experimentally) in terms of the strains, and the stress-strain relations for the material are then derived using Eqs. (9.10).

Once U_0 is known as a function of the strains and the coordinates, we obtain the total strain energy U by simply integrating $U_0\,dV$ throughout the volume of the body:

$$U = \iiint_V U_0\,dV \tag{9.16}$$

Finally, the total potential energy Π is defined as the sum of the external potential energy and the strain energy:

$$\Pi = H + U \tag{9.17}$$

9.3 Complementary energy. Other quantities with the dimensions of work are associated with the deformation process. Consider, for example, the two identical prismatic bars shown in Fig. 9.3. The end displacement of the bar in Fig. 9.3*a* is prescribed to be of magnitude Δ, and a reactive force R is developed as a consequence. The end force of the bar in Fig. 9.3*b* is prescribed to be of magnitude P. If $P = R$, this bar elongates an amount Δ. Thus, the final configuration, stress distribution, and overall behavior of each bar are precisely the same, the only difference being the manner in which we interpreted this behavior to be stimulated. It appears that external work was performed in both cases.

According to the definition presented in the previous article, however, the potential energy of the external forces acting on the bar in Fig. 9.3*a*

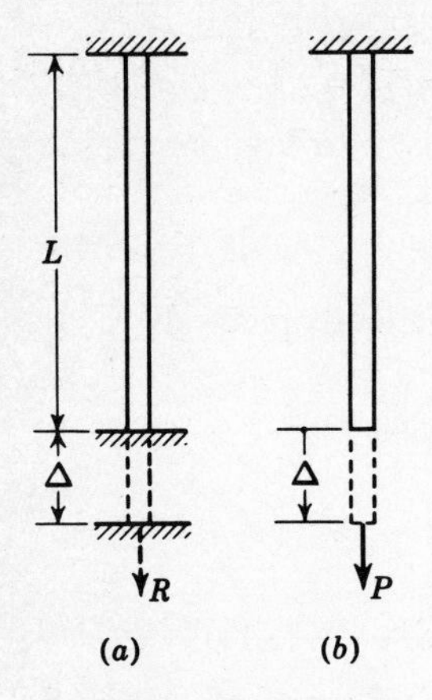

FIGURE 9.3 Bars with prescribed end displacements (*a*) and prescribed end forces (*b*).

is zero, since on no portion S_1 of the surface area are any nonzero external forces prescribed. Furthermore, P was possibly constant throughout the deformation of the bar in Fig. 9.3*b*, while for the bar in Fig. 9.3*a*, R was initially zero and reached its final magnitude only after the total prescribed displacement was realized.

These differences in the interpretation of the bar's behavior suggest the importance of a second type of external potential "energy" called the complementary external energy, which we will denote H^*. If u is the axial displacement of the bar in Fig. 9.3*b*, the bar in general possesses external potential energy

$$H = -\int_0^{\Delta} P\, du \tag{9.18}$$

which may be interpreted as minus the area under the P versus u curve in Fig. 9.4*a*. The complement of this area is (minus) the complementary external potential energy of the bar in Fig. 9.3*a*:

$$H^* = -\int_0^{P} \Delta\, dR \tag{9.19}$$

To interpret H^* in another way, we define

$$W_e^* = \int_0^{P} \Delta\, dR \tag{9.20}$$

as the external complementary work. Then, if W_e^* is independent of the "path," $\Delta\, dR$ is the exact differential of a function H^* which we call the external complementary energy. Note that, in general, neither H nor H^* depends upon the material properties of the bar.

We can easily generalize these ideas. For example, in the case of a three-dimensional body, we define

$$H^* = \iiint_V \left[\int_0^F (u\, dX_b + v\, dY_b + w\, dZ_b)\right] dV - \iint_{S_2} \left[\int_0^F (u\, dX_s + v\, dY_s + w\, dZ_s)\right] dS \tag{9.21}$$

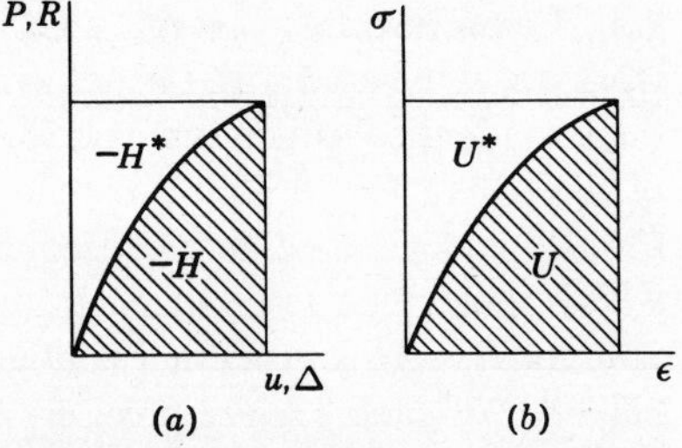

FIGURE 9.4 (*a*) External potential and complementary potential energy; (*b*) strain and complementary strain energy in an axially stressed straight bar.

where the integrals inside the brackets signify that the integration is to be carried from the initial to the final stressed state of the body and S_2 is the portion of the surface area on which the displacements are prescribed.

In general, the volume integral in Eq. (9.21) is zero, since the body forces are usually prescribed. Also, the prescribed displacements are rarely given as functions of the reactive forces. Thus, in most cases,

$$H^* = -\iint_{S_2} (uX_s + vY_s + wZ_s)\, dS \tag{9.22}$$

Similarly, if displacements $\Delta_1, \Delta_2, \ldots, \Delta_n$ are prescribed at certain points on the boundary and if, as a consequence, corresponding reactive forces $R_1, R_2, \ldots, R_n$ are developed,

$$H^* = -(R_1\Delta_1 + R_2\Delta_2 + \cdots + R_n\Delta_n) \tag{9.23}$$

Here the displacements Δ_i are regarded as being independent of the forces R_i. Clearly, Eq. (9.23) is a special case of Eqs. (9.22) and (9.21).

The strain energy U stored in the bar of Fig. 9.3*b* may be interpreted as the area under the stress-strain curve in Fig. 9.4*b*:

$$U = \int_0^{\epsilon^p} \sigma\, d\epsilon \tag{9.24}$$

where ϵ^p is the final strain due to P. In analogy with H^*, we refer to the complement of this area as U^*, the *complementary strain energy:*

$$U^* = \int_0^{\sigma^\Delta} \epsilon\, d\sigma = \epsilon^p \sigma^\Delta - U \tag{9.25}$$

where σ^Δ is the final stress due to Δ. To interpret U^* in another way, we define

$$W_i^* = \int_0^{\sigma^\Delta} \epsilon\, d\sigma \tag{9.26}$$

as the internal complementary work. Then, if W_i^* is independent of the path, $\epsilon\, d\sigma$ is the exact differential of a function U^* which we call the complementary strain energy.

Since, physically, an increment in complementary strain energy results from an increment in stress $d\sigma$ "moving through" its corresponding strain,[5]

[5] When we dismiss our assumption of an infinitesimally slow and adiabatic deformation process and consider the actual nonconservative system, U^* acquires a more definite physical significance. During the deformation of the bar in Fig. 9.3*b*, the work done by the external force P is converted into strain energy, heat, and kinetic energy; the system vibrates momentarily until internal dampening brings it to its final equilibrium configuration. If we define the resulting internal energy as the strain energy stored in the bar, we see that it differs from the change in the "potential energy" of the external loads by an amount equal to U^*. Hence, in this case, U^* physically represents the energy lost (dissipated) in the deformation process.

we can easily extend the definition to the case of the three-dimensional body. Thus, in analogy with U_0, we denote by U_0^* the *complementary strain energy density* (or the *complementary strain energy per unit volume*) and define it as follows:

$$U_0^* = \int_0^F (\epsilon_x \, d\sigma_x + \epsilon_y \, d\sigma_y + \epsilon_z \, d\sigma_z + \gamma_{xy} \, d\tau_{xy} + \gamma_{yz} \, d\tau_{yz} + \gamma_{xz} \, d\tau_{xz}) \tag{9.27}$$

in which it is understood that the integration is to be carried out from the initial state to the final state of stress of the body.

Since

$$dU_0^* = \frac{\partial U_0^*}{\partial \sigma_x} d\sigma_x + \frac{\partial U_0^*}{\partial \sigma_y} d\sigma_y + \frac{\partial U_0^*}{\partial \sigma_z} d\sigma_z + \frac{\partial U_0^*}{\partial \tau_{xy}} d\tau_{xy} + \frac{\partial U_0^*}{\partial \tau_{yz}} d\tau_{yz} + \frac{\partial U_0^*}{\partial \tau_{xz}} d\tau_{xz} \tag{9.28}$$

we see by comparison that the complementary strain energy density possesses the property

$$\epsilon_x = \frac{\partial U_0^*}{\partial \sigma_x} \qquad \epsilon_y = \frac{\partial U_0^*}{\partial \sigma_y} \qquad \cdots \qquad \gamma_{xz} = \frac{\partial U_0^*}{\partial \tau_{xz}} \tag{9.29}$$

In the case of a two-dimensional body, we note that

$$U_0^* = \sigma_x \epsilon_x + \sigma_y \epsilon_y + \tau_{xy} \gamma_{xy} - U_0 \tag{9.30}$$

This relation is a simple generalization of Eq. (9.25). If the body is isotropic and linearly elastic (Hookean), U_0 is given by Eq. (9.14), so that

$$\begin{aligned} U_0^* &= \sigma_x \epsilon_x + \sigma_y \epsilon_y + \tau_{xy} \gamma_{xy} - \tfrac{1}{2}(\sigma_x \epsilon_x + \sigma_y \epsilon_y + \tau_{xy} \gamma_{xy}) \\ &= \tfrac{1}{2}(\sigma_x \epsilon_x + \sigma_y \epsilon_y + \tau_{xy} \gamma_{xy}) = U_0 \end{aligned} \tag{9.31}$$

Thus, in the case of an isotropic linearly elastic body, the strain energy density and the complementary strain energy density are equal. This also follows from an inspection of the one-dimensional case indicated in Fig. 9.4*b*; for if the stress-strain relation is linear, the area under the curve is equal to its complement.

We obtain the total complementary strain energy U^* by simply integrating $U_0^* \, dV$ throughout the volume of the body:

$$U^* = \iiint_V U_0^* \, dV \tag{9.32}$$

In analogy with the total potential energy in Eq. (9.17), we define the sum of the complementary potential energy of the external forces and the complementary strain energy as the *total complementary energy*, denoted Π^*. Thus

$$\Pi^* = H^* + U^* \tag{9.33}$$

THE PRINCIPLES OF POTENTIAL ENERGY

9.4 The principle of stationary potential energy. Upon comparing the equations for external potential energy and strain energy [Eqs. (9.4), (9.5), and (9.8)] with those in Chap. 8 for external and internal virtual work [Eqs. (8.16) and (8.17)], we find that they are remarkably similar. This close similarity is no mere coincidence; in fact, it is the basis of several important principles of structural mechanics.

We recall from Chap. 8 that external and internal virtual work could be interpreted as the work done by the true forces and stresses in moving through *variations* in the displacements and strains. In view of the form of the equations for potential energy, it now appears that virtual work may also be interpreted as being variations in energy resulting from variations in displacements and strains. For example, consider an element of unit volume in a one-dimensional body subjected solely to the stress and strain components σ_x and ϵ_x. The internal virtual work due to a virtual strain is, by definition,

$$\delta U = \sigma_x \, \delta\epsilon_x$$

while, from elementary calculus, the differential change in strain energy in the body due to an increment $d\epsilon_x$ in strain is

$$dU = \frac{\partial U}{\partial \epsilon_x} \, d\epsilon_x = \sigma_x \, d\epsilon_x$$

We see that the symbol δ is more than just a suffix used to indicate a virtual quantity; in fact, it behaves as a *variational operator* which obeys rules of operation very similar to those of d, the first differential operator. If, in analogy, we refer to δ as the *first variation*, we see that internal virtual work may be interpreted as the first variation in the strain energy due to variations in the components of strains. Similarly, if W_e is the work done by external forces in a conservative system

$$-W_e = H$$

and

$$-\delta W_e = \delta H$$

Thus, external virtual work is equivalent to minus the first variation in the potential energy due to variations in the displacements.

From the principle of virtual displacements [Eq. (8.19)],

$$\delta U = -\delta H$$

and

$$\delta U + \delta H = 0$$

which, in view of the above interpretation of δ, may also be written

$$\delta(U + H) = 0$$

Finally, substituting Eq. (9.17) into this result, we get

$$\delta\Pi = 0 \tag{9.34}$$

Thus, we may also state the principle of virtual displacements[6] as follows:

> A deformable system is in equilibrium if the first variation in the total potential energy of the system is zero for every virtual displacement consistent with the constraints.
>
> (THEOREM VII)

Let us now examine the mathematical significance of Eq. (9.34). For simplicity, let us consider a deformable system in which it is possible to express the strains and the displacements of prescribed external forces in terms of only two independent generalized displacements u and v. The truss in Fig. 8.10 is an example of such a system. In this case Π can also be expressed as a function of u and v, so that $\partial\Pi/\partial u$ and $\partial\Pi/\partial v$ are the rates of change of Π with respect to u and v. Therefore, if first u is given a variation δu and then v a variation δv we know from Eq. (9.34) that

$$\delta\Pi = \frac{\partial\Pi}{\partial u}\,\delta u = 0$$

and

$$\delta\Pi = \frac{\partial\Pi}{\partial v}\,\delta v = 0 \tag{9.35}$$

if the structure is in equilibrium. Since the magnitude of these virtual displacements is arbitrary, we see that

$$\frac{\partial\Pi}{\partial u} = \frac{\partial\Pi}{\partial v} = 0 \tag{9.36}$$

which means that for equilibrium to exist,

$$d\Pi = \frac{\partial\Pi}{\partial u}\,du + \frac{\partial\Pi}{\partial v}\,dv = 0 \tag{9.37}$$

A similar result is obtained for the general deformable system, and we conclude that the vanishing of $\delta\Pi$ ensures (or is equivalent to) the vanishing of $d\Pi$.

Now from the calculus we recall that the total differential of a function vanishes at so-called *critical points* of the function. The critical points of a function may be points at which the function is a relative maximum or minimum, or they may be saddle points at which the function is said

[6] See Theorem IV, Chap. 8.

to be a minimax. At such points, the function is said to assume a *stationary value*. Observing that Eq. (9.37) is valid only at the equilibrium configurations of a given system, we arrive at the principle of *stationary potential energy*:

> If a structural system is in static equilibrium, the total potential energy of the system has a stationary value.
>
> (THEOREM VIII)

Equation (9.37) can be interpreted in another way if we think of Π as a continuous function of displacement patterns which are required to be consistent with the constraints. $d\Pi$ is zero only for those displacements corresponding to the equilibrium configurations. Thus, we restate the above principle as follows:

> Of all the possible displacements which satisfy the boundary conditions of a structural system, those corresponding to equilibrium configurations make the total potential energy assume a stationary value.
>
> (THEOREM IX)

9.5 The principle of minimum potential energy. We now demonstrate that for stable equilibrium, Π is a relative minimum.

An equilibrium configuration is stable, we recall, if the system returns to its original configuration after being given a small disturbance. Otherwise, the system is in unstable equilibrium. Consider, for example, the simple pendulum of weight Q in Fig. 9.5. Clearly positions A and B are both equilibrium configurations since in each there is a balance of forces. Configuration A is one of unstable equilibrium. The force Q does positive work during a small angular displacement; consequently, the potential energy decreases and, according to Eq. (9.3), the kinetic energy increases. Configuration B, on the other hand, is one of stable equilibrium. During a small angular displacement, negative work is done; the potential energy increases and creates only an infinitesimal change in kinetic energy. This example demonstrates a property of equilibrium configurations in general; namely, that arbitrary, small, virtual displacements of

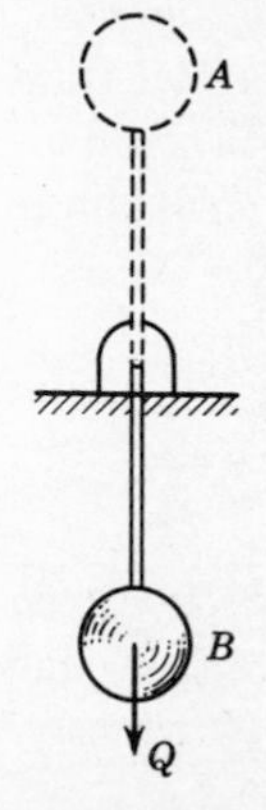

FIGURE 9.5
Simple pendulum.

a system in stable equilibrium increase the potential energy of the system. If the system is in unstable equilibrium, virtual displacements decrease the potential energy and cause the kinetic energy to increase continuously. If small displacements cause no change in potential energy, the system is said to be in neutral equilibrium. According to our definition, neutral equilibrium is also unstable equilibrium.

The classic example of the motion of marbles along the smooth contour in Fig. 9.6 is often used to demonstrate various types of equilibrium.

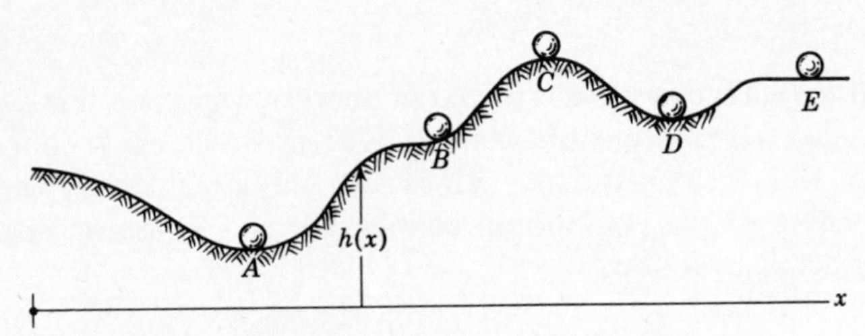

FIGURE 9.6 Marbles resting on a smooth rigid contour.

If the marbles are assumed to be rigid and of weight Q, U is zero and the total potential energy of a given marble is[7]

$$\Pi = H = Qh(x)$$

where $h(x)$ is the height of the contour relative to an arbitrary datum.

In general, the least number of independent variables required to specify the configuration of a system are called *generalized coordinates*. The coordinate x of a single marble of Fig. 9.6, for example, is the generalized coordinate of a system consisting of only one marble. Furthermore, all virtual displacements in a system which are consistent with the constraints can be expressed as a function of the variations in the generalized coordinates. Thus, a consistent virtual displacement of one of the marbles (that is, a virtual displacement which follows the rigid contour) is brought about by giving x a small variation δx, which, in turn, produces a variation in Π. Since each marble is in equilibrium,

$$\delta\Pi = Q\frac{dh}{dx}\,\delta x = 0$$

if

$$\frac{dh}{dx} = 0 \tag{a}$$

which is true at points A, B, C, D, and E. Therefore, these points are possible equilibrium configurations.

Obviously, a small virtual displacement increases the potential energy

[7] The sign of H is determined by noting that Q does negative work during an increase in h. Negative work, we recall, produces a positive change in potential energy.

of the marbles at A and D. Hence, they are in stable equilibrium. Virtual displacements decrease the potential energy of the marble at C and cause no change in that of the marble at E. Position C is one of unstable equilibrium; E is a position of neutral equilibrium. The potential energy of the marble at B is either increased or decreased depending on the sign (direction) of the virtual displacement. Since Π must increase in stable configurations owing to *arbitrary* small virtual displacements, we conclude that the marble at B is in unstable equilibrium.

Now if $\Pi(x)$ corresponds to an equilibrium configuration, $\Pi(x + \delta x)$ is the potential energy of a configuration in the "neighborhood" of $\Pi(x)$ if δx is sufficiently small. According to Taylor's formula,

$$\Pi(x + \delta x) = \Pi(x) + \frac{d\Pi(x)}{dx}\delta x + \frac{1}{2}\frac{d^2\Pi(x)}{dx^2}(\delta x)^2 + \cdots \tag{b}$$

For equilibrium configurations, $d\Pi/dx$ is zero, according to Eq. (a). Noting that $\Delta\Pi = \Pi(x + \delta x) - \Pi(x)$ is the change in Π due to δx, we see that

$$\Delta\Pi = \frac{1}{2}\frac{d^2\Pi}{dx^2}(\delta x)^2 + \frac{1}{3}\frac{d^3\Pi}{dx^3}(\delta x)^3 + \cdots \tag{9.38}$$

or, for the systems in Fig. 9.6,

$$\Delta\Pi = Q\left[\frac{1}{2}\frac{d^2h}{dx^2}(\delta x)^2 + \frac{1}{3}\frac{d^3h}{dx^3}(\delta x)^3 + \cdots\right] \tag{c}$$

Thus, a first-order variation in the displacements at an equilibrium configuration causes a change in potential energy of order $(\delta x)^2$.

Now the sign of the first nonvanishing term in Eq. (9.38) [or Eq. (c)] determines the sign of the change in Π, and, consequently, it indicates the type of equilibrium. If the first nonvanishing derivative is of even order, the sign of $\Delta\Pi$ is independent of the sign of the virtual displacements and Π is either a relative maximum or a relative minimum. For example, in the case of the marbles on the rigid contour in Fig. 9.6, d^2h/dx^2 is positive at A and D and negative at C. Thus, small virtual displacements of the marbles at A and D will result in an increase in their potential energy, whereas a virtual displacement of the marble at C will decrease its potential energy. Π is a relative minimum at A and D and a relative maximum at C. Clearly, the marbles at points A and D are in stable equilibrium and the marble at C is in unstable equilibrium. At position B, the first nonvanishing derivative is of odd order and the sign of $\Delta\Pi$ depends upon the sign of δx. At this point Π is a minimax and the marble is in unstable equilibrium. At E, the position of neutral equilibrium, all derivatives of Π vanish and $\Delta\Pi$ is zero for small virtual displacements.

We conclude from these observations that if Π is a relative minimum the system is in stable equilibrium. In fact, this result is called the *principle of minimum potential energy.* We state it in more meaningful terms as follows:

> Of all the displacements which satisfy the boundary conditions of a structural system, those corresponding to stable equilibrium configurations make the total potential energy a relative minimum.
>
> (THEOREM X)

Consider, for example, the rigid cylinder of weight Q in Fig. 9.7*a* which is in contact with a rigid cylindrical surface. A rigid bar of negligible weight is connected by frictionless hinges to the axes of the cylinder and the cylindrical foundation. Angular motion of the bar is resisted by a rotational spring which develops an internal moment proportional to the cube of the rotation. Thus, if the bar rotates through an angle θ, the spring develops a resisting moment of $K\theta^3$, K being a constant. We assume that the cylinder rolls without slipping and that the internal moment developed in the spring is zero when the bar is in its initial vertical position ($\theta = 0$).

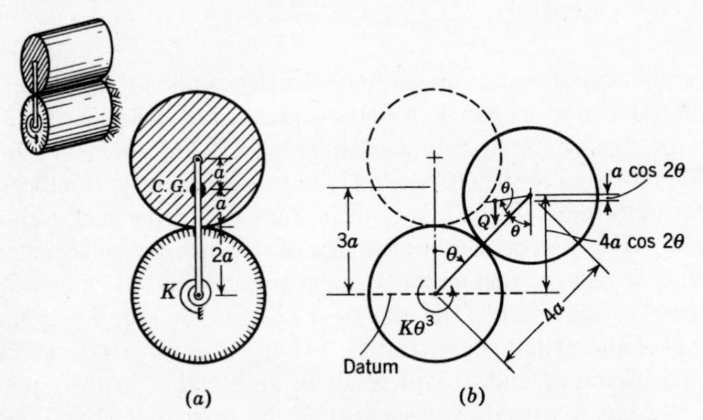

FIGURE 9.7 (*a*) Rigid cylinder of weight Q on a rigid cylindrical surface restrained by a nonlinear rotational spring; (*b*) general configuration of the system as defined by the angular coordinate θ.

To calculate the potential energy of this system, we determine the amount of work done in carrying the system from its original configuration to an arbitrary configuration. To this end, we take the initial vertical position of the bar as a reference and then give the bar an angular displacement θ as indicated in Fig. 9.7*b*. When $\theta = 0$, the weight Q is located a distance $3a$ above a horizontal plane through the axis of the lower cylinder. For an arbitrary value of θ, we find from the geometry of Fig. 9.7*b* that Q is a distance $4a \cos \theta - a \cos 2\theta$ above this plane. Thus

$$H = -Qa[3 - (4 \cos \theta - \cos 2\theta)] \tag{d}$$

The strain energy stored in the spring is given by the formula

$$U = \int_0^\theta K\theta^3 \, d\theta = \frac{K\theta^4}{4} \tag{e}$$

Thus, the total potential energy of the system is

$$\Pi = \frac{K\theta^4}{4} + Qa(4 \cos \theta - \cos 2\theta) - 3Qa \tag{f}$$

Note that the constant term $-3Qa$ depends upon the choice of datum, and since we are to consider only changes in Π it does not enter our calculations.

We now examine the change in the potential energy due to a variation $\delta\theta$ in the generalized coordinate θ:

$$\Delta\Pi = \frac{\partial \Pi}{\partial \theta} \delta\theta + \frac{1}{2} \frac{\partial^2 \Pi}{\partial \theta^2} (\delta\theta)^2 + \frac{1}{3!} \frac{\partial^3 \Pi}{\partial \theta^3} (\delta\theta)^3 + \cdots \tag{g}$$

From Eq. (*f*) we find that

$$\begin{aligned}
\frac{\partial \Pi}{\partial \theta} &= K\theta^3 - 4Qa \sin \theta + 2Qa \sin 2\theta \\
\frac{\partial^2 \Pi}{\partial \theta^2} &= 3K\theta^2 - 4Qa \cos \theta + 4Qa \cos 2\theta \\
\frac{\partial^3 \Pi}{\partial \theta^3} &= 6K\theta + 4Qa \sin \theta - 8Qa \sin 2\theta \\
\frac{\partial^4 \Pi}{\partial \theta^4} &= 6K + 4Qa \cos \theta - 16Qa \cos 2\theta \\
&\cdots\cdots\cdots\cdots\cdots\cdots
\end{aligned} \tag{h}$$

According to the principle of stationary potential energy, the system is in equilibrium when the first variation in Π vanishes. Thus, the equilibrium condition is

$$\delta\Pi = \frac{\partial \Pi}{\partial \theta} \delta\theta = (K\theta^3 - 4Qa \sin \theta + 2Qa \sin 2\theta) \, \delta\theta = 0$$

or, since $\delta\theta$ is arbitrary,

$$K\theta^3 - 4Qa \sin \theta + 2Qa \sin 2\theta = 0 \tag{i}$$

An obvious solution to this equation is $\theta = 0$. If $\Delta\Pi$ is positive at this configuration, the system is in stable equilibrium at $\theta = 0$. Setting θ equal to zero in Eqs. (*h*), we find

$$\frac{\partial\Pi(0)}{\partial\theta} = \frac{\partial^2\Pi(0)}{\partial\theta^2} = \frac{\partial^3\Pi(0)}{\partial\theta^3} = 0 \qquad (j)$$

and

$$\frac{\partial^4\Pi(0)}{\partial\theta^4} = 6K - 12Qa \qquad (k)$$

Thus, the fourth derivative is the lowest-order nonzero derivative of Π at $\theta = 0$. It follows that the sign of the fourth derivative determines the sign of $\Delta\Pi$ and, consequently, the type of equilibrium. Clearly, if $6K > 12Qa$, $\Delta\Pi$ is positive; and if $6K < 12Qa$, $\Delta\Pi$ is negative. Therefore, if $6K > 12Qa$, the system is in stable equilibrium and if

$$6K \leq 12Qa \qquad (l)$$

the system is in unstable equilibrium at $\theta = 0$. Note that if

$$Q = \frac{K}{2a} \qquad (m)$$

$\Delta\Pi$ is zero and the system is in neutral equilibrium. This value of Q is called the *critical load* of the system.

For illustration purposes, let us assume that $Q = \pi K/2a$. Then Q is greater than the critical load and the system is in unstable equilibrium at $\theta = 0$. In this case, a slight disturbance will cause the cylinder to rotate and to seek a position of stable equilibrium. Equation (*i*) becomes

$$\theta^3 - 2\pi \sin\theta + \pi \sin 2\theta = 0 \qquad (n)$$

We solve this equation by trial and error and find that the system is also in equilibrium at $\theta = \pm 115.8°$. Introducing these values of θ into Eqs. (*h*), we find that they lead to a positive change in Π and, therefore, correspond to stable equilibrium configurations. A plot of the potential energy versus θ for this system is given in Fig. 9.8.

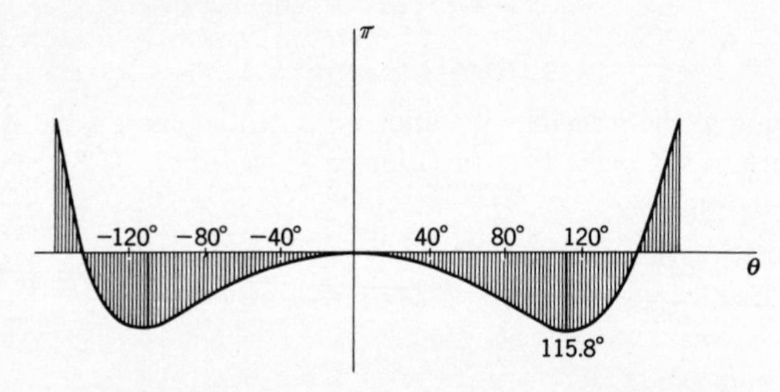

FIGURE 9.8 Variation in the potential energy of the system shown in Fig. 9.7 for the case in which $Q = \pi K/2a$.

9.6 The second variation in the total potential energy. It is important to note that the change in potential energy can also be expressed in the form

$$\Delta\Pi = \delta\Pi + \tfrac{1}{2}\delta^2\Pi + \text{terms of higher order} \tag{9.39}$$

where $\delta^2\Pi$ is called the *second variation* in Π, in analogy with $\delta\Pi$. For one-dimensional systems such as those in Fig. 9.6, we find that

$$\delta^2\Pi = \frac{d^2\Pi}{dx^2}(\delta x)^2 \tag{9.40}$$

Since in equilibrium configurations $\delta\Pi$ is zero, the sign of $\Delta\Pi$ is often determined by the sign of $\delta^2\Pi$. For this reason, the second variation often plays an important role in the study of the stability of structural systems.

From Eq. (9.40) we see that the second variation is quadratic in δx since it is a function of $(\delta x)^2$. In more general systems, we find that $\delta^2\Pi$ is expressible as a *quadratic form* in the virtual displacements (variations of the generalized coordinates). For example, if Π is a function of two generalized coordinates x and y and if the point (a,b) corresponds to an equilibrium configuration, Taylor's formula for functions of several variables gives

$$\begin{aligned}\Delta\Pi &= \Pi(a+\delta x, b+\delta y) - \Pi(a,b)\\ &= \frac{\partial\Pi(a,b)}{\partial x}\delta x + \frac{\partial\Pi(a,b)}{\partial y}\delta y + \frac{1}{2}\left[\frac{\partial^2\Pi(a,b)}{\partial x^2}(\delta x)^2\right.\\ &\qquad \left. + 2\frac{\partial^2\Pi(a,b)}{\partial x\,\partial y}\delta x\,\delta y + \frac{\partial^2\Pi(a,b)}{\partial y^2}(\delta y)^2\right]\\ &\qquad + \text{terms of higher order}\end{aligned} \tag{9.41}$$

Again

$$\delta\Pi = \frac{\partial\Pi(a,b)}{\partial x}\delta x + \frac{\partial\Pi(a,b)}{\partial y}\delta y = 0$$

and, in this case,

$$\delta^2\Pi = [a_{11}(\delta x)^2 + a_{12}\,\delta x\,\delta y + a_{21}\,\delta x\,\delta y + a_{22}(\delta y)^2] \tag{9.42}$$

where

$$a_{11} = \frac{\partial^2\Pi(a,b)}{\partial x^2} \qquad a_{12} = a_{21} = \frac{\partial^2\Pi(a,b)}{\partial x\,\partial y} \qquad a_{22} = \frac{\partial^2\Pi(a,b)}{\partial y^2} \tag{9.43}$$

Equation (9.42) is a quadratic form in δx and δy. The sign of $\Delta\Pi$ clearly depends upon the coefficients a_{11}, a_{12}, a_{21}, and a_{22}, which are functions of the applied loads.

In general, there are five types of quadratic forms:[8] positive definite, positive semidefinite, negative definite, negative semidefinite, and indefinite; the coefficients $a_{11}, a_{12}, \ldots, a_{nn}$ determine the type of form under consideration. If $\delta^2\Pi$ is a positive definite quadratic form, it can be

[8] See Ref. 29, pp. 255–260.

shown[9] that $\Delta\Pi$ is positive and, hence, that Π is a relative minimum. Further, if $\delta^2\Pi$ is negative definite, negative semidefinite, or indefinite, $\Delta\Pi$ is negative and the equilibrium is unstable. If $\delta^2\Pi$ is positive semidefinite or zero, higher variations must be considered.

In the majority of practical cases, it is seldom necessary to determine the type of quadratic form of $\delta^2\Pi$. If in the analysis of linearly elastic systems we introduce the assumption of small displacements, Π can be expressed as a quadratic function of the generalized coordinates. In these cases we are usually faced with the problem of determining the values of the external forces which will cause a system to cease being stable. When such systems are stable, $\delta^2\Pi$ is positive definite and the determinant

$$D = \begin{vmatrix} a_{11} & a_{12} \\ a_{21} & a_{22} \end{vmatrix}$$

is positive. If the applied loads are increased, D decreases. When a *critical* loading system is reached, D vanishes, $\delta^2\Pi$ ceases to be positive definite, and the system becomes unstable and buckles. Thus, for mechanical systems of this type, the stability criterion is

$$\delta^2\Pi = D = 0 \tag{9.44}$$

9.7 Stability analysis of simple structures. To clarify the ideas presented in the previous article, we now investigate the stability of some simple structural systems.

As a first example, let us consider the mechanism shown in Fig. 9.9*a*. The structure consists of two rigid bars of length a hinged at ends A, B, and C. Movement of the system is elastically restrained by linear springs of modulus k_1 at B and by a rotational spring of modulus k_α at B. A unit elongation of a linear spring develops an internal force of magnitude k_1, and a unit rotation of the angular spring develops an internal moment of magnitude k_α. We assume that the springs are in their unstretched positions when the bars are vertical, as in Fig. 9.9*a*.

To obtain the total potential energy of the system, we examine the work done in moving it to some general configuration, as shown in Fig. 9.9*b*. We find that all configurations can be described by specifying only two independent variables, the angular coordinates θ and ψ. Thus, θ and ψ are the generalized coordinates.

For positive values of θ and ψ, the applied load P is displaced a distance Δ, where, from the geometry of the figure,

$$\Delta = 2a - a\cos\theta - a\cos\psi$$

Therefore, the total potential energy of the external force P is

$$H = -P\Delta$$

[9] See Ref. 36.

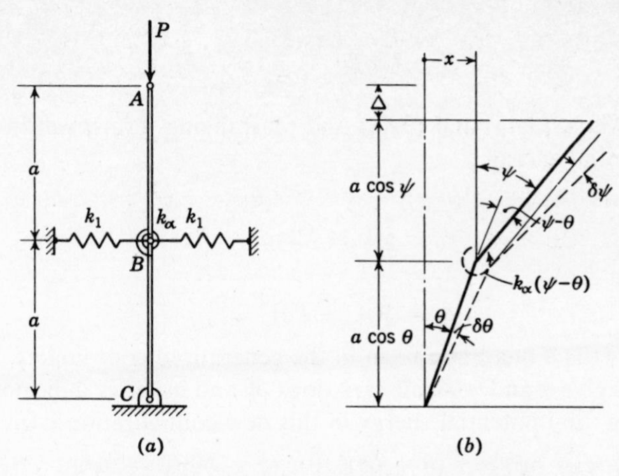

FIGURE 9.9 (*a*) Column composed of elastically restrained rigid bars; (*b*) geometry of deformation.

Now the initial configuration of the system ($\theta = \psi = 0$) is obviously an equilibrium configuration since the load P is balanced by a reactive force at C. For certain values of P, the initial configuration will be one of stable equilibrium. We are interested in determining the value of P for which this configuration becomes unstable and the structure buckles. For this purpose, we need consider only small values of the generalized coordinates θ and ψ. Recalling that $\cos\theta = 1 - \theta^2/2 + \cdots$, etc., we neglect powers of θ and ψ of order three and higher so that potential energy of the external force P is approximated by the formula

$$H = -\frac{Pa}{2}(\theta^2 + \psi^2) \tag{9.45}$$

Since internal forces and moments are developed in the springs, the system also possesses strain energy. From the definition of strain energy,

$$U = 2\int_0^x k_1 x\,dx + \int_0^\alpha k_\alpha \alpha\,d\alpha$$

or

$$U = 2\left(\frac{k_1 x^2}{2}\right) + \frac{k_\alpha \alpha^2}{2} \tag{9.46}$$

where x is the change in length of the linear spring and α is the angle of rotation of the rotational spring.

From the geometry of Fig. 9.9*b*, we find

$$\begin{aligned} x &= a\sin\theta \\ \alpha &= \psi - \theta \end{aligned} \tag{9.47a}$$

or, to the same degree of approximation as Eq. (9.45),

$$x = a\theta \tag{9.47b}$$

Thus
$$U = k_1 a^2 \theta^2 + \frac{k_\alpha}{2}(\psi - \theta)^2 \tag{9.48}$$

Adding Eqs. (9.45) and (9.48) and rearranging terms, we find for the total potential energy

$$\Pi(\theta,\psi) = c_0\theta^2 + c_1\theta\psi + c_2\psi^2 \tag{9.49a}$$

where
$$\begin{aligned} c_0 &= \tfrac{1}{2}(k_\alpha + 2k_1 a^2 - Pa) \\ c_1 &= -k_\alpha \\ c_2 &= \tfrac{1}{2}(k_\alpha - Pa) \end{aligned} \tag{9.49b}$$

Note that Π is a quadratic form in the generalized coordinates.

We now give θ and ψ small variations $\delta\theta$ and $\delta\psi$ as is indicated in Fig. 9.9*b*. The total potential energy in this new configuration is given by

$$\Pi(\theta + \delta\theta, \psi + \delta\psi) = c_0(\theta + \delta\theta)^2 + c_1(\theta + \delta\theta)(\psi + \delta\psi) + c_2(\psi + \delta\psi)^2$$

Expanding and collecting terms, we find

$$\begin{aligned} \Pi(\theta + \delta\theta, \psi + \delta\psi) = {} & (c_0\theta^2 + c_1\theta\psi + c_2\psi^2) \\ & + [(2c_0\theta + c_1\psi)\,\delta\theta + (c_1\theta + 2c_2\psi)\,\delta\psi] \\ & + [c_0(\delta\theta)^2 + c_1\,\delta\theta\,\delta\psi + c_2(\delta\psi)^2] \end{aligned} \tag{9.50}$$

Referring to Eq. (9.49*a*), we recognize the quantity in the first set of parentheses on the right side of Eq. (9.50) as Π. Since

$$\Delta\Pi = \Pi(\theta + \delta\theta, \psi + \delta\psi) - \Pi(\theta,\psi)$$

we have

$$\begin{aligned} \Delta\Pi = {} & [(2c_0\theta + c_1\psi)\,\delta\theta + (c_1\theta + 2c_2\psi)\,\delta\psi] \\ & + [c_0(\delta\theta)^2 + c_1\,\delta\theta\,\delta\psi + c_2(\delta\psi)^2] \end{aligned} \tag{9.51}$$

It is easily verified from the geometry of Fig. 9.9*b* that the quantity within the first set of brackets on the right side of Eq. (9.51) is the virtual work done during the virtual displacements. It is a linear function of $\delta\theta$ and $\delta\psi$. This quantity is the first variation in Π:

$$\delta\Pi = \frac{\partial\Pi}{\partial\theta}\,\delta\theta + \frac{\partial\Pi}{\partial\psi}\,\delta\psi = (2c_0\theta + c_1\psi)\,\delta\theta + (c_1\theta + 2c_2\psi)\,\delta\psi \tag{9.52}$$

In equilibrium configurations, $\delta\Pi$ is zero. Furthermore, since $\delta\theta$ and $\delta\psi$ are arbitrary and independent, the coefficients in parentheses in (9.52) must vanish independently when $\delta\Pi$ is zero. We conclude that the conditions for equilibrium are

$$\begin{aligned} \frac{\partial\Pi}{\partial\theta} &= 0 = 2c_0\theta + c_1\psi \\ \frac{\partial\Pi}{\partial\psi} &= 0 = c_1\theta + 2c_2\psi \end{aligned} \tag{9.53}$$

Obviously, $\theta = \psi = 0$ is an equilibrium configuration.

The quantity in the second set of brackets in Eq. (9.51) is a quadratic form in $\delta\theta$ and $\delta\psi$. This term is (half of) the second variation in Π:

$$\tfrac{1}{2}\delta^2\Pi = c_0(\delta\theta)^2 + c_1\,\delta\theta\,\delta\psi + c_2(\delta\psi)^2 \tag{9.54}$$

Note that

$$\frac{\partial^2\Pi}{\partial\theta^2} = c_0 \qquad \frac{\partial^2\Pi}{\partial\theta\,\partial\psi} = c_1 \qquad \frac{\partial^2\Pi}{\partial\psi^2} = c_2 \tag{9.55}$$

in agreement with Eqs. (9.43) of the preceding article. For stable equilibrium (and positive definite $\delta^2\Pi$), the determinant

$$\begin{vmatrix} c_0 & c_1/2 \\ c_1/2 & c_2 \end{vmatrix}$$

is greater than zero. When P reaches the critical load, the determinant vanishes and the system becomes unstable. Thus, the buckling criterion is

$$c_0c_2 - \tfrac{1}{4}c_1^2 = 0 \tag{9.56}$$

In view of the assumption of small displacements and the fact that $\theta = \psi = 0$ is a stable equilibrium configuration, we can obtain Eq. (9.56) using a different argument. The trivial solution to Eqs. (9.53) defines the only stable equilibrium configuration. Since these equations are homogeneous, they possess a nontrivial solution only if the determinant of the coefficients vanishes. Equation (9.56) is then obtained by equating this determinant to zero. If the assumption of small displacements is removed, however, this argument is no longer valid, since several equilibrium configurations may then be possible.

Introducing Eqs. (9.49*b*) into Eq. (9.56), we obtain a polynomial in P:

$$(k_\alpha - Pa)(k_\alpha + 2k_1a^2 - Pa) - k_\alpha^2 = 0$$

which has two real roots

$$P_1, P_2 = \frac{1}{a}(k_\alpha + k_1a^2 \pm \sqrt{k_\alpha^2 + k_1^2a^4}) \tag{9.57}$$

For illustration purposes, let us assume that $k_\alpha = k_1a^2 = k$. Then

$$P_1 = 0.586\frac{k}{a} \tag{9.58a}$$

$$P_2 = 3.414\frac{k}{a} \tag{9.58b}$$

which means that $0.586k/a$ is the critical load for the system.

If we introduce Eq. (9.58*a*) into Eqs. (9.53) we find that $\psi = 2.414\theta$. This shape is shown in Fig. 9.10*a*. If we introduce Eq. (9.58*b*) into Eqs.

(9.53) we find $\psi = -0.414\theta$. Thus, for $P = 3.414k/a$ the buckled shape of the structure is as indicated in Fig. 9.10*b*. These two possibilities are called the *buckling modes* of the system.

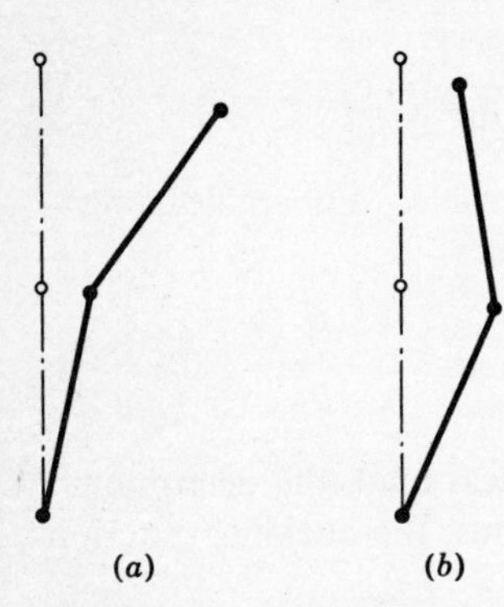

FIGURE 9.10 Buckling modes of the structure in Fig. 9.9.

It is important to note that the total potential energy was expressed in terms of only two independent variables (generalized coordinates) since the system has only two degrees of freedom. We would obtain erroneous results if U in Eq. (9.46) were left as a function of x, since by virtue of Eq. (9.47*b*) x is not independent of θ and ψ.

In many cases, however, it is not convenient to express Π in terms of the least number of independent variables. In these cases, the problem becomes one of minimizing a function whose variables are *constrained* by some side relationships. Such problems of *constrained maxima* and *minima* are easily handled through the use of *Lagrange multipliers*.[10]

Suppose, for example, that Eq. (9.47*b*) is not introduced into Eq. (9.46), so that the problem becomes one of minimizing the function

$$\Pi(\theta,\psi,x) = \tfrac{1}{2}(k_\alpha - Pa)\theta^2 - k_\alpha\theta\psi + k_1x^2 + \tfrac{1}{2}(k_\alpha - Pa)\psi^2 \tag{9.59a}$$

with the additional condition

$$x - a\theta = 0 \tag{9.59b}$$

We form the new function

$$\bar{\Pi} = \Pi(\theta,\psi,x) + \lambda(x - a\theta) \tag{9.60}$$

[10] In general when the variables $u_1, u_2, \ldots, u_n$ of a function $G(u_1, u_2, \ldots, u_n)$ to be minimized must also satisfy m additional conditions of the form

$$g_1(u_1, u_2, \ldots, u_n) = 0 \qquad \cdots \qquad g_m(u_1, u_2, \ldots, u_n) = 0$$

a new function $\bar{G}$ is formed, where

$$\bar{G} = G + \lambda_1 g_1 + \cdots + \lambda_m g_m = G + \sum_{i=1}^{m} \lambda_m g_m$$

The constants $\lambda_1, \ldots, \lambda_m$ are the Lagrange multipliers. When G has a stationary value, $\bar{G}$ must be such that

$$\frac{\partial \bar{G}}{\partial u_1} = 0 \qquad \frac{\partial \bar{G}}{\partial u_2} = 0 \qquad \cdots \qquad \frac{\partial \bar{G}}{\partial u_n} = 0$$

These n conditions plus the m conditions $g_1 = g_2 = \cdots = g_m = 0$ provide $m + n$ independent equations from which the $m + n$ unknowns $u_1, u_2, \ldots, u_n, \lambda_1, \lambda_2, \ldots, \lambda_m$ can be determined. See Ref. 59, p. 254, for more details.

where λ is the Lagrange multiplier. To be a minimum, $\bar{\Pi}$ must satisfy the conditions

$$\frac{\partial \bar{\Pi}}{\partial \theta} = (k_\alpha - Pa)\theta - k_\alpha \psi - a\lambda = 0$$
$$\frac{\partial \bar{\Pi}}{\partial \psi} = -k_\alpha \theta + (k_\alpha - Pa)\psi = 0 \tag{9.61}$$
$$\frac{\partial \bar{\Pi}}{\partial x} = 2k_1 x + \lambda = 0$$

From the last condition we find that $\lambda = -2k_1 x$. Since Eq. (9.59*b*) must also be satisfied,

$$\lambda = -2k_1 a\theta \tag{9.62}$$

Finally, introducing this result into the first two conditions in Eq. (9.61), we obtain Eqs. (9.53), as before.

As another example, let us consider the symmetrical linearly elastic simply supported beam in Fig. 9.11. The beam is subjected to axial forces P and to a sinusoidal transverse loading $p_0 \sin(\pi x/L)$. For simplicity, we neglect shear deformation and assume that the normal stress is given by the elementary formula

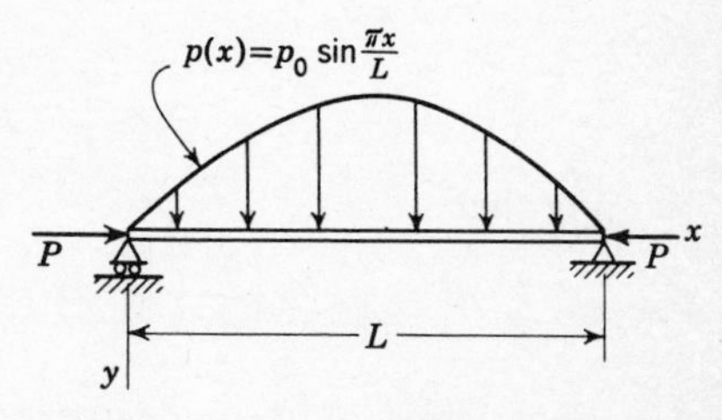

FIGURE 9.11 Simply supported beam column under axial load.

$$\sigma_x = \frac{N}{A} + \frac{My}{I}$$

If σ_y and σ_z are also negligible, $\epsilon_x = E\sigma_x$. According to Eqs. (9.8) and (9.16), the strain energy in this system is of the form

$$U = \iiint_V \left(\int \sigma_x \frac{d\sigma_x}{E} \right) dV = \iiint_V \frac{\sigma_x^2}{2E} \, dA \, dx$$

Thus
$$U = \frac{1}{2E} \int_0^L \left[\int_A \left(\frac{N^2}{A^2} + 2\frac{NM}{AI} y + \frac{M^2 y^2}{I^2} \right) dA \right] dx$$

and since $\int y \, dA = 0$ and $\int y^2 \, dA = I$,

$$U = \int_0^L \left(\frac{N^2}{2AE} + \frac{M^2}{2EI} \right) dx \tag{9.63}$$

The first term in the integrand represents the energy stored in the beam due to the shortening of its centroidal axis; it is independent of the transverse deflection of the beam. If we are to examine just the transverse displacements and the stability of the beam, we need consider only the energy due to changes in curvature. Thus, for the present, we ignore the contribution of the axial load to the strain energy and use, instead of Eq. (9.63),

$$U = \int_0^L \frac{M^2}{2EI}\,dx = \frac{EI}{2}\int_0^L \left(\frac{d^2v}{dx^2}\right)^2 dx \tag{9.64}$$

where v is the transverse displacement.

Now the total potential energy of the external forces is

$$H = -\int_0^L p(x)v\,dx - P\Delta \tag{9.65}$$

in which Δ is the displacement of P due to change in curvature of the beam:

$$\Delta \cong \int_0^L (ds - dx) \tag{9.66a}$$

Noting that

$$ds = \left[1 + \left(\frac{dv}{dx}\right)^2\right]^{\frac{1}{2}} dx = \left[1 + \frac{1}{2}\left(\frac{dv}{dx}\right)^2 + \cdots\right] dx$$

we neglect terms of third degree and higher in the derivative of the displacements and obtain the approximate relation

$$\Delta = \frac{1}{2}\int_0^L \left(\frac{dv}{dx}\right)^2 dx \tag{9.66b}$$

Therefore

$$H = -\int_0^L \left[\frac{P}{2}\left(\frac{dv}{dx}\right)^2 + p(x)v\right] dx \tag{9.67}$$

Again, we obtain the total potential energy due to changes in curvature by adding Eqs. (9.64) and (9.67):

$$\Pi = \int_0^L \left[\frac{EI}{2}\left(\frac{d^2v}{dx^2}\right)^2 - \frac{P}{2}\left(\frac{dv}{dx}\right)^2 - pv\right] dx \tag{9.68}$$

The elastic curve of the beam in Fig. 9.11 is of the form

$$v = C \sin\frac{\pi x}{L} \tag{9.69}$$

where C is an undetermined constant. This function satisfies all of the kinematic boundary conditions. If it is the true elastic curve, C must be such that v also satisfies the equilibrium conditions. We note that the magnitude of the displacement at any point depends upon the magnitude of C. Thus, C behaves as a generalized coordinate.

Introducing Eq. (9.69) into Eq. (9.68) and performing the indicated integration, we find

$$\Pi = \frac{L}{4}\left[\frac{C^2\pi^2}{L^2}\left(\frac{EI\pi^2}{L^2} - P\right) - 2p_0C\right]$$

If the system is in equilibrium,

$$\frac{d\Pi}{dC} = \frac{L}{4}\left[\frac{2C\pi^2}{L^2}\left(\frac{EI\pi^2}{L^2} - P\right) - 2p_0\right] = 0$$

and

$$C = \frac{p_0L^4}{\pi^2(EI\pi^2 - PL^2)} \tag{9.70}$$

so that

$$v = \frac{p_0L^4}{\pi^2(EI\pi^2 - PL^2)}\sin\frac{\pi x}{L} \tag{9.71}$$

Since we have only one generalized coordinate, $\delta^2\Pi$ is positive and the beam is in stable equilibrium if

$$\frac{d^2\Pi}{dC^2} = \frac{\pi^2}{2L}\left(\frac{EI\pi^2}{L^2} - P\right) > 0$$

When P reaches a value which makes $\delta^2\Pi = 0$, however, the beam is no longer in stable equilibrium. Thus, the stability criterion

$$\frac{d^2\Pi}{dC^2} = 0 \tag{9.72a}$$

is satisfied when

$$P = P_{cr} = \frac{\pi^2EI}{L^2} \tag{9.72b}$$

which we recognize as the Euler buckling load for a pinned-end column.

9.8 The Rayleigh-Ritz method. In general, a continuously distributed deformable body consists of an infinity of material points, and, therefore, it has infinitely many degrees of freedom. The Rayleigh-Ritz method[11] is an approximate procedure by which such continuous systems are reduced to systems with a finite number of degrees of freedom. This method is applicable to analyses of deformations, stability, nonlinear behavior, and even vibrations of complex structural systems; it is perhaps one of the most important methods of structural analysis.

Briefly, in the Rayleigh-Ritz method we approximate the components of displacement u, v, and w by functions containing a finite number of independent parameters. We then determine these parameters so that the total potential energy computed on the basis of the approximate

[11] This method was first presented in 1877 by Lord Rayleigh (John William Strutt) and was refined and extended in 1909 by W. Ritz. See Refs. 47 and 50.

displacements is a minimum. For example, suppose that for a given structural system, we assume that u, v, and w are of the form

$$\begin{aligned} u &= a_1\phi_1(x,y,z) + a_2\phi_2(x,y,z) + \cdots + a_n\phi_n(x,y,z) \\ v &= b_1\psi_1(x,y,z) + b_2\psi_2(x,y,z) + \cdots + b_n\psi_n(x,y,z) \\ w &= c_1\eta_1(x,y,z) + c_2\eta_2(x,y,z) + \cdots + c_n\eta_n(x,y,z) \end{aligned} \tag{9.73}$$

where $a_1, \ldots, a_n$, $b_1, \ldots, b_n$, $c_1, \ldots, c_n$ are $3n$ linearly independent parameters yet to be determined and $\phi_1, \ldots, \phi_n$, $\psi_1, \ldots, \psi_n$, $\eta_1, \ldots, \eta_n$ are continuous functions of the coordinates x, y, and z. We select the functions $\phi_1, \ldots, \eta_n$ so that they satisfy all of the kinematic boundary conditions for all values of the constant parameters $a_1, \ldots, c_n$; but they do not necessarily satisfy the static boundary conditions. We recall that such functions are referred to as kinematically admissible functions. Since the components of displacement are now defined in terms of only $3n$ independent quantities, the parameters $a_1, \ldots, c_n$ behave as generalized coordinates and, in effect, the system has only $3n$ degrees of freedom. Using Eqs. (9.73), we compute approximate strains which we then use to evaluate the total potential energy of the system:

$$\Pi(u,v,w) = \Pi(x,y,z,a_1,\ldots,a_n, b_1,\ldots,b_n, c_1,\ldots,c_n) \tag{9.74}$$

Since $\delta u = \sum_{i=1}^{n} \phi_i \, \delta a_i \qquad \delta v = \sum_{i=1}^{n} \psi_i \, \delta b_i \qquad$ and $\qquad \delta w = \sum_{i=1}^{n} \eta_i \, \delta c_i$

variations in the parameters a_i, b_i, and c_i lead to variations in the total potential energy of

$$\delta\Pi = \sum_{i=1}^{n} \left(\frac{\partial \Pi}{\partial a_i} \, \delta a_i + \frac{\partial \Pi}{\partial b_i} \, \delta b_i + \frac{\partial \Pi}{\partial c_i} \, \delta c_i \right) \tag{9.75}$$

Thus, if the system is in equilibrium,

$$\sum_{i=1}^{n} \left(\frac{\partial \Pi}{\partial a_i} \, \delta a_i + \frac{\partial \Pi}{\partial b_i} \, \delta b_i + \frac{\partial \Pi}{\partial c_i} \, \delta c_i \right) = 0 \tag{9.76}$$

for arbitrary values of δa_i, δb_i, and δc_i. It follows that Eq. (9.76) is satisfied only if

$$\begin{array}{cccc} \dfrac{\partial \Pi}{\partial a_1} = 0 & \dfrac{\partial \Pi}{\partial a_2} = 0 & \cdots & \dfrac{\partial \Pi}{\partial a_n} = 0 \\ \dfrac{\partial \Pi}{\partial b_1} = 0 & \dfrac{\partial \Pi}{\partial b_2} = 0 & \cdots & \dfrac{\partial \Pi}{\partial b_n} = 0 \\ \dfrac{\partial \Pi}{\partial c_1} = 0 & \dfrac{\partial \Pi}{\partial c_2} = 0 & \cdots & \dfrac{\partial \Pi}{\partial c_n} = 0 \end{array} \tag{9.77}$$

Equations (9.77) represent a system of $3n$ linearly independent simultaneous equations in the unknown parameters a_i, b_i, and c_i. Once we solve these equations, we introduce the results into Eqs. (9.73) to obtain the approximate components of displacements. We then evaluate strains and stresses and the analysis is complete. In the case of stability analyses, Eqs. (9.77) are homogeneous. We can then compute approximate values of the buckling loads by equating to zero the determinant of the coefficients.

Some important characteristics of the Rayleigh-Ritz method are listed as follows:

1. Ordinarily, the accuracy of the assumed displacement is increased with an increase in the number of parameters used.[12]

2. Although the Rayleigh-Ritz method may lead to fairly accurate expressions for displacements, the corresponding stresses may differ significantly from their exact values. This is due to the fact that stresses generally depend upon derivatives of the displacements. Obviously, the derivatives of approximate functions are usually less accurate approximations than the functions themselves.

3. The differential equations of equilibrium do not enter the analysis. Equilibrium is satisfied in an average sense through minimization of the total potential energy. Thus, stresses computed on the basis of approximate displacements do not, in general, satisfy the equilibrium equations.

4. Since the Rayleigh-Ritz method approximates systems having infinitely many degrees of freedom with systems having a finite number of degrees of freedom, the approximate system is less flexible than the actual system. Hence, the method usually overestimates the stiffness of a structure. It follows that buckling loads computed by the Rayleigh-Ritz method are always greater than (or equal to) the exact values.

As an illustration of the method, let us assume that the deflection curve of the beam in Fig. 9.12 is of the form

$$v = a \sin \frac{\pi x}{L} \tag{a}$$

[12] Although this is often the case, it is possible that successive solutions may not converge to the exact solution. For more information on the convergence of the Rayleigh-Ritz method, consult Refs. 13 and 30.

where a is an undetermined constant. Note that $\sin(\pi x/L)$ satisfies the kinematic boundary conditions at $x = 0$ and $x = L$. If we consider only strain energy due to bending, we find from Eq. (9.68) that

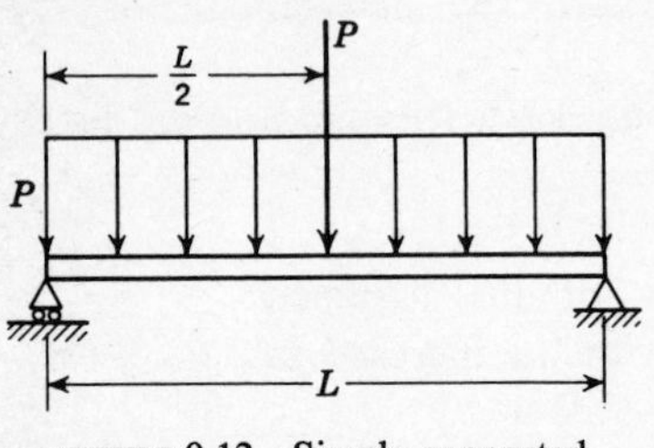

FIGURE 9.12 Simply supported beam.

$$\Pi = \frac{EI}{2}\int_0^L \left(\frac{d^2v}{dx^2}\right)^2 dx - \int_0^L p(x)v\,dx - Pv\left(\frac{L}{2}\right) \tag{b}$$

Substituting Eq. (a) into this equation and performing the integration, we get

$$\Pi = \frac{EI\pi^4}{4L^4}a^2 - 2p_0\frac{L}{\pi}a - Pa$$

We now choose a so as to minimize Π:

$$\frac{\partial \Pi}{\partial a} = 0 = \frac{EI\pi^4}{2L^4}a - \frac{2p_0 L}{\pi} - P$$

Therefore

$$a = \frac{4p_0 L^4}{\pi^5 EI} + \frac{2PL^3}{\pi^4 EI} \tag{c}$$

and

$$v = \frac{2L^3(2p_0 + \pi P)}{\pi^5 EI}\sin\frac{\pi x}{L} \tag{d}$$

Evaluating Eq. (d) at $x = L/2$, we find for the maximum deflection

$$v\left(\frac{L}{2}\right) = a = \frac{4p_0L^4}{\pi^5 EI} + \frac{2PL^3}{\pi^4 EI} = \frac{p_0L^4}{76.5EI} + \frac{PL^3}{48.7EI}$$

whereas the exact value is

$$v\left(\frac{L}{2}\right) = \frac{p_0L^4}{76.8EI} + \frac{PL^3}{48EI}$$

Thus, by using only one parameter we obtain a maximum deflection which is only 0.39 percent in error in the case of a uniform load and 1.46 percent in error in the case of a concentrated load applied at the center line. However, the approximate deflection curve gives a bending moment at $x = L/2$ of

$$M\left(\frac{L}{2}\right) = -EI\frac{d^2v(L/2)}{dx^2} = \frac{4p_0L^2}{\pi^3} + \frac{2PL}{\pi^2} = \frac{p_0L^2}{7.75} + \frac{PL}{4.93}$$

The first term is 3.15 percent in error and the second term is 23.37 percent in error. Note that normal stresses are proportional to the bending moments and, hence, are in error by the same percentages.

To obtain more accurate results, let us now use the two-parameter approximation

$$v = a \sin \frac{\pi x}{L} + b \sin \frac{3\pi x}{L} \tag{e}$$

Note that this function also satisfies the kinematic boundary conditions. Proceeding as before, we use Eq. (*b*) to compute the total potential energy and then from the conditions

$$\frac{\partial \Pi}{\partial a} = 0 \qquad \text{and} \qquad \frac{\partial \Pi}{\partial b} = 0$$

we find that a is the same as that given in Eq. (*c*) and

$$b = \frac{4p_0L^4}{243\pi^5EI} - \frac{2PL^3}{81\pi^4EI} \tag{f}$$

In this case we find for the maximum deflection

$$v\left(\frac{L}{2}\right) = \frac{p_0L^4}{76.8EI} + \frac{PL^3}{48.1EI}$$

which practically coincides with the exact value. The two-parameter approximation gives a bending moment at $x = L/2$ of

$$M = \frac{p_0L^4}{8.05} + \frac{PL}{4.44}$$

The first term is now only 0.63 percent in error, while the error in the second term has been reduced to 11.00 percent. We can further reduce these errors, of course, by increasing the number of parameters in the assumed displacement function.

As a second example, consider the fixed-end column shown in Fig. 9.13. Let us assume that the transverse deflection of the column is of the form

$$v = a(x^3 - 3xL^2 + 2L^3) + b(x - L)^2 \tag{g}$$

where a and b are the undetermined parameters. Note that $v(L) = dv(L)/dx = 0$, so that this function satisfies the kinematic boundary conditions.

If we again consider only strain energy due to bending, then according to Eq. (9.64)

$$U = \frac{EI}{2}\int_0^L \left(\frac{d^2v}{dx^2}\right)^2 dx$$

According to Eqs. (9.65) and (9.66*b*), the potential

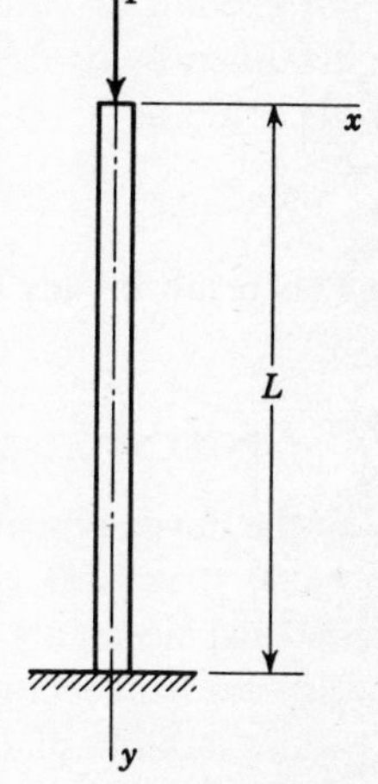

FIGURE 9.13 Fixed-end column.

energy of the external force P is

$$H = -\frac{P}{2}\int_0^L \left(\frac{dv}{dx}\right)^2 dx$$

Thus

$$\Pi = \frac{EI}{2}\int_0^L \left(\frac{d^2v}{dx^2}\right)^2 dx - \frac{P}{2}\int_0^L \left(\frac{dv}{dx}\right)^2 dx \tag{h}$$

Introducing Eq. (g) into Eq. (h), we get

$$\Pi = 6a^2L^3(EI - \tfrac{2}{5}PL^2) + abL^2(6EI - \tfrac{5}{2}PL^2) + 2b^2L(EI - \tfrac{1}{3}PL^2)$$

If a and b are to correspond to a stationary value of Π,

$$\begin{aligned} \frac{\partial \Pi}{\partial a} &= 12L^3(EI - \tfrac{2}{5}PL^2)a + L^2(6EI - \tfrac{5}{2}PL^2)b = 0 \\ \frac{\partial \Pi}{\partial b} &= L^2(6EI - \tfrac{5}{2}PL^2)a + 4L(EI - \tfrac{1}{3}PL^2)b = 0 \end{aligned} \tag{i}$$

A nontrivial solution to this system of homogeneous equations exists only if the determinant of the coefficients vanishes. Equating this determinant to zero, we obtain the equation

$$3P^2 - 104\frac{EI}{L^2}P + 240\frac{(EI)^2}{L^4} = 0 \tag{j}$$

The roots of this equation are

$$P_1 = 2.486\frac{EI}{L^2} \qquad \text{and} \qquad P_2 = 32.181\frac{EI}{L^2} \tag{k}$$

Since these values of P provide nontrivial solutions to Eqs. (i), they correspond to equilibrium positions other than the stable configuration identified by $a = b = 0$. Thus, the smallest root is the buckling load for the column:

$$P_{cr} = 2.486\frac{EI}{L^2} \tag{l}$$

This result is only 0.75 percent larger than the exact value $\pi^2 EI/4L^2$.

POTENTIAL-ENERGY THEOREMS OF STRUCTURAL ANALYSIS

In the developments to follow, we confine our attention to the analysis of stable structural systems at rest. With few exceptions, we assume that the displacements and their derivatives are small and that the strain-displacement relations are linear. In this class of structures, the question of the stability of an equilibrium configuration does not enter the analysis, and, consequently, only the first variation of the potential energy need be examined.

9.9 Castigliano's first theorem.[13] Consider the general three-dimensional body in Fig. 9.14, which is in equilibrium under the action of a system of n forces and moments. If $\Delta_1, \Delta_2, \ldots, \Delta_i, \ldots, \Delta_n$ are the displacements corresponding to these forces (that is, Δ_i is the displacement of P_i), we recall from Eq. (9.6) that the potential energy of the external forces is

$$H = -\sum_{i=1}^{n} P_i \Delta_i \tag{9.78}$$

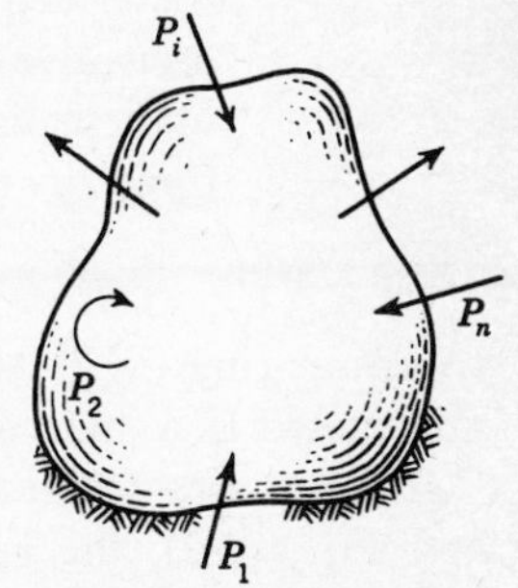

FIGURE 9.14 Three-dimensional body subjected to n concentrated forces and moments.

We assume that it is possible to express the displacement components $u(x,y,z)$, $v(x,y,z)$, $w(x,y,z)$ in terms of the displacements Δ_i of the prescribed surface forces. When this is done, the stresses and strains and the strain energy of the body become functions of $\Delta_1, \Delta_2, \ldots, \Delta_n$. Hence, the displacements Δ_i serve as the generalized coordinates of the problem. The total potential energy is therefore of the form

$$\Pi(\Delta_1, \Delta_2, \ldots, \Delta_n) = U - \sum_{i=1}^{n} P_i \Delta_i \tag{9.79}$$

If we now give each displacement Δ_i a small variation $\delta\Delta_i$, there results a variation in the total potential energy $\delta\Pi$, which must vanish since the structure is in equilibrium. Thus

$$\delta\Pi = 0 = \frac{\partial U}{\partial \Delta_1}\,\delta\Delta_1 + \frac{\partial U}{\partial \Delta_2}\,\delta\Delta_2 + \cdots + \frac{\partial U}{\partial \Delta_n}\,\delta\Delta_n - P_1\,\delta\Delta_1 - P_2\,\delta\Delta_2 - \cdots - P_n\,\delta\Delta_n$$

Rearranging terms, we have

$$\left(\frac{\partial U}{\partial \Delta_1} - P_1\right)\delta\Delta_1 + \left(\frac{\partial U}{\partial \Delta_2} - P_2\right)\delta\Delta_2 + \cdots + \left(\frac{\partial U}{\partial \Delta_n} - P_n\right)\delta\Delta_n = 0$$

Now since the variations $\delta\Delta_1, \delta\Delta_2, \ldots, \delta\Delta_n$ are completely arbitrary, each term in parentheses must vanish independently. That is,

$$\frac{\partial U}{\partial \Delta_1} - P_1 = 0 \qquad \frac{\partial U}{\partial \Delta_2} - P_2 = 0 \qquad \cdots \qquad \frac{\partial U}{\partial \Delta_n} - P_n = 0$$

or, in general,

$$\frac{\partial U}{\partial \Delta_i} = P_i \tag{9.80}$$

[13] First presented by Alberto Castigliano in 1873 (Ref. 10). This form of the theorem is also referred to as Castigliano's theorem, Part I. Castigliano's second theorem (or Castigliano's theorem, Part II) is discussed in Art. 9.14.

Equation (9.80) is a mathematical statement of *Castigliano's first theorem:*

> If the strain energy of a structural system is expressed in terms of the n independent displacements corresponding to a system of n prescribed forces $P_1, P_2, \ldots, P_n$, the first partial derivative of the strain energy with respect to any of these displacements Δ_i at point i is equal to the force P_i at i in the direction of Δ_i.
>
> (THEOREM XI)

By comparing Eq. (9.80) with Eqs. (9.10), we see that Eqs. (9.10) can be interpreted as a "microscopic" form of Castigliano's theorem.

It is important that we note the identity of Castigliano's first theorem and Eq. (8.22), the general equation of the unit-dummy-displacement method. *The internal virtual work due to a virtual displacement* $\delta\Delta_i$, we recall, *is equal to the first variation in the strain energy due to a variation* $\delta\Delta_i$ *in the true displacement* Δ_i. Thus, the principle of virtual displacements [Eq. (8.19)]

$$\delta W_e = P_i\,\delta\Delta_i = \delta U = \int_V \sigma\,\delta\epsilon^{\Delta}\,dV$$

gives Eq. (9.80) immediately, since

$$\delta U = \frac{\partial U}{\partial \Delta_i}\,\delta\Delta_i$$

Since the two concepts are identical, we can now interpret more generally the virtual strains $\delta\epsilon^{\Delta}$:

$$\frac{\partial U}{\partial \Delta_i}\,\delta\Delta_i = \int_V \left(\frac{\partial U_0}{\partial \Delta_i}\right)\delta\Delta_i\,dV = \int_V \left(\frac{\partial U_0}{\partial \epsilon_x}\frac{\partial \epsilon_x}{\partial \Delta_i} + \cdots + \frac{\partial U_0}{\partial \gamma_{xz}}\frac{\partial \gamma_{xz}}{\partial \Delta_i}\right)\delta\Delta_i\,dV$$

Introducing Eqs. (9.10),

$$\delta U = \frac{\partial U}{\partial \Delta_i}\,\delta\Delta_i = \int_V \left(\sigma_x \frac{\partial \epsilon_x}{\partial \Delta_i}\,\delta\Delta_i + \sigma_y \frac{\partial \epsilon_y}{\partial \Delta_i}\,\delta\Delta_i + \cdots + \tau_{xz}\frac{\partial \gamma_{xz}}{\partial \Delta_i}\,\delta\Delta_i\right) dV$$

$$= \int_V (\sigma_x\,\delta\epsilon_x{}^{\Delta_i} + \sigma_y\,\delta\epsilon_y{}^{\Delta_i} + \cdots + \tau_{xz}\,\delta\gamma_{xz}{}^{\Delta_i})\,dV$$

from which it follows that

$$\delta\epsilon_x{}^{\Delta_i} = \frac{\partial \epsilon_x}{\partial \Delta_i}\,\delta\Delta_i \qquad \delta\epsilon_y{}^{\Delta_i} = \frac{\partial \epsilon_y}{\partial \Delta_i}\,\delta\Delta_i \qquad \cdots \qquad \delta\gamma_{xz}{}^{\Delta_i} = \frac{\partial \gamma_{xz}}{\partial \Delta_i}\,\delta\Delta_i \quad (9.81)$$

Since Castigliano's first theorem is equivalent to the principle of virtual displacements, its application will obviously lead to equilibrium conditions. Consider, for example, the structure in Fig. 9.15, which consists of two linearly elastic hinged bars acted upon by a concentrated force P at the center hinge. Although the material is linear, the structure is *geometrically nonlinear*, in that equilibrium conditions necessarily depend upon the geometry of the deformed structure.

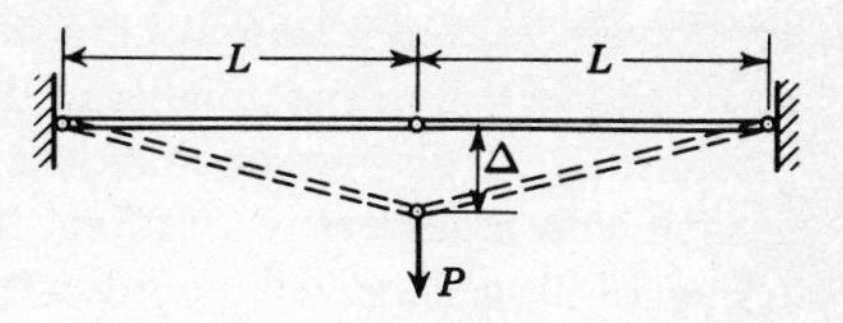

FIGURE 9.15 Geometrically nonlinear structure.

If Δ is the displacement of P, the strain in each bar is

$$\epsilon = \frac{1}{L}(\sqrt{L^2 + \Delta^2} - L) = \left[1 + \left(\frac{\Delta}{L}\right)^2\right]^{\frac{1}{2}} - 1 = 1 + \frac{1}{2}\left(\frac{\Delta}{L}\right)^2 + \cdots - 1$$

or, neglecting powers of Δ/L greater than 2,

$$\epsilon \approx \frac{1}{2}\left(\frac{\Delta}{L}\right)^2$$

With this approximation, the total strain energy is

$$U = 2AL\frac{E\epsilon^2}{2} = \frac{AE}{4L^3}\Delta^4$$

where A is the area of each bar. Further, if N is the internal axial force developed in each bar,

$$N = AE\epsilon = \frac{AE}{2L^2}\Delta^2$$

Finally, we have from Castigliano's first theorem

$$\frac{\partial U}{\partial \Delta} = \frac{AE}{L^3}\Delta^3 = P \qquad (a)$$

or

$$P = \sqrt{\frac{8N^3}{AE}} \qquad (b)$$

Note that Eqs. (*a*) and (*b*) are equilibrium conditions.

9.10 Structural analysis by Castigliano's first theorem. Although Eq. (9.80) is applicable to nonlinearly elastic structures, the majority of applications of Castigliano's first theorem are concerned with analysis of linearly elastic structures which undergo small displacements. Since the theorem parallels the unit-dummy-displacement method, displacements are taken as unknowns and a number of independent equilibrium conditions equal to the number of degrees of freedom of the structure are established. This means that when the theorem is used to analyze a system with n degrees of freedom, the strain energy must be expressed in terms of n generalized coordinates $\Delta_1, \Delta_2, \ldots, \Delta_n$, even though the actual number of external forces may not be n.

Consider, for example, the pin-connected truss in Fig. 8.10, in which the displacements u and v correspond to the external forces $P\cos\theta$ and $P\sin\theta$. The strain energy in bar m is

$$U_m = \int_V \left(\int_0^F \sigma_m \, d\epsilon_m \right) dV = \left(\int_0^F E\epsilon_m \, d\epsilon_m \right) A_m L_m = EA_m L_m \frac{\epsilon_m^{\,2}}{2}$$

Hence, the total strain energy is

$$U = \frac{E}{2} \sum_{m=1}^{n} \epsilon_m^{\,2} A_m L_m$$

The strain ϵ_m is given in terms of the generalized displacements u and v by Eq. (a) of Art. 8.7. Introducing this relation into the expression for strain energy gives

$$U = \frac{E}{2} \sum_{m=1}^{n} \frac{Am}{Lm} (u^2 \cos^2 \alpha_m + 2uv \cos \alpha_m \sin \alpha_m + v^2 \sin^2 \alpha_m) \qquad (a)$$

Applying Castigliano's first theorem, we find

$$\begin{aligned} \frac{\partial U}{\partial u} &= P\cos\theta = u \sum_{m=1}^{n} \frac{EA_m}{L_m} \cos^2 \alpha_m + v \sum_{m=1}^{n} \frac{EA_m}{L_m} \cos \alpha_m \sin \alpha_m \\ \frac{\partial U}{\partial v} &= P\sin\theta = u \sum_{m=1}^{n} \frac{EA_m}{L_m} \cos \alpha_m \sin \alpha_m + v \sum_{m=1}^{n} \frac{EA_m}{L_m} \sin^2 \alpha_m \end{aligned} \qquad (b)$$

which we recognize as Eqs. (e) and (f) of Art. 8.7. The remaining steps in the analysis are identical to those used in the virtual work procedure. Note that for $\delta u = \delta v = 1$,

$$\delta \epsilon_m^{\,u} = \frac{\cos \alpha_m}{L_m} = \frac{\partial \epsilon_m}{\partial u}$$

and

$$\delta \epsilon_m^{\,v} = \frac{\sin \alpha_m}{L_m} = \frac{\partial \epsilon_m}{\partial v}$$

in agreement with Eqs. (9.81).

It is important to note that Eqs. (b) are still valid when $\theta = 0$, that is, when the external force corresponding to v is zero. If such is the case, we must still formulate the problem in terms of two displacements u and v, even though no external force exists in the direction of u. Again, this is because the system has two degrees of freedom and, therefore, two independent variables are required to define the strains in the structure.

Let us now extend these ideas and consider the application of Castigliano's first theorem to more general linear structures. In the analysis of discrete linear structures, it is always possible to express the components of strain as linear functions of the generalized coordinates. Thus, for a system with n degrees of freedom, we may write

$$\begin{aligned} \epsilon_x &= a_{11}\Delta_1 + a_{12}\Delta_2 + \cdots + a_{1n}\Delta_n \\ \epsilon_y &= a_{21}\Delta_1 + a_{22}\Delta_2 + \cdots + a_{2n}\Delta_n \\ &\cdots\cdots\cdots\cdots\cdots\cdots\cdots\cdots \\ \gamma_{xz} &= a_{61}\Delta_1 + a_{62}\Delta_2 + \cdots + a_{6n}\Delta_n \end{aligned} \tag{9.82}$$

where the quantities $a_{11}, a_{12}, \ldots, a_{6n}$ are known functions of the coordinates x, y, z. Equation (a) of Art. 8.7 is clearly a special case of Eqs. (9.82). We also note that for a Hookean material the strain energy density is a quadratic form in the strain components. Thus, it is also a quadratic form in the displacements Δ_i.

For simplicity, let us temporarily confine our attention to structures in which the stress and strain distributions in each member are adequately described by stress components σ_x, τ_{xy} and strain components ϵ_x, γ_{xy}; the remaining components are assumed to be negligible. Such distributions exist, for example, in coplanar beams, frames, and trusses. Equations (9.82) then reduce to

$$\begin{aligned} \epsilon_x &= a_{11}\Delta_1 + a_{12}\Delta_2 + \cdots + a_{1n}\Delta_n \\ \gamma_{xy} &= a_{41}\Delta_1 + a_{42}\Delta_2 + \cdots + a_{4n}\Delta_n \end{aligned} \tag{9.83}$$

and, since $\sigma_x = E\epsilon_x$ and $\tau_{xy} = G\gamma_{xy}$, the strain energy becomes

$$U = \int_V \left(\int_0^F E\epsilon_x \, d\epsilon_x + G\gamma_{xy} \, d\gamma_{xy} \right) dV = \int_V \left(\frac{E\epsilon_x^{\,2}}{2} + \frac{G\gamma_{xy}^{\,2}}{2} \right) dV \tag{9.84}$$

From Castigliano's first theorem,

$$P_i = \frac{\partial U}{\partial \Delta_i} = \int_V \left(E\epsilon_x \frac{\partial \epsilon_x}{\partial \Delta_i} + G\gamma_{xy} \frac{\partial \gamma_{xy}}{\partial \Delta_i} \right) dV$$

and from Eqs. (9.83),

$$\frac{\partial \epsilon_x}{\partial \Delta_i} = a_{1i} \qquad \frac{\partial \gamma_{xy}}{\partial \Delta_i} = a_{4i}$$

Thus

$$P_i = \int_V [E(a_{11}\Delta_1 + \cdots + a_{1n}\Delta_n)a_{1i} + G(a_{41}\Delta_1 + \cdots + a_{4n}\Delta_n)a_{4i}] \, dV$$

or

$$\begin{aligned} P_i = \Delta_1 \int_V (Ea_{11}a_{1i} + Ga_{41}a_{4i}) \, dV + \cdots + \Delta_i \int_V (Ea_{1i}^{\,2} + Ga_{4i}^{\,2}) \, dV \\ + \cdots + \Delta_n \int_V (Ea_{1n}a_{1i} + Ga_{4n}a_{4i}) \, dV \end{aligned} \tag{9.85}$$

By differentiating U with respect to each of the displacements Δ_1, $\Delta_2, \ldots, \Delta_n$, we obtain the set of n independent equations

$$\begin{aligned} P_1 &= k_{11}\Delta_1 + k_{12}\Delta_2 + \cdots + k_{1i}\Delta_i + \cdots + k_{1n}\Delta_n \\ P_2 &= k_{21}\Delta_1 + k_{22}\Delta_2 + \cdots + k_{2i}\Delta_i + \cdots + k_{2n}\Delta_n \\ &\cdots\cdots\cdots\cdots\cdots\cdots\cdots\cdots\cdots\cdots \\ P_i &= k_{i1}\Delta_1 + k_{i2}\Delta_2 + \cdots + k_{ii}\Delta_i + \cdots + k_{in}\Delta_n \\ &\cdots\cdots\cdots\cdots\cdots\cdots\cdots\cdots\cdots\cdots \\ P_n &= k_{n1}\Delta_1 + k_{n2}\Delta_2 + \cdots + k_{ni}\Delta_i + \cdots + k_{nn}\Delta_n \end{aligned} \tag{9.86}$$

where
$$k_{ij} = \int_V (Ea_{1j}a_{1i} + Ga_{4j}a_{4i})\, dV \tag{9.87}$$

The forces $P_1, P_2, \ldots, P_n$ are known quantities. Hence, we may determine the displacements Δ_i by simply solving Eqs. (9.86). We then calculate the strains and, subsequently, the stresses by substituting the Δ's into Eqs. (9.83), and the analysis is completed.

The constants k_{ij} are called *stiffnesses* of the structure.[14] If all of the displacements are zero except Δ_j, and Δ_j is unity, the ith member of Eqs. (9.86) becomes

$$P_i = 0 + 0 + \cdots + k_{ij}1 + \cdots + 0 = k_{ij}$$

Thus, physically, k_{ij} *is the force at i in the direction i due to a unit displacement at j in the direction j.* Noting that $\epsilon_x = a_{1j}$ and $\gamma_{xy} = a_{4j}$ when Δ_j is unity and all other displacements are zero, we see that the quantities Ea_{1j} and Ga_{4j} in Eq. (9.87) are the stresses $(\sigma_x)_j$ and $(\tau_{xy})_j$ due to a unit displacement at j ($\Delta_i = 0$, $i \neq j$). Note that in the case of linearly elastic systems $k_{ij} = k_{ji}$.

The stiffness at point i of a general *nonlinearly elastic* three-dimensional body is

$$k_{ij} = \int_V \left[(\sigma_x)_j \frac{\partial \epsilon_x}{\partial \Delta_i} + (\sigma_y)_j \frac{\partial \epsilon_y}{\partial \Delta_i} + \cdots + (\tau_{xz})_j \frac{\partial \gamma_{xz}}{\partial \Delta_i} \right] dV \tag{9.88}$$

where $(\sigma_x)_j, (\sigma_y)_j, \ldots, (\tau_{xz})_j$ are the stresses produced by a unit displacement of point j, $\Delta_j = 1$. We express Eq. (9.88) more concisely in the symbolic form

$$k_{ij} = \int_V (\sigma)_j \frac{\partial \epsilon}{\partial \Delta_i}\, dV \tag{9.89}$$

Since $\partial\epsilon/\partial\Delta_i = \delta\epsilon^{\Delta_i}$, it follows that, in general, k_{ij} is the internal virtual work done by the true stresses produced by a unit displacement at j in moving through the virtual strains produced by a unit virtual displacement at i.

[14] See page 250.

To illustrate these ideas, let us consider the coplanar rigid frame in Fig. 9.16*a*. To obtain the strain energy of the frame in terms of the generalized coordinates, it is convenient to first calculate the energy stored in giving a straight bar completely general deformations, as is indicated in Fig. 9.16*b*.

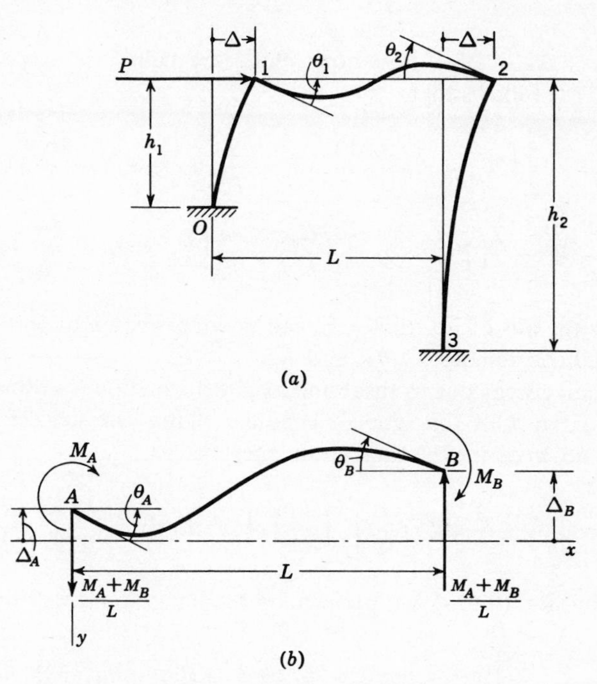

FIGURE 9.16 (*a*) A coplanar frame with rigid joints; (*b*) general deformations of a straight beam.

From elementary beam theory we recall that

$$\epsilon_x = -\frac{y}{\rho} = \frac{d^2v}{dx^2}y \tag{9.90}$$

where, for the beam in Fig. 9.16*b*,

$$\frac{d^2v}{dx^2} = -\frac{M_z}{EI} = \frac{-1}{EI}\left(M_A - \frac{M_A + M_B}{L}x\right)$$

Integrating this equation twice and using the conditions $v(0) = -\Delta_A$ and $dv(0)/dx = \theta_A$, we obtain

$$v = -\frac{1}{EI}\left(M_A\frac{x^2}{2} - \frac{M_A + M_B}{L}\frac{x^3}{6}\right) + \theta_A x - \Delta_A$$

From the conditions $v(L) = -\Delta_B$ and $dv(L)/dx = \theta_B$, we find

$$M_A = \frac{4EI}{L}\theta_A + \frac{2EI}{L}\theta_B + \frac{6EI}{L^2}\Delta_y$$
$$M_B = \frac{2EI}{L}\theta_A + \frac{4EI}{L}\theta_B + \frac{6EI}{L^2}\Delta_y \tag{9.91}$$

where $\Delta_y = \Delta_B - \Delta_A$. Equations (9.91) are called the *slope-deflection equations.* It follows that

$$v = \theta_A x - \frac{1}{L}\left(2\theta_A + \theta_B + \frac{3\Delta_y}{L}\right)x^2 + \frac{1}{L^2}\left(\theta_A + \theta_B + \frac{2\Delta_y}{L}\right)x^3 \tag{9.92a}$$

and
$$\frac{d^2v}{dx^2} = -\frac{2}{L}\left(2\theta_A + \theta_B + \frac{3\Delta_y}{L}\right) + \frac{6}{L^2}\left(\theta_A + \theta_B + \frac{2\Delta_y}{L}\right)x \tag{9.92b}$$

and, with the aid of Eq. (9.90), ϵ_x can now be written in terms of three generalized coordinates θ_A, θ_B, and Δ_y.

The strain energy due to shear and axial deformation is ordinarily small in comparison with that due to bending. Thus, we neglect it for the present, and write for the total strain energy

$$U = \int_V \frac{E(\epsilon_m)^2}{2}\,dV = \frac{E}{2}\int_0^L \left(\frac{d^2v}{dx^2}\right)^2 \int_A y^2\,dA\,dx = \frac{EI}{2}\int_0^L \left(\frac{d^2v}{dx^2}\right)^2 dx$$

Introducing Eq. (9.92*b*) and performing the integration, we obtain

$$U = \frac{2EI}{L}\left[\theta_A{}^2 + \theta_A\theta_B + \theta_B{}^2 + 3\frac{\Delta_y}{L}(\theta_A + \theta_B) + 3\frac{\Delta_y{}^2}{L^2}\right] \tag{9.93}$$

We now return to the analysis of the frame in Fig. 9.16*a*. Owing to our assumption of small deformations, the horizontal displacements of joints 1 and 2 are essentially the same. This is in keeping with elementary beam theory in which $ds = dx$; hence, the distances between end points in the deformed and undeformed beam are the same. This is also consistent with our assumption that the energy due to axial deformations is negligible compared with that due to bending. In view of Eq. (9.92*a*), the displacements (and rotations) of all points between joints of the frame are known in terms of the displacements and rotations of the joints. This equation plays the role of an infinity of conditions of constraint and, in effect, reduces the frame to a system with a finite number of degrees of freedom. Since there are only three unknown joint displacements (θ_1, θ_2, and Δ) in the case of the frame in Fig. 9.16*a*, this structure is three times kinematically indeterminate. Its deformation is thus characterized by only three generalized coordinates.

Assuming that EI is the same for all members, we use Eq. (9.93) to obtain the strain energy in members 01, 12, and 23:

$$U_{01} = \frac{2EI}{h_1}\left(\theta_1^2 - 3\frac{\Delta}{h_1}\theta_1 + 3\frac{\Delta^2}{h_1^2}\right)$$

$$U_{12} = \frac{2EI}{L}(\theta_1^2 + \theta_1\theta_2 + \theta_2^2)$$

$$U_{23} = \frac{2EI}{h_2}\left(\theta_2^2 - 3\frac{\Delta}{h_2}\theta_2 + 3\frac{\Delta^2}{h_2^2}\right)$$

The total strain energy is

$$U = U_{01} + U_{12} + U_{23}$$

Finally, from Castigliano's first theorem, we obtain the equations

$$\begin{aligned}
\frac{\partial U}{\partial \theta_1} &= 0 = \frac{4EI}{h_1}\left(1 + \frac{h_1}{L}\right)\theta_1 + \frac{2EI}{L}\theta_2 - \frac{6EI}{h_1^2}\Delta \\
\frac{\partial U}{\partial \theta_2} &= 0 = \frac{2EI}{L}\theta_1 - \frac{4EI}{h_2}\left(1 + \frac{h_2}{L}\right)\theta_2 - \frac{6EI}{h_2^2}\Delta \\
\frac{\partial U}{\partial \Delta} &= P = -\frac{6EI}{h_1^2}\theta_1 - \frac{6EI}{h_2^2}\theta_2 + \frac{12EI}{h_1^3h_2^3}(h_1^3 + h_2^3)\Delta
\end{aligned} \tag{9.94}$$

We now solve these equations for the displacements. Taking for illustration purposes $h_1 = h_2 = L$, we find

$$\theta_1 = \theta_2 = \frac{PL^2}{28EI} \qquad \text{and} \qquad \Delta = \frac{PL^3}{84EI}$$

We now evaluate end moments and displacements of any point by simply substituting these results into the appropriate relation given by Eqs. (9.91) and (9.92*a*).

On comparing Eqs. (9.94) with Eqs. (9.86), we see that the coefficients of θ_1, θ_2, and Δ are stiffnesses of the structure. The quantity $2EI/L$ in the first equation, for example, is the moment developed at joint 1 due to a unit rotation of joint 2, etc. Physical interpretations of these coefficients are indicated in Fig. 9.17. Also note that Eqs. (9.94) are equilibrium conditions. The first member of Eqs. (9.94), for example, can also be obtained by writing the condition for equilibrium of moments at joint 1 ($M_{10} + M_{12} = 0$) in terms of θ_1, θ_2, and Δ with the aid of Eqs. (9.91). We are again reminded that in equilibrium (stiffness) methods of structural analysis, displacements are taken as unknowns, compatibility is taken for granted at the outset, a number of independent conditions equal to the number of degrees of freedom (kinematic indeterminacy) are established [Eqs. (9.94)], and final displacements are chosen so as to provide equilibrium.

It is interesting to note that the above procedure can also be used to obtain forces and moments at points at which the displacements and

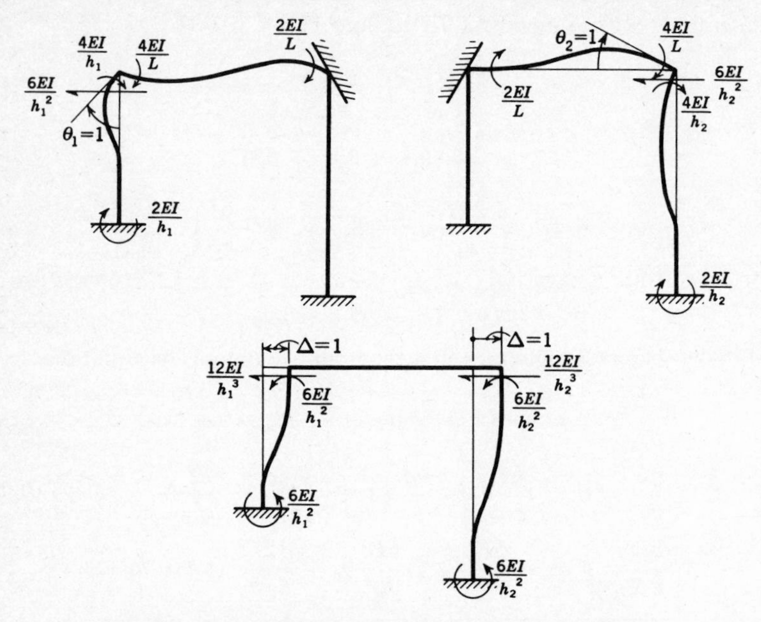

FIGURE 9.17 Physical interpretation of the stiffness coefficients in Eqs. (9.94) corresponding to the frame in Fig. 9.16*a*. Physically, a stiffness coefficient is the force (or moment) developed due to a unit displacement (or rotation) when all other generalized displacements are zero.

rotations are zero (prescribed). Suppose, for example, that we wish to determine the force P_i at a point i whose corresponding displacement Δ_i is zero. To accomplish this, we first treat Δ_i as if it were not zero and compute the strain energy as a function of Δ_i. We then apply Castigliano's first theorem and evaluate P_i by taking the limit as Δ_i approaches zero:

$$P_i = \lim_{\Delta_i \to 0} \frac{\partial U}{\partial \Delta_i} \tag{9.95}$$

For example, to evaluate the end moment at joint 3 of the frame in Fig. 9.16*a*, we calculate U on the assumption that θ_3 is not zero. Then

$$M_{31} = \lim_{\theta_3 \to 0} \frac{\partial U}{\partial \theta_3}$$

Such calculations are often unnecessary, however, because the support moments and forces can usually be obtained from statics once the nonzero generalized displacements and their corresponding forces have been determined.

9.11 The principle of minimum strain energy. If we choose only those virtual displacements which are zero at the points of application of the applied loads, then

$$\delta H = -(P_1 \cdot 0 + P_2 \cdot 0 + \cdots + P_n \cdot 0) = 0$$

Hence, in this case $\delta\Pi$ equals δU and the equilibrium condition becomes

$$\delta U = \frac{\partial U}{\partial \Delta_i} \delta\Delta_i = 0 \tag{9.96}$$

Since the present discussion pertains to stable structures, $\delta^2\Pi$ and, hence, $\delta^2 U$ are positive definite and U is a minimum when the system is in equilibrium. This result is called the *principle of minimum strain energy:*[15]

> If the constraints in a structural system are such that the external forces perform no work (or, if only such virtual displacements are allowed, that no external work is done), then of all the compatible deformations the structure can undergo, those which correspond to stable equilibrium configurations make the strain energy a minimum.
>
> (THEOREM XII)

This principle amounts to a special case of the theorem of minimum potential energy and Castigliano's first theorem. We mention it only for the sake of completeness, and, for our purposes, we need not consider it further.[16]

THE COMPLEMENTARY ENERGY PRINCIPLES

In Chap. 8 we found that virtual work could be performed in a system in two ways: first, by true forces moving through virtual displacements and, second, by virtual forces moving through true displacements. Later, we found that the ideas of virtual displacements formed the basis for the important principles of minimum potential energy. The possibility of formulating the principle of virtual forces in terms of energy is now quite obvious. In fact, we shall find that the translation of the principle into one dealing with variations in the total complementary energy of a system provides us with some of the most useful theorems in structural mechanics.

9.12 The principle of stationary complementary energy. Upon comparing the equations for the total complementary energy [Eqs. (9.21), (9.32), and (9.33)] with those in Chap. 8 for the complementary external and internal

[15] This principle is also referred to as the first principle of least work and the first theorem of minimum strain energy.

[16] For a more detailed discussion of this principle, see Ref. 2, p. 353.

virtual work [Eqs. (8.30) and (8.31)], we find that they are remarkably similar. Consider, for example, an element in a one-dimensional body subjected solely to the stress and strain components σ_x and ϵ_x. The internal virtual work done due to a virtual stress is, by definition,

$$\delta U^* = \epsilon_x \, \delta\sigma_x$$

Recalling from Eqs. (9.29) that

$$\epsilon_x = \frac{\partial U^*}{\partial \sigma_x}$$

we see that the complementary internal virtual work can also be written

$$\delta U^* = \frac{\partial U^*}{\partial \sigma_x} \, \delta\sigma_x$$

which, according to the calculus, represents *a change in U^* due to a variation $\delta\sigma_x$ in the true stresses.* Similarly, since

$$-H^* = W_e^*$$

we may write $-\delta H^*$ for the external complementary virtual work. Thus, the principle of virtual forces may be written in the form

$$\delta U^* + \delta H^* = 0$$

or, in view of the distributive property of δ,

$$\delta(U^* + H^*) = 0$$

Finally, introducing Eq. (9.33) into this result, we get

$$\delta\Pi^* = 0 \tag{9.97}$$

Therefore, we see that the principle of virtual forces[17] can also be stated as follows:

> The displacements in a deformable system are compatible and consistent with the constraints if the first variation in the total complementary energy is zero for every virtual force system satisfying the equilibrium conditions.
>
> (THEOREM XIII)

To translate this theorem into a more meaningful form, let us consider a structural system in which the complementary energy is given in terms of only two stress components σ_x and σ_y. By giving σ_x and σ_y variations $\delta\sigma_x$ and $\delta\sigma_y$, which are themselves in equilibrium, we obtain the complementary energy in neighboring equilibrium states. The first variation in Π^* due to these virtual stresses,

$$\delta\Pi^* = \frac{\partial \Pi^*}{\partial \sigma_x} \, \delta\sigma_x + \frac{\partial \Pi^*}{\partial \sigma_y} \, \delta\sigma_y$$

[17] See Theorem VI, Chap. 8.

vanishes in those particular equilibrium states which are also compatible. Since $\delta\sigma_x$ and $\delta\sigma_y$ are arbitrary (except for satisfying the equilibrium conditions), the vanishing of $\delta\Pi^*$ implies the vanishing of $\partial\Pi^*/\partial\sigma_x$ and $\partial\Pi^*/\partial\sigma_y$. Hence, when $\delta\Pi^*$ is zero,

$$d\Pi^* = \frac{\partial\Pi^*}{\partial\sigma_x}\,d\sigma_x + \frac{\partial\Pi^*}{\partial\sigma_y}\,d\sigma_y = 0$$

which means that Π^* assumes a stationary value.

These observations are summarized by the *principle of stationary complementary energy:*

> Of all the stresses and forces which satisfy the equilibrium conditions of a structural system, those corresponding to the true compatible state of deformation make the total complementary energy assume a stationary value.
>
> (THEOREM XIV)

Furthermore, in certain cases we may also argue that Π^* is a minimum. For example, if $\Pi^*(\sigma_x + \delta\sigma_x, \sigma_y + \delta\sigma_y)$ is the complementary energy in an equilibrium state neighboring the true compatible one $\Pi^*(\sigma_x,\sigma_y)$, the change in Π^* due to $\delta\sigma_x$ and $\delta\sigma_y$ is obtained from Taylor's formula:

$$\Delta\Pi^* = \delta\Pi^* + \tfrac{1}{2}\,\delta^2\Pi^* + \cdots$$

in which $\delta\Pi^*$ is zero, from Eq. (9.97), and

$$\delta^2\Pi^* = \frac{\partial^2\Pi^*}{\partial\sigma_x^{\,2}}\,(\delta\sigma_x)^2 + 2\,\frac{\partial^2\Pi^*}{\partial\sigma_x\,\partial\sigma_y}\,(\delta\sigma_x)(\delta\sigma_y) + \frac{\partial^2\Pi^*}{\partial\sigma_y^{\,2}}\,(\delta\sigma_y)^2$$

is the second variation in Π^*. Thus, $\delta^2\Pi^*$ is a quadratic form in the virtual stresses (or forces). In the compatible stable equilibrium configurations of certain structures, it can be shown that $\delta^2\Pi^*$ is positive definite,[18] in which case $\Delta\Pi^*$ is positive and, therefore, Π^* is a minimum.

It is important to appreciate the relationship between the principles of stationary potential energy and stationary complementary energy. The potential-energy principle was derived from the principle of virtual displacements and, hence, provides equilibrium conditions. The complementary-energy principle was derived from the principle of virtual forces and, hence, provides compatibility conditions.

To fix ideas, let us consider the simple system in Fig. 9.18, which consists of a nonlinear spring and an external force F. The material

[18] Consider, in particular, the case of a linearly elastic structure undergoing small displacements for which the matrix of the quadratic form $\delta^2\Pi^*$ is the inverse of that for $\delta^2\Pi$. For stable equilibrium, we recall that $\delta^2\Pi$ is positive definite. But the inverse of a real symmetric positive definite matrix is also positive definite (see Ref. 29, p. 257). Therefore, $\delta^2\Pi^*$ is also positive definite and Π^* is a minimum.

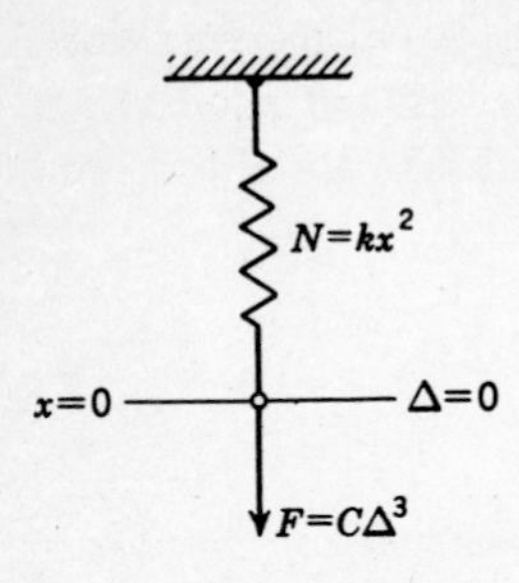

FIGURE 9.18 Nonlinear spring.

properties of the spring are such that an elongation x produces an internal force $N = kx^2$, k being a known constant. The force F is assumed to be proportional to the cube of its displacement Δ; that is, $F = C\Delta^3$, where C is a known constant. The "energies" of the system are as follows:

$$U = \int_0^x N\,dx = \int_0^x kx^2\,dx = \frac{k}{3}x^3 \qquad (a)$$

$$U^* = \int_0^N x\,dN = \int_0^N \left(\frac{N}{k}\right)^{\frac{1}{2}} dN = \frac{2}{3\sqrt{k}}N^{\frac{3}{2}} \qquad (b)$$

$$H = -\int_0^\Delta F\,d\Delta = -\int_0^\Delta C\Delta^3\,d\Delta = -\frac{C}{4}\Delta^4 \qquad (c)$$

$$H^* = -\int_0^F \Delta\,dF = -\int_0^F \left(\frac{F}{C}\right)^{\frac{1}{3}} dF = -\frac{3}{4C^{\frac{1}{3}}}F^{\frac{4}{3}} \qquad (d)$$

To compare the two concepts, let us now apply both energy principles simultaneously:

Stationary potential energy

We express the total potential energy in terms of displacements,

$$\Pi = \frac{k}{3}x^3 - \frac{C}{4}\Delta^4$$

and assume compatibility at the outset,

$$x = \Delta$$

From the principle of stationary potential energy,

$$\frac{\partial \Pi}{\partial \Delta} = 0 = kx^2\frac{\partial x}{\partial \Delta} - C\Delta^3$$

since $\quad \delta\Pi = \dfrac{\partial \Pi}{\partial \Delta}\delta\Delta = 0$

but $\quad \dfrac{\partial x}{\partial \Delta} = 1$

because the deformations are compatible. Therefore

$$kx^2 = C\Delta^3$$

or $\quad N = F$

which we recognize as the equilibrium condition.

Stationary complementary energy

We express the total complementary energy in terms of forces,

$$\Pi^* = \frac{2}{3\sqrt{k}}N^{\frac{3}{2}} - \frac{3}{4C^{\frac{1}{3}}}F^{\frac{4}{3}}$$

and assume equilibrium at the outset,

$$N = F$$

From the principle of stationary complementary energy,

$$\frac{\partial \Pi^*}{\partial F} = 0 = \frac{1}{\sqrt{k}}N^{\frac{1}{2}}\frac{\partial N}{\partial F} - \frac{1}{C^{\frac{1}{3}}}F^{\frac{1}{3}}$$

since $\quad \delta\Pi^* = \dfrac{\partial \Pi^*}{\partial F}\delta F = 0$

but $\quad \dfrac{\partial N}{\partial F} = 1$

because the forces are in equilibrium. Therefore

$$\left(\frac{N}{k}\right)^{\frac{1}{2}} = \left(\frac{F}{C}\right)^{\frac{1}{3}}$$

or $\quad x = \Delta$

which we recognize as the compatibility condition.

Note that

$$\frac{\partial U}{\partial \Delta} = kx^2 \frac{\partial x}{\partial \Delta} = N \frac{\partial x}{\partial \Delta} = F$$

in agreement with Castigliano's first theorem. It is also interesting to note that

$$\frac{\partial U^*}{\partial F} = \left(\frac{N}{k}\right)^{\frac{1}{2}} \frac{\partial N}{\partial F} = x \frac{\partial N}{\partial F} = \Delta$$

a result which will have more significance after the following article.

9.13 Engesser's first theorem.[19] Consider, once again, the nonlinearly elastic three-dimensional body of general shape in Fig. 9.14. We assume that the body is homogeneous and isotropic, that displacements are small[20] and compatible, and that the body is in a state of stable equilibrium under the action of the system of n forces and moments P_1, $P_2, \ldots, P_n$. Again, let $\Delta_1, \Delta_2, \ldots, \Delta_n$ be the displacements corresponding to these forces. If we now regard $P_1, P_2, \ldots, P_n$ as being reactions produced by a set of prescribed displacements $\Delta_1, \Delta_2, \ldots, \Delta_n$, and if the Δ's are independent of these forces, we can express the external complementary energy in the form [see Eq. (9.23)]

$$H^* = -\sum_{i=1}^{n} P_i \Delta_i \tag{9.98}$$

If the stresses are expressed in terms of the forces P_i, the total complementary energy of the body becomes a function of $P_1, P_2, \ldots, P_n$:

$$\Pi^*(P_1, P_2, \ldots, P_n) = U^* - \sum_{i=1}^{n} P_i \Delta_i \tag{9.99}$$

Hence, forces P_i serve as the "generalized" forces of the system. If we now give the force system small variations δP_i which, themselves, satisfy the equilibrium conditions, there results a variation in the total

[19] First presented in 1889 by F. Engesser (Ref. 19). The theorem is often referred to as the generalized form of Castigliano's second theorem. See Art. 9.14.

[20] With certain modifications, these principles can be applied to systems undergoing large displacements, but U^* is no longer the sum of the complementary energies of each part of the structure and Π^* is no longer given by Eq. (9.33) unless H^* and U^* are appropriately redefined. See for example, Ref. 36, p. 136. The developments presented in the remainder of this chapter are based on the assumption of small displacements.

complementary energy $\delta\Pi^*$, which must vanish since the configuration of the body is a compatible one. Thus,

$$\delta\Pi^* = 0 = \frac{\partial U^*}{\partial P_1}\,\delta P_1 + \frac{\partial U^*}{\partial P_2}\,\delta P_2 + \cdots + \frac{\partial U^*}{\partial P_n}\,\delta P_n - \Delta_1\,\delta P_1 - \Delta_2\,\delta P_2 - \cdots - \Delta_n\,\delta P_n$$

Rearranging terms, we have

$$\left(\frac{\partial U^*}{\partial P_1} - \Delta_1\right)\delta P_1 + \left(\frac{\partial U^*}{\partial P_2} - \Delta_2\right)\delta P_2 + \cdots + \left(\frac{\partial U^*}{\partial P_n} - \Delta_n\right)\delta P_n = 0$$

Since, except for satisfying the equilibrium conditions, the variations $\delta P_1, \delta P_2, \ldots, \delta P_n$ are completely arbitrary, each term in parentheses must vanish independently. That is,

$$\frac{\partial U^*}{\partial P_1} - \Delta_1 = 0 \qquad \frac{\partial U^*}{\partial P_2} - \Delta_2 = 0 \qquad \cdots \qquad \frac{\partial U^*}{\partial P_n} - \Delta_n = 0$$

or, in general,

$$\frac{\partial U^*}{\partial P_i} = \Delta_i \tag{9.100}$$

Equation (9.100) is a mathematical statement of *Engesser's first theorem:*

> If the complementary strain energy of a structural system undergoing small displacements is expressed in terms of the n independent forces corresponding to a system of n prescribed displacements $\Delta_1, \Delta_2, \ldots, \Delta_n$, the first partial derivative of the complementary strain energy with respect to any force P_i at point i is equal to the displacement Δ_i at i in the direction of P_i.
>
> (THEOREM XV)

By comparing Eq. (9.100) with Eqs. (9.29), we see that Eqs. (9.29) may be interpreted as a "microscopic" form of Eq. (9.100). It is also interesting to compare the above theorem with Castigliano's first theorem.[21]

It is important to note the identity of Engesser's first theorem and Eq. (8.45) of the general unit-dummy-load method. *The internal complementary virtual work due to a virtual force* δP_i, we recall, *is equal to the first variation in the complementary strain energy due to a variation* δP_i,

[21] See Theorem XI, Art. 9.9.

in the true force P_i, provided δP_i satisfies the equilibrium conditions. Thus, from the principle of virtual forces [Eq. (8.33)]

$$\delta W_e^* = \Delta_i\, \delta P_i = \delta U^* = \int_V (\epsilon\, \delta\sigma^p)\, dV = \frac{\partial U^*}{\partial P_i}\, \delta P_i$$

gives Eq. (9.100) immediately. In view of this identity in results, we may interpret the virtual stresses $\delta\sigma^P$ in the above equation:

$$\delta U^* = \frac{\partial U^*}{\partial P_i}\, \delta P_i = \int_V \left(\frac{\partial U_0^*}{\partial \sigma_x} \frac{\partial \sigma_x}{\partial P_i} + \cdots + \frac{\partial U_0^*}{\partial \tau_{xz}} \frac{\partial \tau_{xz}}{\partial P_i} \right) \delta P_i\, dV$$

Introducing Eqs. (9.29)

$$\delta U^* = \int_V \left(\epsilon_x \frac{\partial \sigma_x}{\partial P_i}\, \delta P_i + \epsilon_y \frac{\partial \sigma_y}{\partial P_i}\, \delta P_i + \cdots + \gamma_{xz} \frac{\partial \tau_{xz}}{\partial P_i}\, \delta P_i \right) dV$$

$$= \int_V (\epsilon_x\, \delta\sigma_x^{P_i} + \epsilon_y\, \delta\sigma_y^{P_i} + \cdots + \gamma_{xz}\, \delta\tau_{xz}^{P_i})\, dV$$

from which it follows that

$$\delta\sigma_x^{P_i} = \frac{\partial \sigma_x}{\partial P_i}\, \delta P_i \qquad \delta\sigma_y^{P_i} = \frac{\partial \sigma_y}{\partial P_i}\, \delta P_i \qquad \cdots \qquad \delta\tau_{xz}^{P_i} = \frac{\partial \tau_{xz}}{\partial P_i}\, \delta P_i \tag{9.101}$$

9.14 Castigliano's second theorem.[22] A special case of Engesser's first theorem which is applicable only to linearly elastic structures is called *Castigliano's second theorem.* According to Eq. (9.31), in such structures the strain energy is equal to the complementary strain energy,

$$U = U^*$$

Therefore, Eq. (9.100) can be written

$$\frac{\partial U}{\partial P_i} = \Delta_i \tag{9.102}$$

in which it is understood that U is to be expressed in terms of the forces P_i.

Equation (9.102) is a statement of *Castigliano's second theorem.* We emphasize that it is applicable only to linearly elastic systems.

9.15 Deflections of simple structures. Engesser's first theorem and Castigliano's second theorem provide powerful tools for calculating deflections of structural systems. We now illustrate their utility by considering a number of examples.

[22] This theorem is also referred to as Castigliano's theorem, Part II.

THREE-DIMENSIONAL BARS. Let us first consider the three-dimensional cantilevered bar of general shape in Fig. 9.19*a*. The bar is subjected to a general system of transverse loads and is assumed to be linearly elastic. A typical element of the bar is shown in Fig. 9.19*b*. Cross sections normal to the bar's geometric axis are located with respect to the fixed end by the coordinate s, indicated in the figure, and cross-sectional coordinates x, y, and z are established at the centroid of each section. It is required to determine the displacement at a point i on the bar in a specified direction n.

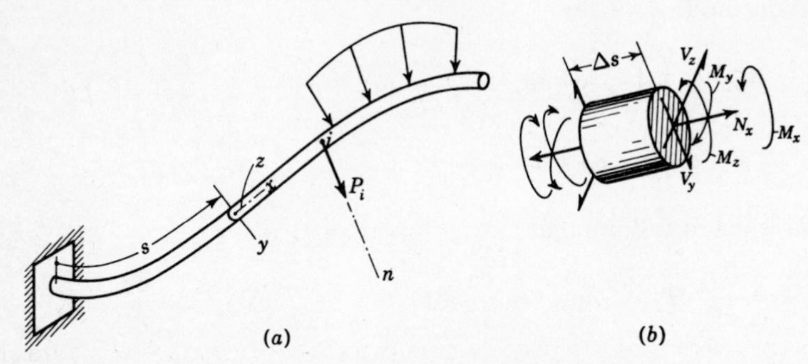

FIGURE 9.19 (*a*) Three-dimensional bar under general loading; (*b*) typical element of the bar.

Assuming that y and z are principal axes and that the effects of restrained warping are negligible, the stresses developed on a typical section are given by the familiar formulas

$$\sigma_s = \frac{N_s}{A} + \frac{M_y z}{I_y} + \frac{M_z y}{I_z}$$

$$\tau_{xy} = \frac{V_y Q_z}{I_z b} + \frac{M_x c_y(y,z)}{J} \tag{9.103}$$

$$\tau_{xz} = \frac{V_z Q_y}{I_y b} + \frac{M_x c_z(y,z)}{J}$$

where M_x is the twisting moment and $c_y(y,z)$ and $c_z(y,z)$ are functions of y and z which depend upon the cross-sectional geometry. All other stress components are assumed to be negligible.

We find from Eqs. (9.15), (9.16), and (9.31) that

$$U^* = U = \iiint \left(\frac{\sigma_x^{\,2}}{2E} + \frac{\tau_{xy}^{\,2}}{2G} + \frac{\tau_{xz}^{\,2}}{2G} \right) dA\, ds \tag{9.104}$$

Substituting Eqs. (9.103) into this relation, we get

$$U^* = \int_0^S \left\{ \int_A \left[\frac{1}{2E}\left(\frac{N_s}{A} + \frac{M_y z}{I_y} + \frac{M_z y}{I_z}\right)^2 + \frac{V_y^{\,2}}{2G}\left(\frac{Q_z}{I_z b}\right)^2 + \frac{V_z^{\,2}}{2G}\left(\frac{Q_y}{I_y b}\right)^2 + \frac{M_x^{\,2}}{2GJ^2}(c_y^{\,2} + c_z^{\,2}) \right] dA \right\} ds \quad (9.105)$$

where S is the total arc length of the bar's geometric axis and A is the cross-sectional area. Noting that

$$\int_A y\, dA = \int_A z\, dA = \int_A yz\, dA = 0$$

$$\int_A y^2\, dA = I_z \qquad \int_A z^2\, dA = I_y$$

and, approximately,

$$\int_A (c_y^{\,2} + c_z^{\,2})\, dA = J$$

and assuming that the cross-sectional dimensions of the bar are constant throughout its length, we integrate Eq. (9.105) over the area and obtain the equation

$$U^* = \int_0^S \left(\frac{N_s^{\,2}}{2EA} + \frac{M_y^{\,2}}{2EI_y} + \frac{M_z^{\,2}}{2EI_z} + \frac{\kappa_z V_y^{\,2}}{2GA} + \frac{\kappa_y V_z^{\,2}}{2GA} + \frac{M_x^{\,2}}{2GJ} \right) ds \quad (9.106)$$

where κ_y and κ_z are the shear correction factors[23]

$$\kappa_y = \frac{A}{I_y^{\,2}} \int_A \left(\frac{Q_y}{b}\right)^2 dA \quad (9.107a)$$

$$\kappa_z = \frac{A}{I_z^{\,2}} \int_A \left(\frac{Q_z}{b}\right)^2 dA \quad (9.107b)$$

Note that for rectangular sections such as that in Fig. 9.20,

$$\kappa_y = \kappa_z = \frac{bh}{bh^3/12} \int_{-h/2}^{+h/2} \frac{1}{b^2} \left[\frac{b}{2}\left(\frac{h^2}{4} - y^2\right) \right]^2 b\, dy = \frac{6}{5} = 1.20 \quad (9.108)$$

For thin-walled sections, we find that these factors are close to unity.

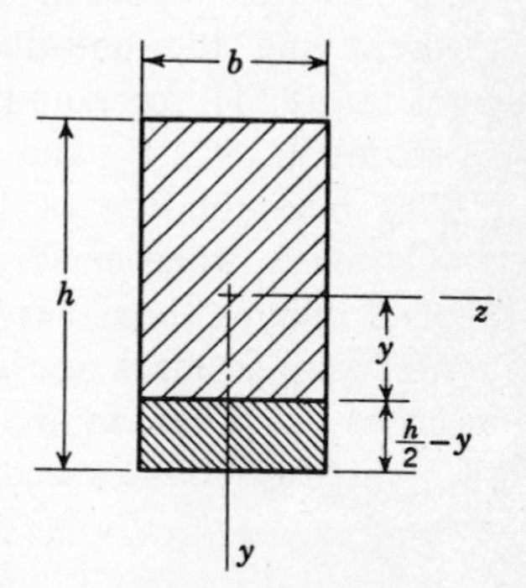

FIGURE 9.20 Rectangular cross section.

[23] See Eq. (6.23).

Since the bar in Fig. 9.19 is statically determinate, we can use simple statics to express the stress resultants in Eq. (9.106) in terms of the force P_i. Then, according to Eq. (9.100),

$$\Delta_i = \frac{\partial U^*}{\partial P_i} = \int_0^S \left(\frac{N_s \, \partial N_s/\partial P_i}{EA} + \frac{M_y \, \partial M_y/\partial P_i}{EI_y} + \frac{M_z \, \partial M_z/\partial P_i}{EI_z} + \kappa_z \frac{V_y \, \partial V_y/\partial P_i}{GA} + \kappa_y \frac{V_z \, \partial V_z/\partial P_i}{GA} + \frac{M_x \, \partial M_x/\partial P_i}{GJ} \right) ds \quad (9.109)$$

Since the structure is linearly elastic, the stress resultants in Eq. (9.109) are linear in the applied loads. Doubling the load, for example, doubles the stress resultants, etc. It follows that the partial derivatives in this equation represent physically the values of the corresponding stress resultant due to a unit (virtual) force at i in direction n. Thus, we can rewrite Eq. (9.109) in the form

$$\Delta_i = \int_0^S \left(\frac{N_s n_s}{EA} + \frac{M_y m_y}{EI_y} + \frac{M_z m_z}{EI_z} + \kappa_z \frac{V_y v_y}{GA} + \kappa_y \frac{V_z v_z}{GA} + \frac{M_x m_x}{GJ} \right) ds \quad (9.110)$$

where

$$n_s = \frac{\partial N_s}{\partial P_i} \qquad m_y = \frac{\partial M_y}{\partial P_i} \qquad m_z = \frac{\partial M_z}{\partial P_i}$$
$$v_y = \frac{\partial V_y}{\partial P_i} \qquad v_z = \frac{\partial V_z}{\partial P_i} \qquad m_x = \frac{\partial M_x}{\partial P_i} \quad (9.111)$$

and n_s = the axial force due to a unit virtual force at i in direction n $(P_i = 1)$

m_y = the bending moment due to a unit virtual force at i in direction n $(P_i = 1)$

. . . .

With this interpretation, the relationship between Castigliano's second theorem and the unit-dummy-load method becomes quite obvious. Equation (8.51), for example, can be obtained directly from Eq. (9.110) by setting M_y, V_y, V_z, and M_x equal to zero.

This interpretation of Eq. (9.109) also suggests the procedure for evaluating Δ_i when no force actually exists at i. If such is the case, we apply a dummy (imaginary) force P_i at this point in the direction of the desired displacement and then evaluate the stress resultants and U^* in terms of P_i. Then, in analogy with Eq. (9.95), we apply Engesser's first theorem (Castigliano's second theorem) and take the limit as P_i approaches zero:

$$\Delta_i = \lim_{P_i \to 0} \frac{\partial U^*}{\partial P_i} \quad (9.112)$$

Let us now apply these results to the bar shown in Fig. 9.21. The bar is subjected to a concentrated force P at end A and it is required that we determine the displacement of A in the direction of P. We begin by establishing convenient coordinate systems s, s', and s'', as indicated in the figure. Then, from statics, we find

Member AB:

$$M_z = -Ps \qquad V_y = -P \qquad M_x = 0$$

Member BC:

$$M_z = -Ps' \qquad V_y = -P \qquad M_x = -PL$$

Member CD:

$$M_z = -P(L + s'') \qquad V_y = -P \qquad M_x = PL$$

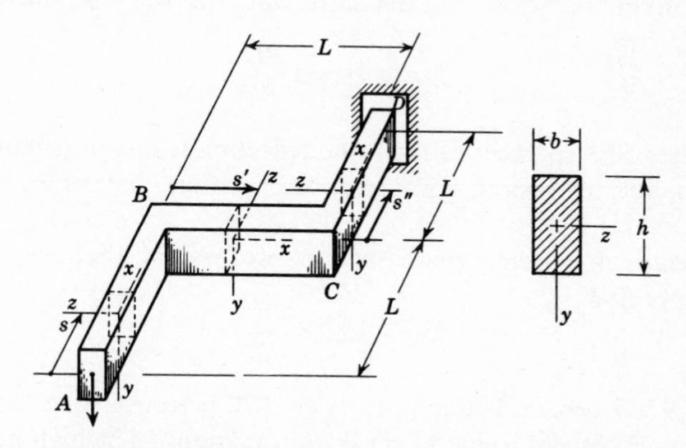

FIGURE 9.21 Cantilevered bar of rectangular cross section.

Clearly N_s is zero and, for our choice of coordinates, M_y and V_z are zero for all members.

Introducing these functions in Eq. (9.106), we find for the complementary strain energy of the bar

$$U^* = \int_0^L \left(\frac{P^2 s^2}{2EI_z} + \frac{1.2P^2}{2GA} \right) ds + \int_0^L \left(\frac{P^2 s'^2}{2EI_z} + \frac{1.2P^2}{2GA} + \frac{P^2L^2}{2GJ} \right) ds'$$
$$+ \int_0^L \left[\frac{P^2(L + s'')^2}{2EI_z} + \frac{1.2P^2}{2GA} + \frac{P^2L^2}{2GJ} \right] ds''$$

Here we have set $\kappa_z = 1.2$ since the cross section is rectangular. We now use Eq. (9.109) to obtain the displacement of point A in the direction of P:

$$\Delta_A = \frac{\partial U}{\partial P} = \int_0^L \left(\frac{Ps^2}{EI_z} + \frac{1.2P}{GA} \right) ds + \int_0^L \left(\frac{Ps'^2}{EI_z} + \frac{1.2P}{GA} + \frac{PL^2}{GJ} \right) ds'$$
$$+ \int_0^L \left[\frac{P(L + s'')^2}{EI_z} + \frac{1.2P}{GA} + \frac{PL^2}{GJ} \right] ds''$$

or

$$\Delta_A = \frac{3PL^3}{EI_z} + \frac{3.6PL}{GA} + \frac{2PL^3}{GJ}$$

For illustration purposes, let us assume that $E/G = 2.5$ and $b/h = 1$. Then

$$I_z = \frac{h^4}{12} \qquad A = h^2 \qquad \text{and}^{24} \qquad J = 0.141h^4$$

and

$$\Delta_A = \frac{PL^3}{Eh^4}\left[36 + 9\left(\frac{h}{L}\right)^2 + 35.461\right]$$

The first term within the brackets indicates the deflection due to bending, the second term is due to shear deformation, and the third term is due to torsion. Note that in this case the deflection due to bending is practically the same as that due to torsion whereas that due to shear deformation depends upon the square of the ratio h/L. If we take, for example, $h/L = \frac{1}{10}$, then

$$\Delta_A = 71.551\,\frac{PL^3}{Eh^4}$$

In this case, 50.31 percent of the total deflection is due to bending, 49.56 percent is due to torsion, and shear deformation contributes only 0.13 percent.

If, instead of a long square section, we assume that $b/h = \frac{1}{10}$ and $h/L = \frac{1}{4}$, we find

$$\Delta_A = 16{,}392\,\frac{PL^3}{Eh^4}$$

of which 97.77 percent is due to torsion, 2.20 percent is due to bending, and only 0.03 percent is due to shear deformation. The high percentage of torsional deflection, of course, is due to the very low torsional stiffness of thin narrow sections. Note that no appreciable error is introduced by completely neglecting shear deformations.

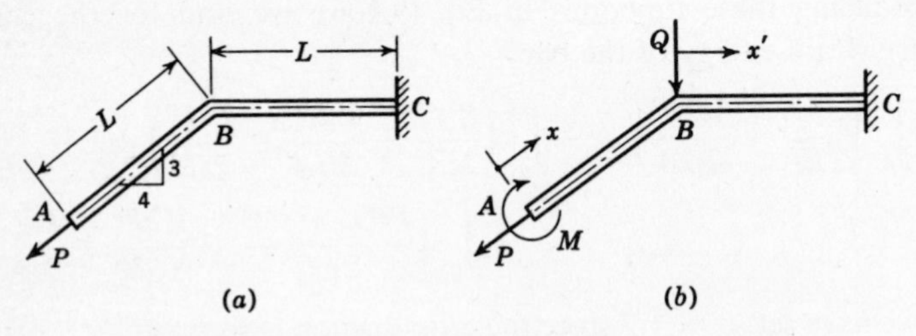

FIGURE 9.22 (*a*) Coplanar bar subjected to an end force P; (*b*) dummy moment and vertical force introduced in order to evaluate the rotation of end A and the vertical displacement of point B.

DEFLECTIONS OF UNLOADED POINTS. Let us now demonstrate the procedure to be used to calculate displacements of points at which no forces are applied. Consider, for example, the coplanar bar in Fig. 9.22*a* and

[24] See Eqs. (3.33) and Table 3.1.

suppose that we wish to find the rotation of end A and the vertical displacement of point B. Since P is only applied force acting on the structure, we must introduce dummy forces and moments corresponding to the deformations which we wish to evaluate. In the present case, this means that we must apply imaginary moment M at point A and an imaginary vertical force Q at point B, as indicated in Fig. 9.22*b*.

From statics, we find

Member AB:

$$M_z = M \qquad N_s = P \qquad V_y = 0$$

Member BC:

$$M_z = M - (\tfrac{3}{5}P + Q)x' \qquad N_s = \tfrac{4}{5}P \qquad V_y = -(\tfrac{3}{5}P + Q)$$

where x' is the coordinate indicated in the figure. Thus,

$$U^* = \int_0^L \left(\frac{M^2}{2EI} + \frac{P^2}{2EA} \right) dx + \int_0^L \left\{ \frac{[M - (\tfrac{3}{5}P + Qx')]^2}{2EI} + \frac{16P^2}{50EA} + \kappa \frac{(\tfrac{3}{5}P + Q)^2}{2GA} \right\} dx'$$

or

$$U^* = \frac{M^2L}{2EI} - \frac{3MP}{5EI} - \frac{MQL^2}{2EI} + \frac{9P^2}{50EI} + \frac{3PQL^2}{10EI} + \frac{Q^2L^3}{6EI} + \frac{16P^2L}{50EA} + \kappa \frac{(\tfrac{3}{5}P + Q)^2}{2GA}$$

According to Eq. (9.112), the rotation of end A is obtained by differentiating U^* with respect to M and taking the limit as M and Q approach zero:

$$\theta_A = \lim_{M,Q \to 0} \frac{\partial U^*}{\partial M} = -\frac{3P}{5EI}$$

The minus sign indicates that A rotates in a direction opposite to that of M. Similarly

$$\Delta_B = \lim_{M,Q \to 0} \frac{\partial U^*}{\partial Q} = \frac{3PL^2}{10EI} + \frac{3\kappa P}{5GA}$$

DEFORMATIONS OF TRUSSES. From the above example we note that rotations (slopes) are obtained by differentiating U^* with respect to the corresponding moments and that displacements are obtained by differentiating U^* with respect to the corresponding forces. In fact, practically any type of deformation can be obtained in this manner if we formulate U^* properly.

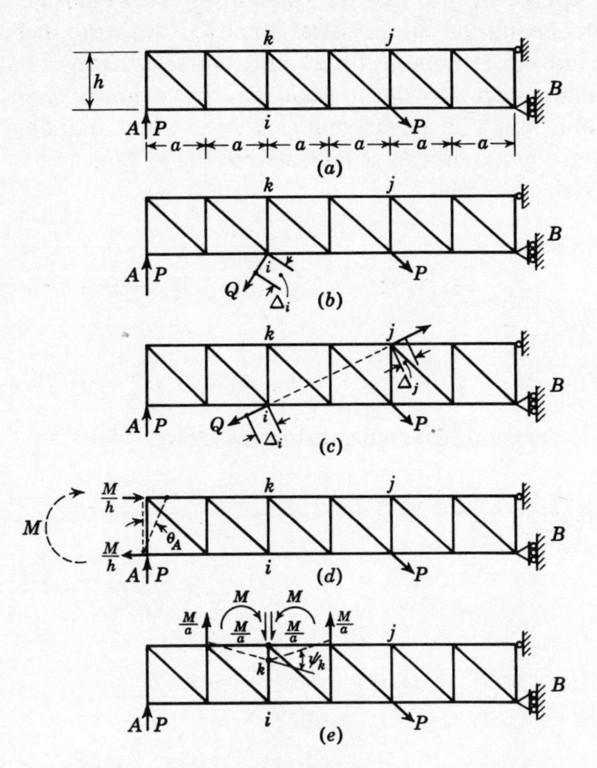

FIGURE 9.23 Illustration of procedures to be used to calculate various types of deformation of a simple coplanar truss. (*a*) The truss; (*b*) displacement of joint i; (*c*) relative displacement of joints i and j; (*d*) rotation of end A; (*e*) change in slope at joint k.

To emphasize this point, consider the coplanar truss shown in Fig. 9.23*a*, which is subjected to a general system of joint forces. To evaluate the displacement of joint i, we simply apply a dummy force Q at i (Fig. 9.23*b*) and compute U^* as a function of Q and the applied loads. Then, as demonstrated previously,

$$\Delta_i = \lim_{Q \to 0} \frac{\partial U^*}{\partial Q} = \sum_{m=1}^{n} N_m \frac{\partial N_m}{\partial Q} \lambda_m \tag{9.113a}$$

where N_m is the axial force in member m, $\lambda_m = L_m/EA_m$, and n is the number of truss members.

Relative displacements of joints can be evaluated in a similar manner. For example, the displacement of joint j relative to i is obtained by applying a pair of opposing forces Q at i and j as indicated in Fig. 9.23*c*.

Then

$$\Delta_{ij} = \Delta_i + \Delta_j = \lim_{Q \to 0} \frac{\partial U^*}{\partial Q} = \sum_{m=1}^{n} N_m \frac{\partial N_m}{\partial Q} \lambda_m \tag{9.113b}$$

where Δ_{ij} is the displacement of j relative to i.

The rotation of end A, of course, is obtained by applying a couple M at A as indicated in Fig. 9.23*d*. Then

$$\theta_A = \lim_{M \to 0} \frac{\partial U^*}{\partial M} = \sum_{m=1}^{n} N_m \frac{\partial N_m}{\partial M} \lambda_m \tag{9.113c}$$

If we apply opposing couples at joint k, as indicated in Fig. 9.23*e*, we obtain the change in slope at k:

$$\psi_k = \lim_{M \to 0} \frac{\partial U^*}{\partial M} = \sum_{m=1}^{n} N_m \frac{\partial N_m}{\partial M} \lambda_m \tag{9.113d}$$

NONLINEARLY ELASTIC BEAMS. We previously noted that applications of Engesser's first theorem are not restricted to linearly elastic structural systems. As an illustration of its application to a materially nonlinear system, let us compute the transverse displacement of end A of the nonlinearly elastic beam in Fig. 9.24. For small strains, the material is assumed to exhibit a parabolic stress-strain relation

$$\sigma_x = E(\operatorname{sgn} \epsilon_x)\epsilon_x^{\frac{1}{2}}$$

where E is a material constant and $\operatorname{sgn} \epsilon_x = 1$ if $\epsilon_x > 0$, -1 if $\epsilon_x < 0$, and 0 if $\epsilon_x = 0$. From elementary beam theory

$$\epsilon_x = \frac{y}{\rho} = \left(\frac{\sigma_x}{E}\right)^2$$

where $1/\rho$ is the curvature of the deformed axis of the beam. Thus, if we neglect shear deformations, the complementary strain energy is

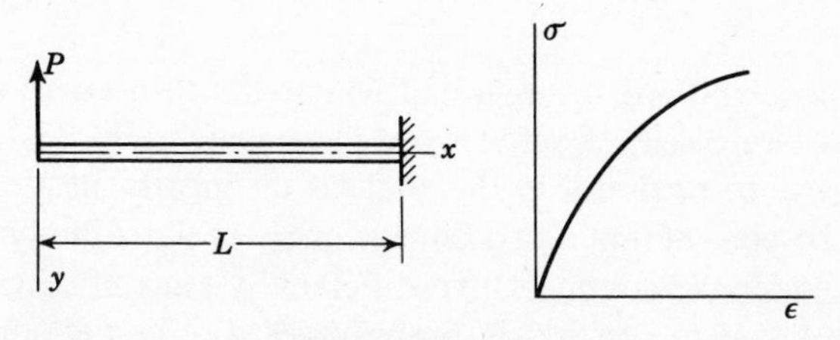

FIGURE 9.24 Nonlinearly elastic cantilever beam. The material is assumed to exhibit a parabolic stress-strain relation $\sigma_x = E(\operatorname{sgn} \epsilon_x)\epsilon_x^{\frac{1}{2}}$.

given by

$$U^* = \iiint_V \left(\int_0^{\sigma_x} \epsilon_x \, d\sigma_x \right) dV$$

$$= \iiint_V \left(\int_0^{\sigma_x} \frac{\sigma_x^{\,2}}{E^2} \, d\sigma_x \right) dV$$

$$= \iiint_V \frac{\sigma_x^{\,3}}{3E^2} \, dV$$

The bending moment developed at each section is, by definition,

$$M_z = \int_A \sigma_x y \, dA = \int_A E \frac{y^{\frac{3}{2}}}{\rho^{\frac{1}{2}}} \, dA = \frac{EI^*}{\rho^{\frac{1}{2}}} = \frac{I^* \sigma_x}{y^{\frac{1}{2}}}$$

where

$$I^* = \int_A y^{\frac{3}{2}} \, dA$$

Therefore

$$\sigma_x = \frac{M_z y^{\frac{1}{2}}}{I^*}$$

and

$$U^* = \int_0^L \int_A \frac{1}{3E} \left(\frac{M_z y^{\frac{1}{2}}}{I^*} \right)^3 dA \, dx$$

or

$$U^* = \int_0^L \frac{M_z^{\,3}}{3(EI^*)^2} \, dx \tag{9.114}$$

For the beam in Fig. 9.24

$$M_z = Px$$

so that

$$U^* = \frac{P^3 L^4}{12(EI^*)^2}$$

Therefore, we obtain for the deflection of end A

$$\Delta_A = \frac{\partial U^*}{\partial P} = \frac{P^2 L^4}{4(EI^*)^2}$$

9.16 The Rayleigh-Betti theorem and Maxwell's theorem of reciprocity. In this article, we discuss briefly two well-known "reciprocal" theorems that often prove to be useful in the analysis of linearly elastic structures.

We begin by considering a two-dimensional Hookean body which is in equilibrium under the action of two different systems of external forces. The first force system consists of body forces X_b, Y_b and surface forces X_s, Y_s which produce displacements u and v; the second force system consists of body forces $\bar{X}_b$, $\bar{Y}_b$ and surface forces $\bar{X}_s$, $\bar{Y}_s$ which produce displacements $\bar{u}$ and $\bar{v}$. By introducing Hooke's law and the strain-displacement relations and by successive applications of Green's theorem, it is not difficult to show that[25]

$$\iint_A (X_b\bar{u} + Y_b\bar{v})\,dx\,dy + \int_C (X_s\bar{u} + Y_s\bar{v})\,ds$$
$$= \iint_A (\bar{X}_b u + \bar{Y}_b v)\,dx\,dy + \int_C (\bar{X}_s u + \bar{Y}_s v)\,ds \qquad (9.115)$$

Thus, we conclude:

> If a linearly elastic body is subjected to two different force systems, the work that would be done by the first system of forces in moving through the displacements produced by the second system of forces is equal to the work that would be done by the second system of forces in moving through the displacements produced by the first system of forces.
>
> (THEOREM XVI)

This is called *the reciprocal theorem of Rayleigh and Betti.*[26]

A special case of the above theorem was discovered somewhat earlier by Clerk Maxwell.[27] Consider, for example, the case in which each of the two force systems consists of a single force of the same magnitude P, as indicated in Fig. 9.25*a*. In the first case P is applied at a point i in a direction n_1 and it causes a displacement $(\Delta_{ji})_{n_2}$ of a point j in a direction n_2. In the second case, P is applied at j in direction n_2 and it causes a displacement $(\Delta_{ij})_{n_1}$ of point i in direction n_1. According to the reciprocal theorem of Rayleigh and Betti,

$$P(\Delta_{ij})_{n_1} = P(\Delta_{ji})_{n_2}$$

or

$$(\Delta_{ij})_{n_1} = (\Delta_{ji})_{n_2} \qquad (9.116)$$

Equation (9.116) is called *Maxwell's reciprocal theorem:*

> For a linearly elastic body, the displacement of point i in direction n_1 due to a force P at point j in direction n_2 is equal to the displacement of point j in direction n_2 due to a force P at point i in direction n_1.
>
> (THEOREM XVII)

[25] The derivation of this relation is straightforward. See, for example, Ref. 72, p. 169.

[26] Refs. 48 and 7.

[27] See Ref. 40.

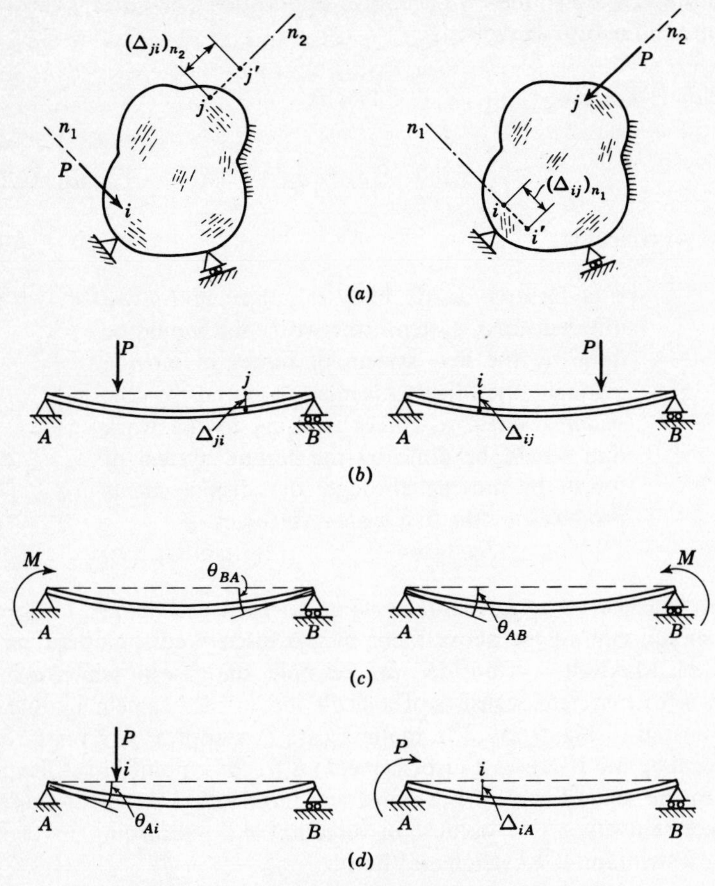

FIGURE 9.25 Illustrations of Maxwell's reciprocal theorem. (*a*) Elastic body subjected to two different force systems each consisting of a single force P; (*b*) reciprocity of displacements; (*c*) slopes; (*d*) displacements and slopes.

Other applications of Maxwell's theorem are illustrated in Fig. 9.25*b*, *c*, and *d*. Accordingly

$$\begin{aligned} \Delta_{ij} &= \Delta_{ji} \\ \theta_{AB} &= \theta_{BA} \\ \theta_{Ai} &= \Delta_{iA} \end{aligned} \tag{9.117}$$

where Δ_{ij}, θ_{AB}, . . . , Δ_{iA} are the deformations indicated in the figure. Note that the moment P in Fig. 9.25*d* has the same numerical value as the force P, but it has, of course, different units. Thus, Δ_{iA} in Eqs. (9.117) might be interpreted as a displacement per unit length.

9.17 Engesser's second theorem, or the theorem of least work.[28] In demonstrating the application of Engesser's first theorem and Castigliano's second theorem in Art. 9.15, all of the structures considered were statically determinate; in every case it was possible to express the total complementary strain energy in terms of the applied loads. Because of our assumption that displacements are small and that the stress-strain law is monotonically increasing, there is only one possible stress distribution that can exist in a statically determinate structure under a given set of loads.

Let us now examine the more general case in which the structure is both externally and internally statically indeterminate. Consider, for example, the structural system in Fig. 9.26*a*, which consists of two deformable bodies under the action of a set of applied forces. Let us assume that the system is stable and statically determinate and that U^* is given in terms of the forces $P_1, P_2, \ldots, P_n$, whose corresponding displacements are, again, $\Delta_1, \Delta_2, \ldots, \Delta_n$.

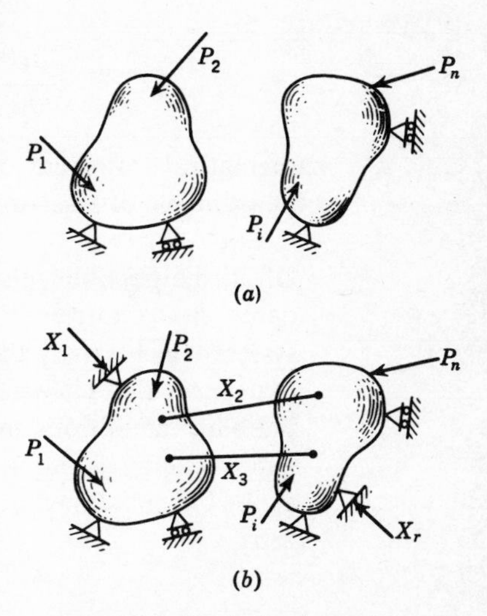

FIGURE 9.26 Statically determinate (*a*) and indeterminate (*b*) structures.

Suppose that we now introduce a number of additional constraints on the boundary and connect the bodies with deformable members so that the statically indeterminate system in Fig. 9.26*b* is obtained. If the number of additional external and internal constraints is r, we have, in effect, introduced a number of redundant forces $X_1, X_2, \ldots, X_r$. Further, if we temporarily assume that the prescribed displacements at these boundary points are zero and that the connecting members fit perfectly before the loads are applied, the redundant forces do not contribute to the external complementary energy. That is, H^* is given solely in terms of $P_1, P_2, \ldots, P_n$ and $\Delta_1, \Delta_2, \ldots, \Delta_n$, although U^* is now a function of both the forces P_i and X_i.

Now since the structure is in stable equilibrium without any redundants present, any set of values of the forces X_i will satisfy the equilibrium

[28] This theorem, among other things, is also referred to as the generalized form of Castigliano's theorem of least work, Engesser's theorem of compatibility, and the theorem of stationary (minimum) complementary strain energy. The general form which we are to consider was first presented by F. Engesser in 1889 (Ref. 19), although a specialized form of the theorem applicable only to linearly elastic systems was presented by A. Castigliano in 1879 (Ref. 10).

conditions. There exists only one set of values, however, that will also satisfy the compatibility conditions; and, according to the complementary energy principle, it is this set which makes Π^* assume a stationary value. Thus, if we give the redundants variations $\delta X_1, \delta X_2, \ldots, \delta X_r$ which themselves satisfy the equilibrium conditions, we obtain r independent compatibility conditions of the form

$$\delta\Pi^* = \frac{\partial U^*}{\partial X_i}\,\delta X_i + \frac{\partial H^*}{\partial X_i}\,\delta X_i = 0 \qquad (9.118)$$

but H^* is independent of X_i since, by hypothesis, the prescribed displacements at the points of action of the redundants are zero. Thus, δH^* is zero, which means X_i must be such that δU^* vanishes. Therefore, Eq. (9.118) reduces to

$$\frac{\partial U^*}{\partial X_i} = 0 \qquad (9.119)$$

This is a mathematical statement of a special form of *Engesser's second theorem*, or the principle of least work.[29] From it, we conclude:

> Of all the possible sets of values of the redundants in a statically indeterminate structural system which satisfy the equilibrium conditions, that particular set which also satisfies the compatibility conditions makes the complementary strain energy a minimum, provided the prescribed displacements of these redundants are zero.
>
> (THEOREM XVIII)

The physical significance of Eq. (9.119) is quite obvious in the case of externally indeterminate structures. The two redundant forces at the interior supports of the beam in Fig. 9.27*a*, for example, clearly satisfy the conditions

$$\frac{\partial U^*}{\partial X_1} = 0 \quad \text{and} \quad \frac{\partial U^*}{\partial X_2} = 0$$

since the displacements of these supports are zero. However, the theorem is also applicable when internal forces are selected as redundants,[30] as in Fig. 9.27*b* and *c*. For example, if the shear X_i at point i in the frame in

[29] This theorem is sometimes referred to as the second theorem of least work, to distinguish it from Theorem XII discussed previously.

[30] We still assume, for the present, that the prescribed displacements of X_i are zero and that internal redundant members have a perfect fit, initially.

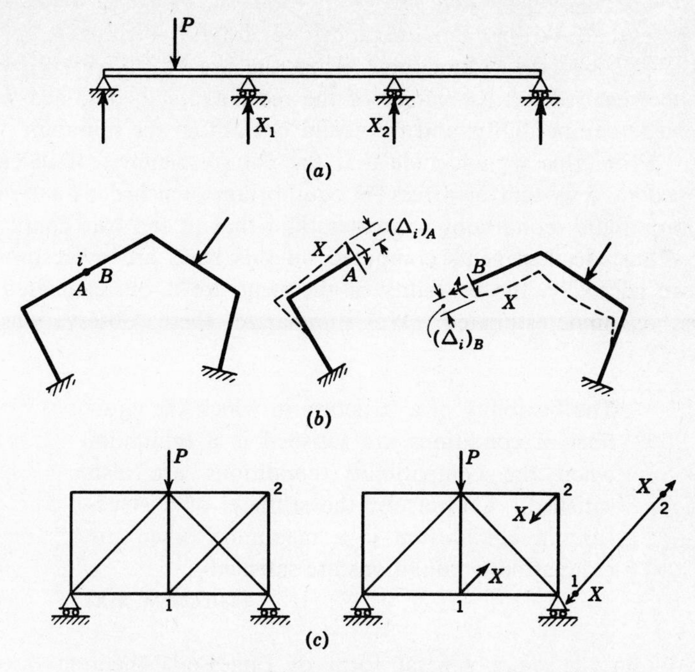

FIGURE 9.27 Statically indeterminate structures.

Fig. 9.27*b* is selected as one of the three redundants, the condition that $\partial U^*/\partial X_i$ be zero is merely the requirement that the displacement $(\Delta_i)_A$ of i on part A of the frame plus that of i of part B, $(\Delta_i)_B$, be zero—that is, the deformations of A are compatible with those of B. Similarly, the redundant force in bar $\overline{12}$ of the internally indeterminate truss in Fig. 9.27*c* must also satisfy Eq. (9.119). To verify this, we isolate the redundant member, as is indicated, and express the total complementary energy as the sum of two parts

$$U^* = \bar{U}^* + U^*_{12}$$

where $\bar{U}^*$ is the complementary energy of the truss obtained by removing bar $\overline{12}$ and U^*_{12} is that contained in the bar $\overline{12}$. Thus

$$\frac{\partial U^*}{\partial X} = \frac{\partial \bar{U}^*}{\partial X} + \frac{\partial U^*_{12}}{\partial X}$$

but, in view of Eq. (9.113*b*) (and Fig. 9.23*c*), $\partial \bar{U}^*/\partial X$ is Δ_{12}, the relative inward displacement of joints 1 and 2 along $\overline{12}$ due to X, and $\partial U_{12}/\partial X$ is e_{12}, the elongation of bar $\overline{12}$. Clearly, $\Delta_{12} = -e_{12}$, if the deformations of both parts of the structure are to be compatible. Hence

$$\frac{\partial U^*}{\partial X} = 0$$

Suppose that, through an oversight, we did not include in U^* the effects of all r of the redundants. Then the use of Eq. (9.119) would yield incorrect results; the values of the redundants so obtained would not provide compatibility, and we would not obtain the minimum value of U^*. From this we conclude that the complementary strain energy developed by a system of forces (in equilibrium) which does *not* satisfy the compatibility conditions is greater than that at the true compatible state. Thus, displacements computed on this basis are overestimated; or, more generally, the flexibility of the structure is overestimated and stiffness is underestimated. We summarize these observations, as follows:[31]

> The flexibility of a structure in which the equilibrium conditions are satisfied is a minimum when the compatibility conditions are also satisfied. Conversely, the stiffness of a structure in equilibrium is a maximum when the compatibility conditions are satisfied.
>
> (THEOREM XIX)

To obtain the more general form of Engesser's theorem of compatibility, let us now consider the case in which the displacements of the boundary points on which the redundant forces act are prescribed and are not zero. In addition any of the connecting members in Fig. 9.26*b* (internal redundants) may have an initial prescribed lack of fit. Then the contribution of the redundant forces to the external complementary energy is

$$-X_1 s_1 - X_2 s_2 - \cdots - X_r s_r$$

where $s_1, s_2, \ldots, s_r$ are the prescribed displacements. Now δH^* is no longer zero and Eq. (9.119) is not applicable. In fact, if we give the redundant forces which satisfy the compatibility conditions variations δX_i which satisfy the equilibrium conditions, we know that

$$\delta\Pi^* = \frac{\partial U^*}{\partial X_i}\,\delta X_i + \frac{\partial H^*}{\partial X_i}\,\delta X_i = \left(\frac{\partial U^*}{\partial X_i} - s_i\right)\delta X_i = 0$$

Hence

$$\frac{\partial U^*}{\partial X_i} = s_i \tag{9.120}$$

For example, suppose that bar $\overline{12}$ of the truss in Fig. 9.27*c* is initially too short by an amount e_0 to fit properly between joints 1 and 2. Then, when

[31] These ideas were discussed from a different viewpoint by D. Williams (Ref. 75) and E. H. Brown (Ref. 9); as stated above, the theorem is similar to that given by J. H. Argyris and S. Kelsey (Ref. 2).

the complete structure reaches the state of both equilibrium and compatibility, the complementary strain energy is such that

$$\frac{\partial U^*}{\partial X} = e_0$$

Note that for linearly elastic structures, U and U^* are equal and Eqs. (9.119) and (9.120) become

$$\frac{\partial U}{\partial X_i} = 0 \tag{9.121}$$

and

$$\frac{\partial U}{\partial X_i} = s_i \tag{9.122}$$

respectively. Equations (9.121) and (9.122) are usually referred to as *Castigliano's theorem of least work or Castigliano's theorem of compatibility.*[32]

It is interesting to note that in the case of prescribed displacements, if we erroneously did not include the influence of all redundants in computing the complementary strain energy (and hence employed an incompatible virtual force system), we would underestimate the true forces by using Eq. (9.120). It follows that for the compatible state, these forces are a maximum. We may state this conclusion as follows:[33]

> If a structure is given a set of prescribed displacements, of all the sets of values of the redundants satisfying the equilibrium conditions, those satisfying the compatibility condition make the complementary strain energy a maximum.
>
> (THEOREM XX)

9.18 Structural analysis by compatibility methods. The theorems presented earlier are extremely useful in the analysis of discrete statically indeterminate structures. To demonstrate this, we confine our attention to linearly elastic systems for which it is always possible to express the stresses as linear functions of the applied loads and the redundant forces. Thus, for a system which is r times statically indeterminate, we may write

$$\begin{aligned} \sigma_x &= b_{11}X_1 + b_{12}X_2 + \cdots + b_{1r}X_r + \sigma_{x0} \\ \sigma_y &= b_{21}X_1 + b_{22}X_2 + \cdots + b_{2r}X_r + \sigma_{y0} \\ &\cdots\cdots\cdots\cdots\cdots\cdots\cdots\cdots \\ \tau_{xz} &= b_{61}X_1 + b_{62}X_2 + \cdots + b_{6r}X_r + \tau_{xz0} \end{aligned} \tag{9.123}$$

[32] This terminology is used by Matheson (Ref. 39).

[33] This theorem was presented for linearly elastic structures by D. Williams (Ref. 75), who referred to "the theorem of maximum strain energy." It was generalized by Argyris and Kelsey (Ref. 2). See also Ref. 9.

where the quantities $b_{11}, b_{12}, \ldots, b_{6r}$ are known functions of the coordinates and $\sigma_{x0}, \sigma_{y0}, \ldots, \tau_{xz0}$ are the known stresses due solely to the applied loads (that is, the stresses developed in the statically determinate structure). We also note that for a Hookean material the complementary strain energy density is a quadratic form in the stress components. Hence, it is also a quadratic form in the forces X_i.

For simplicity, let us temporarily confine our attention to structures in which the stress and strain distribution in each member are adequately described by the stress components σ_x, τ_{xy} and the strain components ϵ_x, γ_{xy}; the remaining components are assumed negligible. Again, such distributions exist in coplanar beams, frames, and trusses. Equations (9.123) then reduce to

$$\begin{aligned} \sigma_x &= b_{11}X_1 + b_{12}X_2 + \cdots + b_{1r}X_r + \sigma_{x0} \\ \tau_{xy} &= b_{41}X_1 + b_{42}X_2 + \cdots + b_{41}X_r + \tau_{xy0} \end{aligned} \qquad (9.124)$$

and since $\epsilon_x = \sigma_x/E$ and $\gamma_{xy} = \tau_{xy}/G$, the complementary strain energy is

$$U^* = \int_V \left(\int_0^F \frac{\sigma_x}{E}\, d\sigma_x + \frac{\tau_{xy}}{G}\, d\tau_{xy} \right) dV = \int_V \left(\frac{\sigma_x^{\;2}}{2E} + \frac{\tau_{xy}^{\;2}}{2G} \right) dV \qquad (9.125)$$

From Engesser's theorem of compatibility

$$\frac{\partial U^*}{\partial X_i} = s_i = \int_V \left(\frac{\sigma_x}{E}\frac{\partial \sigma_x}{\partial X_i} + \frac{\tau_{xy}}{G}\frac{\partial \tau_{xy}}{\partial X_i} \right) dV$$

and from Eqs. (9.124)

$$\frac{\partial \sigma_x}{\partial X_i} = b_{1i} \qquad \frac{\partial \tau_{xy}}{\partial X_i} = b_{4i}$$

Thus

$$\begin{aligned} s_i = \int_V \Bigg[& \frac{1}{E}(b_{11}X_1 + \cdots + b_{1r}X_r + \sigma_{x0})b_{1i} \\ & + \frac{1}{G}(b_{41}X_1 + \cdots + b_{4r}X_r + \tau_{xy0})b_{4i} \Bigg] dV \end{aligned}$$

or

$$\begin{aligned} s_i = {} & X_1 \int_V \left(\frac{b_{11}}{E}\, b_{1i} + \frac{b_{41}}{G}\, b_{4i} \right) dV + \cdots + X_i \int_V \left(\frac{b_{1i}^{\;2}}{E} + \frac{b_{4i}^{\;2}}{G} \right) dV + \cdots \\ & + X_r \int_V \left(\frac{b_{1r}}{E}\, b_{1i} + \frac{b_{4r}}{G}\, b_{4i} \right) dV + \int_V \left(\frac{\sigma_{x0}}{E}\, b_{1i} + \frac{\tau_{xy0}}{G}\, b_{4i} \right) dV \end{aligned} \qquad (9.126)$$

By differentiating U^* with respect to each of the forces $X_1, X_2, \ldots, X_r$, we obtain the set of r independent equations

$$\begin{aligned} s_1 &= f_{11}X_1 + f_{12}X_2 + \cdots + f_{1r}X_r + f_{01} \\ s_2 &= f_{21}X_1 + f_{22}X_2 + \cdots + f_{2r}X_r + f_{02} \\ &\cdots\cdots\cdots\cdots\cdots\cdots\cdots\cdots \\ s_i &= f_{i1}X_1 + f_{i2}X_2 + \cdots + f_{ir}X_r + f_{0i} \\ &\cdots\cdots\cdots\cdots\cdots\cdots\cdots\cdots \\ s_r &= f_{r1}X_1 + f_{r2}X_2 + \cdots + f_{rr}X_r + f_{0r} \end{aligned} \tag{9.127}$$

where

$$f_{ij} = \int_V \left(\frac{b_{1j}}{E} b_{1i} + \frac{b_{4j}}{G} b_{4i} \right) dV \tag{9.128}$$

and

$$f_{0i} = \int_V \left(\frac{\sigma_{x0}}{E} b_{1i} + \frac{\tau_{xy0}}{G} b_{4i} \right) dV \tag{9.129}$$

We may evaluate the forces X_i by solving Eqs. (9.127). We then obtain final stresses and, subsequently, strains by substituting the X's into Eqs. (9.124). This completes the analysis.

Since the quantity on the left side of Eq. (9.126) is a displacement at i in the direction of X_i, each term on the right side must also be a displacement at i in that direction. Hence, the quantity f_{0i} physically represents the displacement of i in the direction of X_i due solely to applied loads. In other words, it is the displacement which point i undergoes in the statically determinate system (when all the X's are zero). Similarly, $f_{ij}X_j$ must be the displacement produced at i solely by X_j.

The quantities f_{ij} are, of course, the flexibilities of the structure.[34] If all of the forces and applied loads are zero except X_j, and X_j is unity, the ith member of Eqs. (9.127) becomes

$$s_i = 0 + 0 + \cdots + f_{ij} \cdot 1 + 0 + \cdots + 0 = f_{ij}$$

Thus, physically, f_{ij} *is the displacement at i in the direction i due to a unit force at j in the direction j.* Note that $f_{ij} = f_{ji}$, in agreement with Maxwell's reciprocal theorem.

According to Eqs. (9.124), $\sigma_x = b_{1j}$ and $\tau_{xy} = b_{4j}$ when X_j is unity and all other forces are zero. It follows that the quantities b_{1j}/E and b_{4j}/G in Eq. (9.128) are the strains $(\epsilon_x)_j$ and $(\gamma_{xy})_j$ due to a unit force at j ($X_i = 0$, $i \neq j$). Thus, we can also write Eq. (9.128) in the form

$$f_{ij} = \int_V \left[(\epsilon_x)_j \frac{\partial \sigma_x}{\partial X_i} + (\gamma_{xy})_j \frac{\partial \tau_{xy}}{\partial X_i} \right] dV \tag{9.130}$$

[34] See Art. 8.12, Eqs. (8.60).

When in this form, Eq. (9.130) also gives the flexibilities for structures which are not linearly elastic. In fact, for the nonlinearly elastic three-dimensional body,

$$f_{ij} = \int_V \left[(\epsilon_x)_j \frac{\partial \sigma_x}{\partial X_i} + (\epsilon_y)_j \frac{\partial \sigma_y}{\partial X_i} + \cdots + (\gamma_{xz})_j \frac{\partial \tau_{xz}}{\partial X_i} \right] dV \quad (9.131)$$

where $(\epsilon_x)_j, (\epsilon_y)_j, \ldots, (\gamma_{xz})_j$ are the strains produced by a unit force at j, $X_j = 1$. We express Eq. (9.131) more concisely in the symbolic form

$$f_{ij} = \int_V \epsilon_j \frac{\partial \sigma}{\partial X_i} \, dV \quad (9.132)$$

Since $\partial \sigma / \partial X_i = \delta \sigma^{X_i}$, it follows that, in general, f_{ij} is the internal complementary virtual work done by the virtual stresses produced by a unit virtual force at i in moving through the true strains produced by a unit force at j.

Through similar arguments, we may also show that, in general,

$$f_{0i} = \int_V \epsilon_0 \frac{\partial \sigma}{\partial X_i} \, dV$$

in which ϵ_0 are the strains due to applied loads in the statically determinate system.

To demonstrate these ideas, consider the linearly elastic truss shown in Fig. 9.28*a* which is twice statically indeterminate. The stress in a typical member m is given by

$$\sigma_m = \frac{N_m}{A_m}$$

where A_m is the area of the mth member and N_m is the axial force in the mth member.

Let us choose the forces in bars 1 and 2 as redundants and denote them X_1 and X_2, respectively, as indicated in Fig. 9.28*b*. Then the force in any member can be expressed in the form

$$N_m = X_1 n'_m + X_2 n''_m + N_{0m} \quad (9.133)$$

where n'_m = the force in member m due to a unit value of the redundant force X_1

n''_m = the force in member m due to a unit value of the redundant force X_2

N_{0m} = the force in member m of the statically determinate structure obtained by removing the redundants X_1 and X_2. Thus, N_{0m} is the force due to the applied loads.

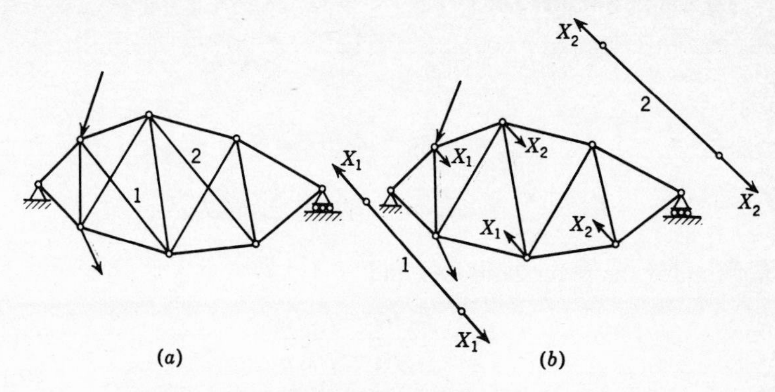

FIGURE 9.28 (a) Statically indeterminate truss; (b) redundant forces.

It follows that

$$\sigma_m = X_1 \frac{n'_m}{A_m} + X_2 \frac{n''_m}{A_m} + \frac{N_{0m}}{A_m} \tag{9.134}$$

Comparing this equation with Eqs. (9.123), we see that in this case

$$b_{11} = \frac{n'_m}{A_m} \qquad b_{12} = \frac{n''_m}{A_m} \qquad \sigma_{0m} = \frac{N_{0m}}{A_m}$$

where $\sigma_{0m} = \sigma_{x0}$ for member m.

The total complementary strain energy in the truss is computed as follows:

$$U^* = \sum_{m=1}^{n} \left(\int_{L_m} \frac{\sigma_m{}^2}{2E} \, dA \, dx \right) = \frac{1}{2} \sum_{m=1}^{n} N_m{}^2 \lambda_m \tag{9.135}$$

where n is the total number of truss members (including the redundant bars) and $\lambda_m = L_m / EA_m$. Thus

$$U^* = \frac{1}{2} \sum_{m=1}^{n} (X_1 n'_m + X_2 n''_m + N_{0m})^2 \lambda_m$$

According to the principle of least work,

$$\frac{\partial U^*}{\partial X_1} = 0 = X_1 \sum_{m=1}^{n} n'^2_m \lambda_m + X_2 \sum_{m=1}^{n} n'_m n''_m \lambda_m + \sum_{m=1}^{n} N_{0m} n'_m \lambda_m \tag{9.136a}$$

$$\frac{\partial U^*}{\partial X_2} = 0 = X_1 \sum_{m=1}^{n} n'_m n''_m \lambda_m + X_2 \sum_{m=1}^{n} n''^2_m \lambda_m + \sum_{m=1}^{n} N_{0m} n''_m \lambda_m \tag{9.136b}$$

Comparing these equations with Eqs. (9.127) to (9.129), we see that in this case the flexibilities are

$$f_{11} = \sum_{m=1}^{n} n'^2_m \lambda_m \qquad f_{22} = \sum_{m=1}^{n} n''^2_m \lambda_m$$

$$f_{12} = f_{21} = \sum_{m=1}^{n} n'_m n''_m \lambda_m$$

$$f_{01} = \sum_{m=1}^{n} N_{0m} n'_m \lambda_m \qquad f_{02} = \sum_{m=1}^{n} N_{0m} n''_m \lambda_m$$

Solving for the redundants, we find

$$X_1 = \frac{f_{01} f_{22} - f_{02} f_{12}}{f_{11} f_{22} - f_{12} f_{21}}$$
$$X_2 = \frac{f_{02} f_{11} - f_{01} f_{21}}{f_{11} f_{22} - f_{12} f_{21}} \qquad (9.137)$$

We can now obtain the final stress in each bar by introducing these values into Eq. (9.134).

As a final example, let us consider the linearly elastic parabolic arch shown in Fig. 9.29*a* which is subjected to a uniformly varying load of maximum intensity p_0. The arch is fixed at end A and hinged at B and is, therefore, two times statically indeterminate. For simplicity, we assume that the moment of inertia of the arch varies according to the secant law $I_s = I_0 \sec \alpha$, where I_s is the moment of inertia of a section an arbitrary arc length s from the fixed end, I_0 is the moment of inertia at the crown (point 0), and

$$\alpha = \tan^{-1} \frac{dy}{dx}$$

x and y being the coordinates indicated in the figure. Then

$$\frac{ds}{I_s} = \frac{dx}{I_0}$$

and integration along the axis of the arch is greatly simplified.

To begin the analysis, we select for redundants the horizontal reactive force and the moment at support A. We denote these by X_1 and X_2, respectively, as indicated in Fig. 9.29*b*. Note that the arch is in equilibrium for any value of X_1 and X_2. We next isolate a typical segment of the arch, as shown in Fig. 9.29*c*, and, from statics, obtain the relations

$$M_s = -X_1 y + X_2 \frac{L - x}{L} + \frac{p_0 x}{6L}(L^2 - x^2)$$
$$N_s = -X_1 \cos \alpha + X_2 \frac{\sin \alpha}{L} - \frac{p_0}{6L}(L^2 - 3x^2) \sin \alpha \qquad (9.138)$$
$$V_s = -X_1 \sin \alpha - X_2 \frac{\cos \alpha}{L} + \frac{p_0}{6L}(L^2 - 3x^2) \cos \alpha$$

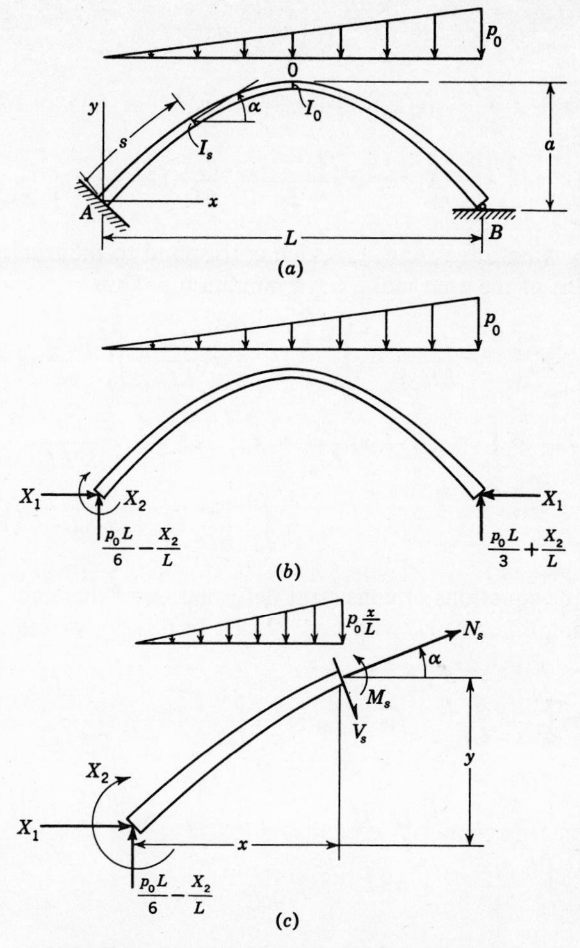

FIGURE 9.29 (*a*) Parabolic arch subjected to a uniformly varying load; (*b*) redundant force and moment acting on the arch; (*c*) free body of a finite segment of the arch.

where M_s, N_s, and V_s are the stress resultants indicated in the figure. According to Eq. (9.106), the complementary strain energy developed in the structure is given by

$$U^* = \int_0^S \left(\frac{N_s^2}{2EA} + \frac{M_s^2}{2EI_s} + \frac{\kappa V_s^2}{2GA} \right) ds$$

Since deformations due to N_s and V_s are ordinarily small compared with those due to bending, we neglect the axial force[35] and shearing force

[35] This assumption is justified except in the case of very low flat arches. See Ref. 65, p. 364.

terms in U^* and use simply

$$U^* = \int_0^S \frac{M_s^2}{2EI_s}\, ds$$

Recalling that $ds/I_s = dx/I_0$ and introducing M_s of Eqs. (9.138), we get

$$U^* = \int_0^L \left[-X_1 y + X_2 \frac{L-x}{L} + \frac{p_0 x}{6L}(L^2 - x^2) \right]^2 \frac{dx}{EI_0}$$

Now the values of X_1 and X_2 which correspond to the true compatible deformations of the arch make U^* a minimum. Thus

$$\frac{\partial U^*}{\partial X_1} = 0 = X_1 \int_0^L y^2 \frac{dx}{EI_0} - X_2 \int_0^L \frac{y}{L}(L-x)\frac{dx}{EI_0} - \int_0^L \frac{p_0 x y}{6L}(L^2 - x^2)\frac{dx}{EI_0}$$

$$\frac{\partial U^*}{\partial X_2} = 0 = -X_1 \int_0^L \frac{y}{L}(L-x)\frac{dx}{EI_0} + X_2 \int_0^L (L-x)^2 \frac{dx}{L^2 EI_0} + \int_0^L \frac{p_0 x}{6L^2}(L+x)(L-x)^2 \frac{dx}{EI_0} \qquad (9.139)$$

These are the equations of consistent deformation of the arch.

Comparing Eqs. (9.139) with Eqs. (9.127) to (9.129), we see that in this case

$$f_{11} = \int_0^L y^2 \frac{dx}{EI_0} \qquad f_{12} = f_{21} = -\int_0^L \frac{y}{L}(L-x)\frac{dx}{EI_0}$$

$$f_{22} = \int_0^L (L-x)^2 \frac{dx}{L^2 EI_0} \qquad f_{01} = -\int_0^L \frac{p_0 x y}{6L}(L^2 - x^2)\frac{dx}{EI_0}$$

$$f_{02} = \int_0^L \frac{p_0 x}{6L^2}(L+x)(L-x)^2 \frac{dx}{EI_0}$$

Physical interpretations of these quantities are indicated in Fig. 9.30.

From the geometry of Fig. 9.29*a*

$$y = \frac{4ax}{L^2}(L-x)$$

where a is the rise of the arch. Introducing this equation into the above integrals and performing the integration, we obtain for the flexibilities of the arch

$$f_{11} = \frac{8a^2 L}{15EI_0} \qquad f_{12} = f_{21} = -\frac{aL}{3EI_0} \qquad f_{22} = \frac{L}{3EI_0}$$

$$f_{01} = -\frac{p_0 a L^3}{30EI_0} \qquad f_{02} = \frac{7p_0 L^3}{360EI_0}$$

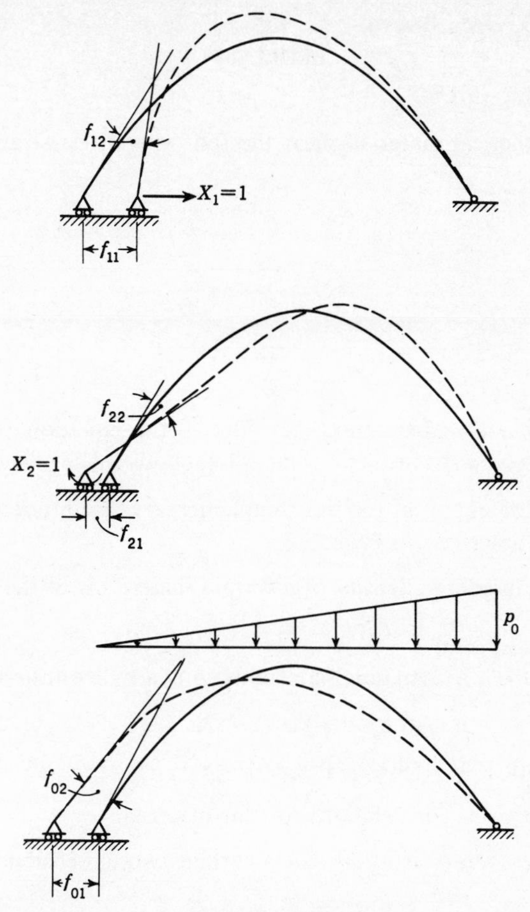

FIGURE 9.30 Physical interpretation of the flexibilities of the arch in Fig. 9.29.

We now substitute these quantities into Eqs. (9.139) and, after some simplifications, obtain

$$16aX_1 - 10X_2 - p_0L^2 = 0$$

$$-120aX_1 + 120X_2 + 7p_0L^2 = 0$$

Thus
$$X_1 = \frac{5p_0L^2}{72a} \qquad X_2 = \frac{p_0L^2}{90} \tag{9.140}$$

Final bending moments, axial forces, and shearing forces are obtained by introducing Eqs. (9.140) into Eq. (9.138).

PROBLEMS

9.1. Derive Eq. (9.15*a*).

9.2. A hypothetical material obeys the following stress-strain law in two dimensions:

$$\epsilon_x = \frac{{\sigma_x}^2}{E} - \nu\frac{\sigma_y}{E}$$

$$\epsilon_y = \frac{{\sigma_y}^2}{E} - \nu\frac{\sigma_x}{E}$$

$$\gamma_{xy} = \frac{{\tau_{xy}}^2}{G}$$

where E, G, and ν are constants. Derive the equation for the strain energy density of a strained two-dimensional body made of such a material.

9.3. Derive the equation for the complementary strain energy density of the material described in Prob. 9.2.

9.4. The strain energy density of a certain material is of the form

$$U_0 = C_1(I_1 - 3) + C_2(I_2 - 3) + pe$$

where C_1 and C_2 are material constants, p is a hydrostatic pressure, and

$$I_1 = 3 + 2(\epsilon_x + \epsilon_y + \epsilon_z) = 3 + 2e$$

$$I_2 = 3 + 4e + 4(\epsilon_x\epsilon_y + \epsilon_y\epsilon_z + \epsilon_x\epsilon_z) - 4(\gamma_{xy}^2 + \gamma_{yz}^2 + \gamma_{xz}^2)$$

Derive the stress-strain relations for this material.

9.5. The stress-strain relations for a certain two-dimensional anisotropic body are

$$\sigma_x = C_{xx}\epsilon_x + C_{xy}\epsilon_y$$

$$\sigma_y = C_{xy}\epsilon_y + C_{yy}\epsilon_y$$

$$\tau_{xy} = D_{xy}\gamma_{xy}$$

where C_{xx}, C_{xy}, C_{yy}, and D_{xy} are material constants. Derive the equation for (*a*) the strain energy density and (*b*) the complementary strain energy density for this material.

9.6. The stresses developed in a thin elastic plate are

$$\sigma_x = \frac{Ez}{1-\nu^2}\left(\frac{\partial^2 w}{\partial x^2} + \nu\frac{\partial^2 w}{\partial y^2}\right)$$

$$\sigma_y = \frac{Ez}{1-\nu^2}\left(\frac{\partial^2 w}{\partial y^2} + \nu\frac{\partial^2 w}{\partial x^2}\right)$$

$$\tau_{xy} = \frac{Ez}{1-\nu}\frac{\partial^2 w}{\partial x\,\partial y}$$

where w is the component of displacement normal to the plane of the plate and z is the coordinate measured from the middle plane of the plate. The energy contributed by the remaining components is assumed to be negligible. Show that the strain energy developed in a thin flat plate is

$$U = \frac{Eh^3}{24(1-\nu^2)} \iint_A \left\{ \left(\frac{\partial^2 w}{\partial x^2} + \frac{\partial^2 w}{\partial y^2} \right)^2 - 2(1-\nu)\left[\frac{\partial^2 w}{\partial x^2}\frac{\partial^2 w}{\partial y^2} - \left(\frac{\partial^2 w}{\partial x \, \partial y} \right)^2 \right] \right\} dx\, dy$$

where h is the thickness of the plate and A is the area of the plate.

9.7. Show that the total potential energy of an elastic beam column on an elastic foundation is

$$\frac{1}{2}\int_0^L \left[EI\left(\frac{d^2 v}{dx^2} \right)^2 - P\left(\frac{dv}{dx} \right)^2 + kv^2 - 2pv \right] dx$$

where P is the axial load, k is the foundation modulus, and p is the transverse load intensity.

9.8. An elastic thin-walled open section is subjected to an eccentric longitudinal force. One section of the structure is restrained against warping. Assuming that y and z are principal cross-sectional axes and that the sectorial pole coincides with the shear center and neglecting shear deformations, show that the strain energy in this structure is given by the formula

$$U = \int_0^L \left(\frac{N_x^2}{2EA} + \frac{M_y^2}{2EI_y} + \frac{M_z^2}{2EI_z} + \frac{W_\omega^2}{2E\Gamma} \right) dx$$

where W_ω is the bimoment and Γ is the warping constant.

9.9. Evaluate the total potential energy of the system shown below. The spring is nonlinear; the internal force developed is proportional to the square of the elongation $N = kx^2$. The external force P is produced by a magnetic field. Thus, the magnitude of P is inversely proportional to the square of the distance of the weight Q from the magnet $P = CQ/d^2$, where C is a constant and d is the distance indicated in the figure.

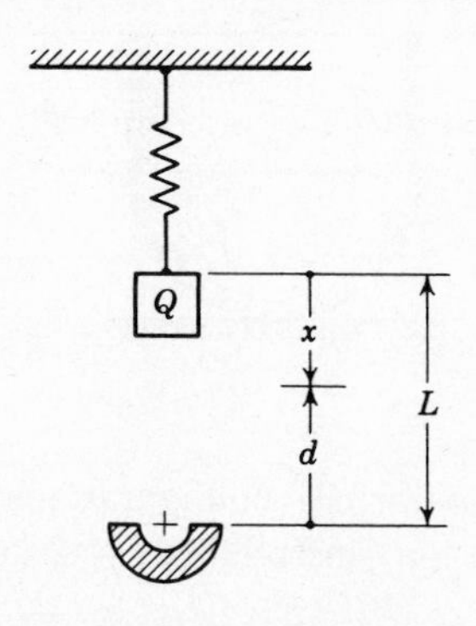

9.10. Evaluate the total complementary energy of the system in Prob. 9.9.

9.11. The total potential energy of a certain system is

$$\Pi = \frac{x^5}{5} - \frac{a}{4}x^4 - \frac{2a}{3}x^3 + a^2$$

where x is the generalized coordinate and a is a constant. Determine all of the equilibrium configurations of this system and indicate which ones are stable and unstable.

9.12. Use the principle of minimum potential energy to determine stable equilibrium configurations of the system shown. The bar is rigid and of negligible weight, and all surfaces are smooth.

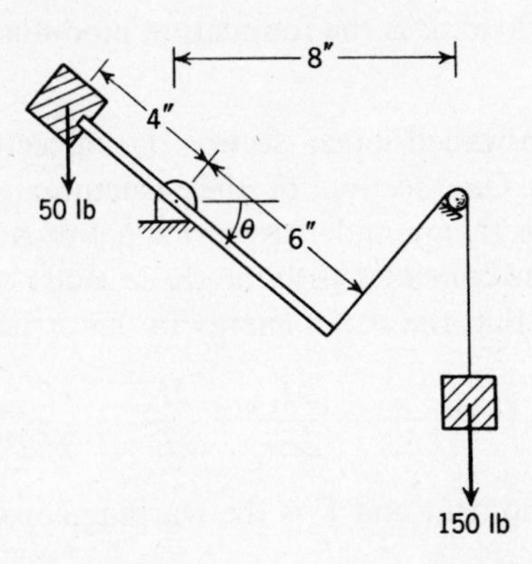

9.13. The homogeneous cylinder of weight Q is attached to a hemispherical base of radius r and weight Q. The hemisphere rests on a smooth surface. Determine the range of h/r for which this configuration is stable.

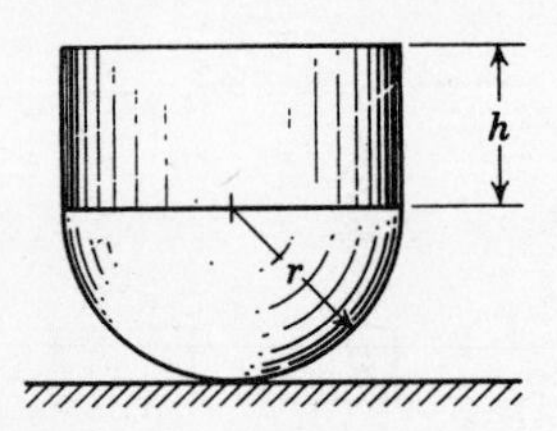

9.14. Determine the equilibrium configurations of the system shown. The linear spring is in its unstretched position when $\theta = 45°$. The rigid bar is of negligible weight and the cord is inextensible. All surfaces are smooth.

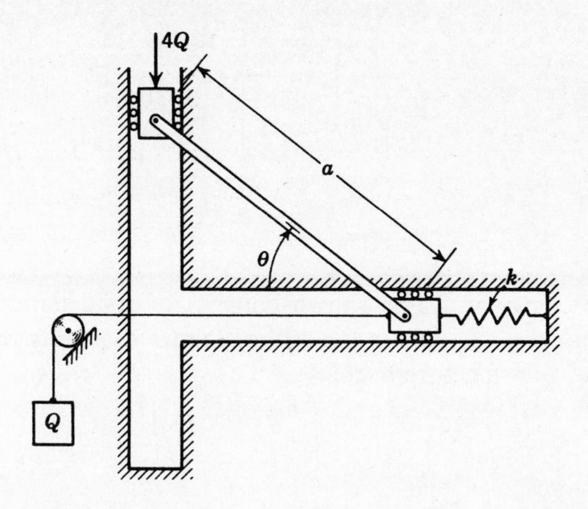

9.15–9.16. In each of the systems shown, the linear springs are in their unstretched positions when $\theta = 30°$. The bars are assumed to be rigid and of negligible weight. Use the principle of minimum potential energy to determine the equilibrium configurations of these systems and indicate which are stable and unstable.

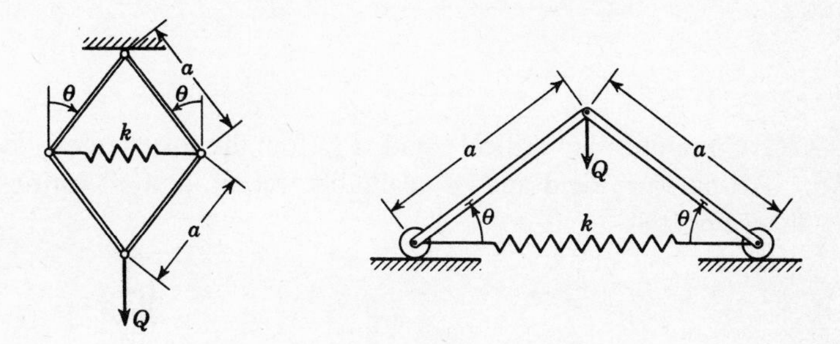

9.17–9.20. Solve Probs. 8.3–8.6 using the principle of minimum potential energy instead of the principle of virtual displacements. Compare and discuss the procedures used in applying both principles to these problems.

9.21. A 400-lb rigid homogeneous circular slab is suspended by three elastic wires as indicated below. The slab is subjected to an eccentric load of 2,000 lb at point D. The cross-sectional areas of the wires at A, B, and C are 0.01, 0.05, and 0.03 sq in., respectively. Determine the stress in each wire at the equilibrium configuration of the system.

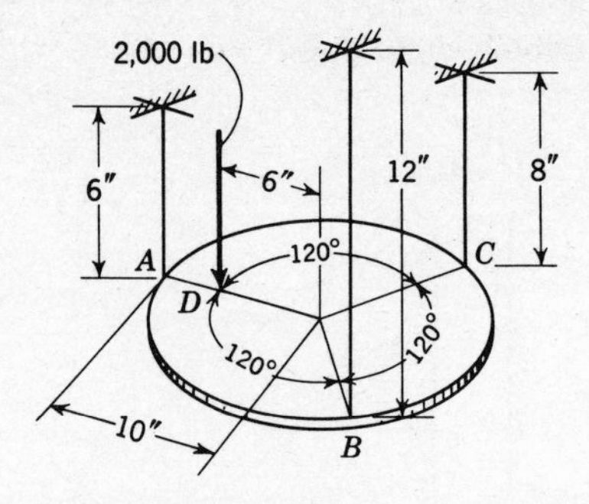

9.22. The rectangular frame shown consists of rigid bars connected by frictionless hinges. Compute the critical load P assuming that P remains vertical when the structure buckles.

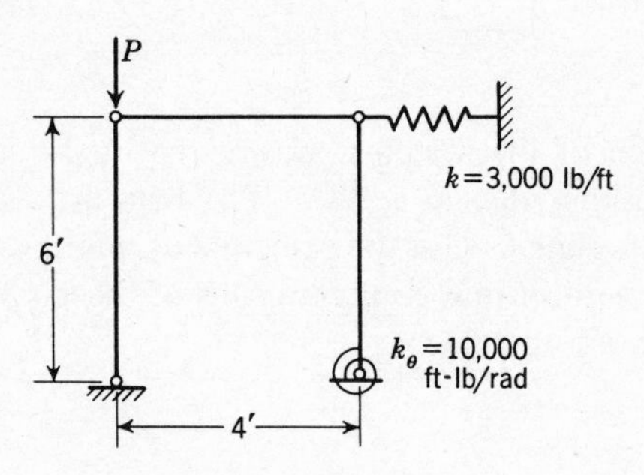

9.23–9.26. Compute the critical load P_{cr} for the mechanical systems shown. All bars are rigid and of negligible weight, and all springs have linear characteristics.

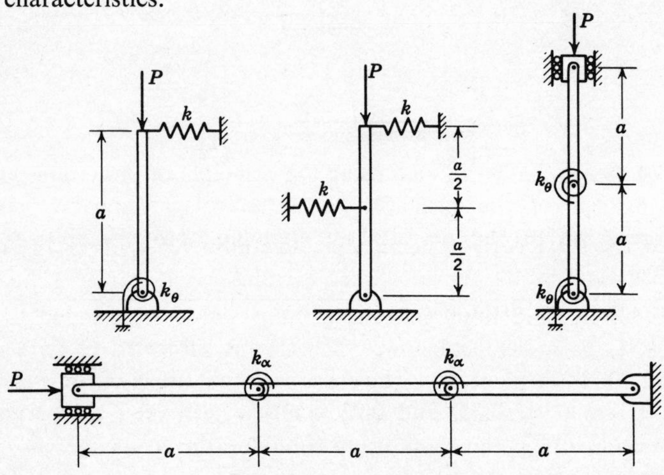

9.27. Compute the critical load for the system shown. The bars are rigid and of negligible weight, and the springs have linear characteristics. Indicate the relationships between various structural properties which will lead to symmetrical or antisymmetrical buckling modes.

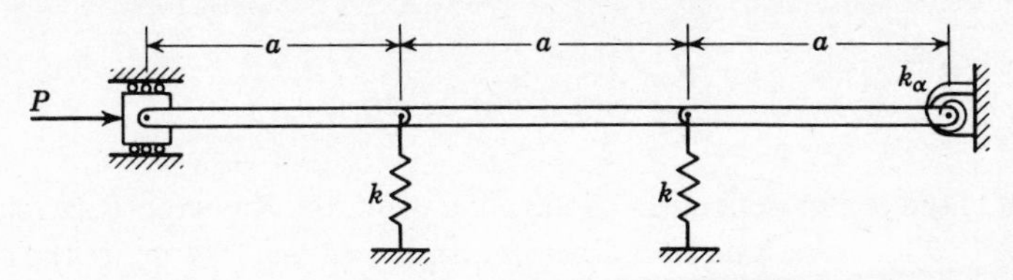

9.28. Use the Rayleigh-Ritz method to determine the transverse displacement of point A of the beam shown.

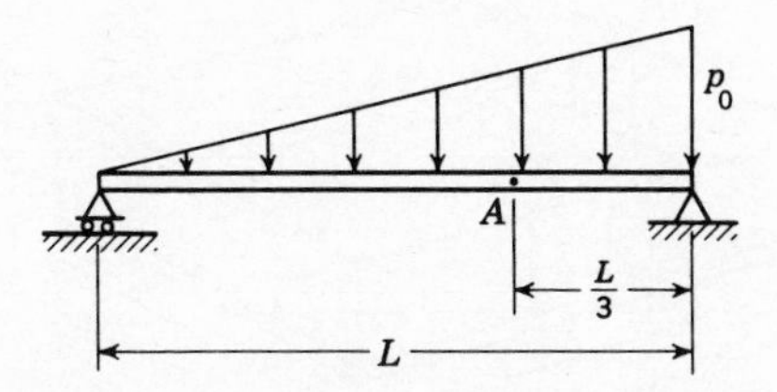

9.29. Use the Rayleigh-Ritz method to determine the critical load of a uniform column with fixed ends, assuming $v = Cx^2(L - x)^2$.

9.30. Compute the midspan deflection of the propped-end beam shown. Assume $v = C[1 - \cos (2\pi x/L)]$. Then compute the critical load.

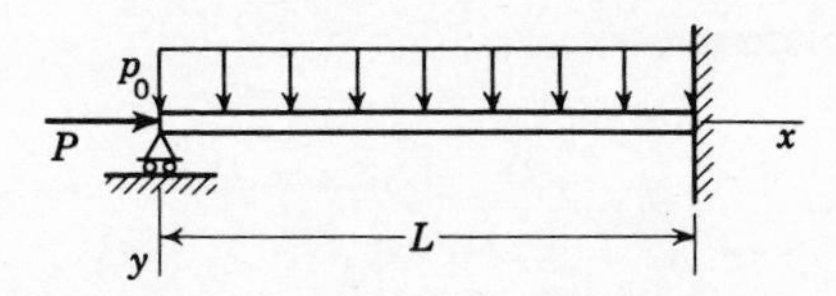

9.31. Use the Rayleigh-Ritz method to determine the buckling load of the simply supported beam on an elastic foundation shown below. Assume $v = a \sin (\pi x/L)$. (See Prob. 9.7.)

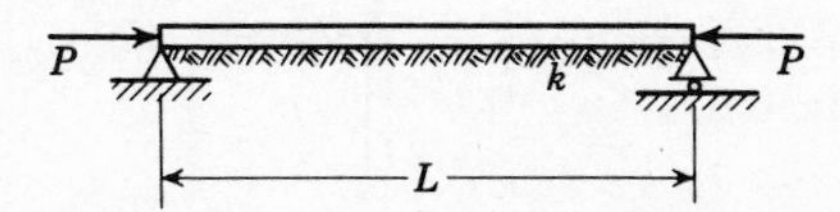

9.32. Use the Rayleigh-Ritz method to determine the midspan deflection of the fixed-end beam on an elastic foundation shown. Assume $v = Cx^2(L - x)^2$. (See Prob. 9.7.)

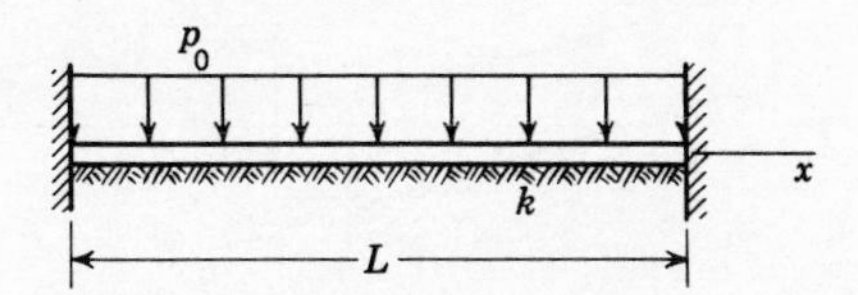

9.33. The kinematic boundary conditions for the uniformly loaded simply supported rectangular plate shown are

$$w = \frac{\partial^2 w}{\partial x^2} = 0 \qquad \text{at } x = 0,\ x = a$$

$$w = \frac{\partial^2 w}{\partial y^2} = 0 \qquad \text{at } y = 0,\ y = b$$

Compute the center deflection of the plate using the Rayleigh-Ritz method (see Prob. 9.6). Assume $w = C \sin (\pi x/a) \sin (\pi y/b)$. Compare the result for the case of a square plate with the exact value $0.0487(1 - \nu^2)p_0 a^4/Eh^3$.

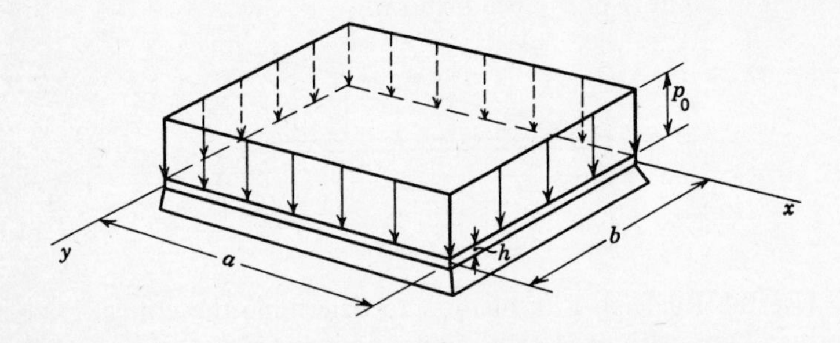

9.34. Repeat Prob. 9.33 using a polynomial approximation of w.

9.35. The kinematic boundary conditions for the uniformly loaded fixed-edge rectangular plate shown are

$$w = \frac{\partial w}{\partial x} = 0 \qquad \text{at } x = 0,\ x = a$$

$$w = \frac{\partial w}{\partial y} = 0 \qquad \text{at } y = 0,\ y = b$$

Compute the center deflection of the plate using the Rayleigh-Ritz method (see Prob. 9.6). Assume $w = C[1 - \cos (2\pi x/a)][1 - \cos (2\pi y/b)]$. Assuming that $\nu = 0.3$, compare the result for the case of a square plate with the exact value $0.01375 p_0 a^4/Eh^3$.

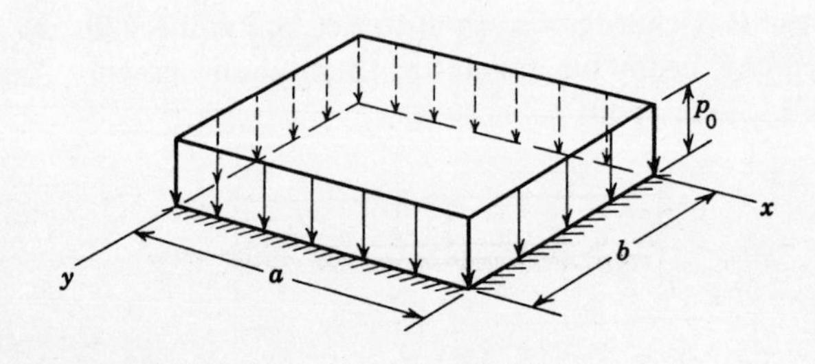

9.36. Repeat Prob. 9.35 using a polynomial approximation of w.

9.37–9.40. Solve Probs. 8.11–8.14 using Castigliano's first theorem.

9.41. Joint 2 of the frame shown below is displaced vertically owing to the applied force P. Considering only energy due to bending, use Castigliano's first theorem to evaluate the end moments in each member.

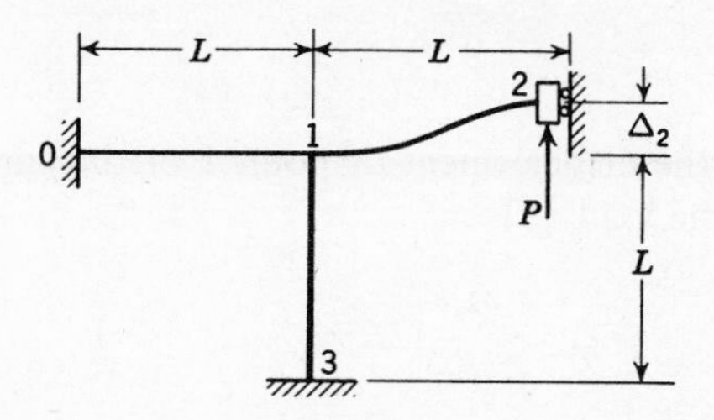

9.42. Using Castigliano's first theorem,

(*a*) Solve for the generalized displacements of the frame shown. Neglect shear and axial deformation.

(*b*) Interpret physically, by means of appropriate sketches, the significance of the coefficients appearing in one of the equations for the unknown displacements.

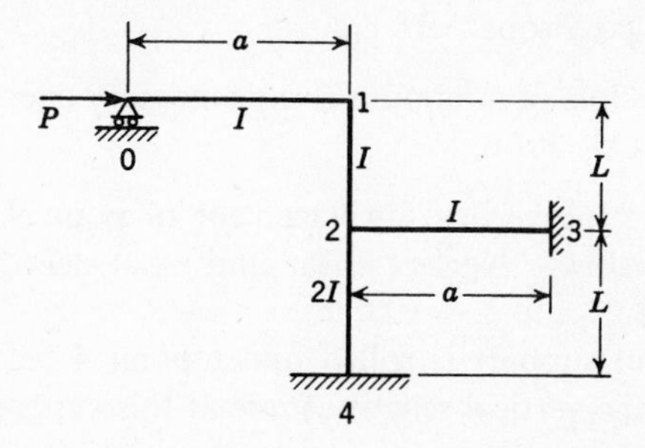

9.43. The coplanar frame shown is linearly elastic, and shear and axial deformations in members $\overline{01}$, $\overline{12}$, and $\overline{23}$ are negligible. The rod $\overline{02}$ is assumed to develop only axial stresses. Units are taken so that the cross-sectional area of the rod is numerically equal to half the moment of inertia of the other members. Using Castigliano's first theorem,

(*a*) Obtain the equations necessary to solve for the generalized displacements.

(*b*) Interpret physically, with the aid of diagrams, the meaning of the coefficients of the displacements in one or more of the equations in part (*a*).

(c) Evaluate the axial force in the rod and the end moment M_{01}.

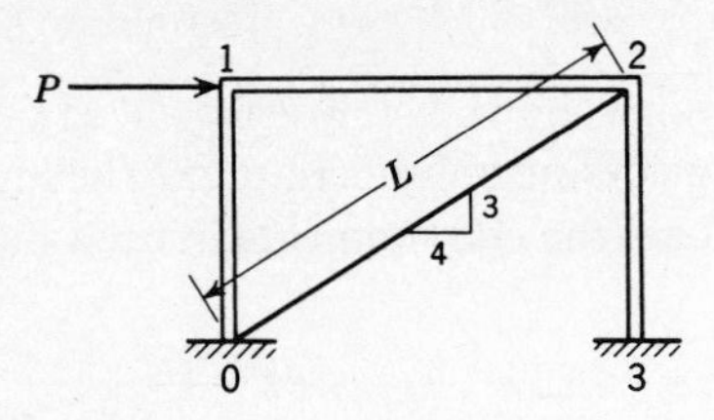

9.44. Show that the displacement of point 1 of the curved bar shown in the direction of the load Q is

$$\Delta = \frac{QR}{E}[2R^2(\pi - 2) + \pi]$$

if $E/G = 2$ and $\kappa = 1$.

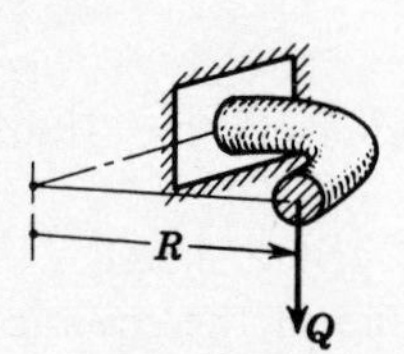

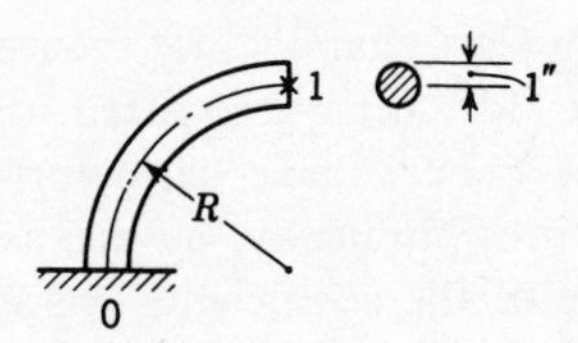

9.45–9.47. Use Castigliano's second theorem instead of the unit-dummy-load method to solve Probs. 8.18-8.20.

9.48. Use the principle of minimum complementary energy (or Engesser's first theorem) to solve Prob. 8.25.

9.49. (a) Compute the vertical displacement of point 4 of the bent bar shown below. Neglect shear and axial deformation and take $E/G = 2.25$ and $a = 10.0$ in.

(b) If a point support is rolled under point 4 before P is applied, what is the vertical reactive force at this support?

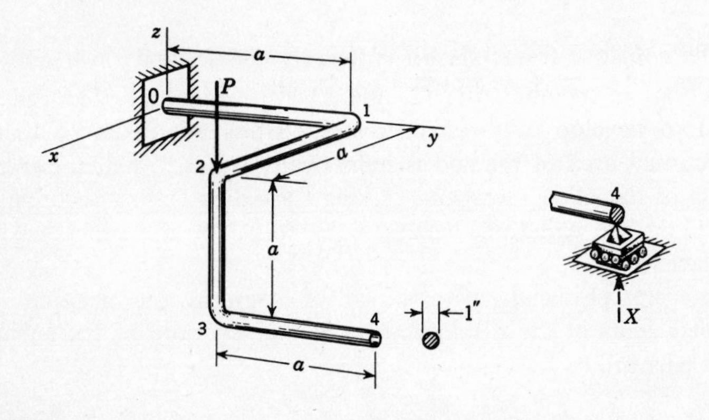

9.50. The cantilevered beam shown is constructed of a nonlinearly elastic material which exhibits a stress-strain relation of the form

$$\sigma_x = K \operatorname{sgn}(\epsilon_x)\epsilon_x^{1/n}$$

where K is a material constant and n is a positive integer. Use Engesser's first theorem to evaluate the transverse deflection of the free end.

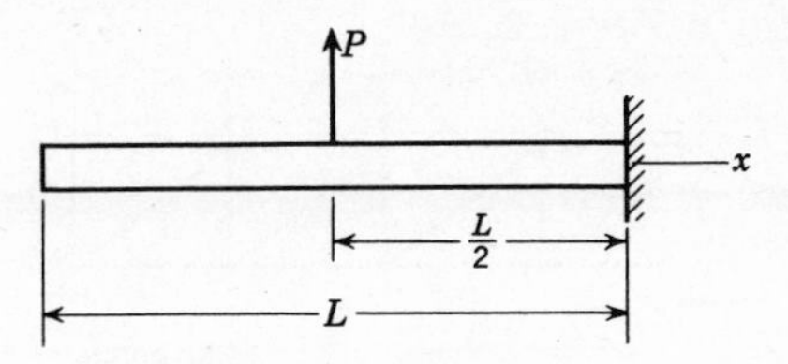

9.51. Evaluate the relative displacement of points A and B of the circular ring shown. Neglect shear deformations.

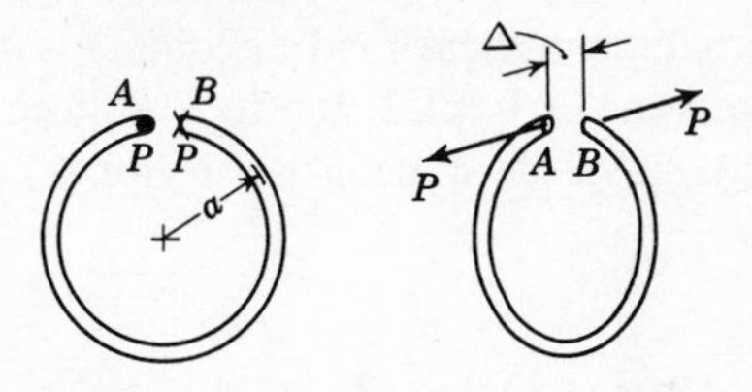

9.52. Evaluate the relative displacements of points A and B of the rectangular frame shown. Neglect shear deformations.

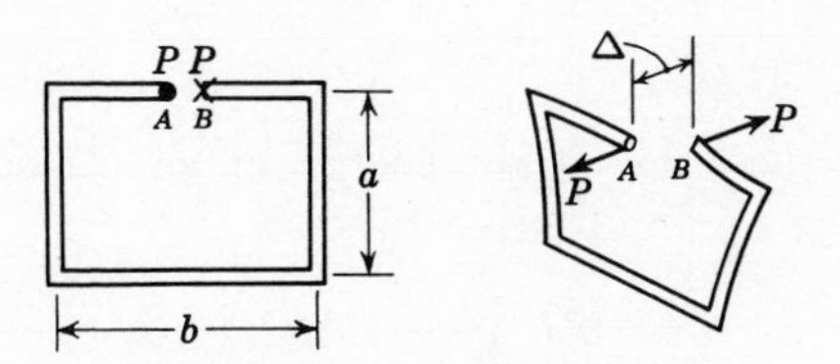

9.53. Compute the rotation of the vertical member AB of the coplanar truss shown. The cross-sectional areas of all members of the truss are 1.0 sq in. and $E = 30 \times 10^6$ psi.

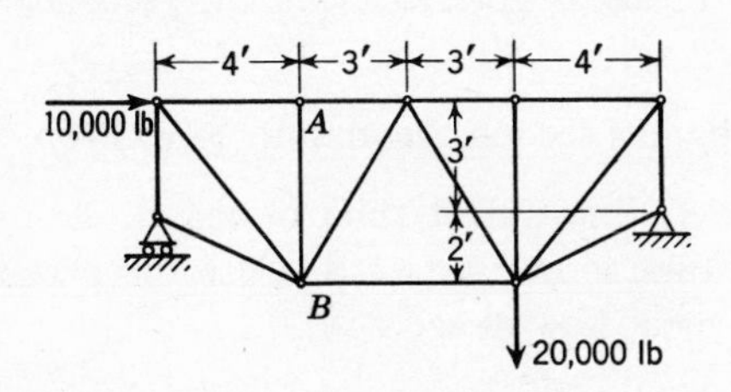

9.54. All members of the coplanar truss shown are 1.0 sq in. in area and $E = 10^7$ psi. Using Castigliano's second theorem, compute

(*a*) the displacement of joint 2 relative to joint 7
(*b*) the change in slope of the top chord at joint 4
(*c*) the rotation of the end vertical member 01
(*d*) the horizontal displacement of joint 9

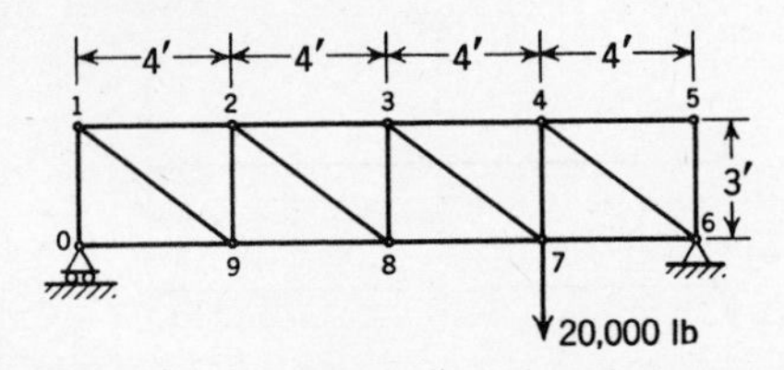

9.55. Verify Maxwell's theorem for a vertical load at joint 3 and an end couple at end *A* of the truss shown. That is, show that the rotation of member 01 due to a unit vertical load at joint 3 is numerically equal to the vertical displacement of joint 3 due to a unit couple at end *A*. All members of the truss have the same cross-sectional area *a*.

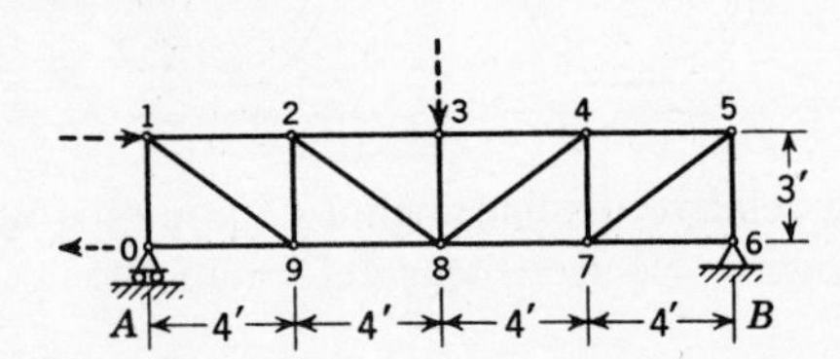

9.56. Show that the total strain energy in an elastic single-cell thin-walled tube in pure torsion is given by

$$U = \frac{q^2 L}{2G} \oint \frac{ds}{t} = \frac{M_t{}^2 L}{8\Omega^2 G} \oint \frac{ds}{t}$$

where L is the length of the tube, q is the shear flow, Ω is the sectorial area of the cell, and M_t is the twisting moment.

9.57–9.60. Solve Probs. 3.12–3.15 using the (second) principle of least work.

9.61. Use Castigliano's second theorem to solve Prob. 5.19.

9.62. A three-celled thin-walled tube of length L is subjected to end torques T. The tube is free to warp. Use the principle of least work to determine the shear flow in each cell.

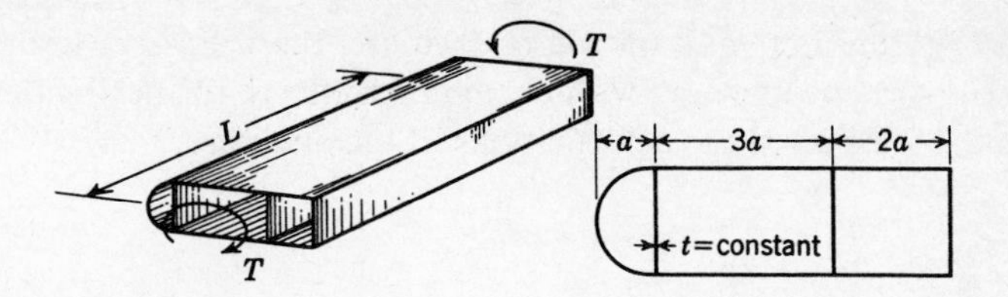

9.63. Using Castigliano's second theorem (Engesser's first theorem),

(*a*) Evaluate the reactive forces developed at the supports of the frame shown.

(*b*) Interpret physically, by means of appropriate sketches, the significance of the flexibility coefficients and the load terms in the compatibility conditions obtained in part (*a*).

The moments of inertia and the areas of all members are equal.

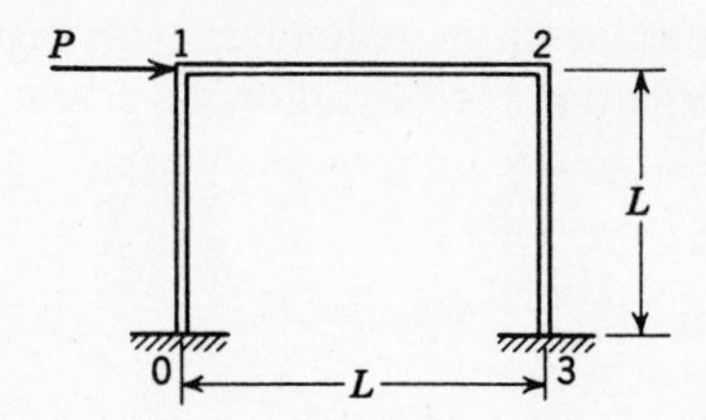

9.64. Find the stress in each member of the truss shown below. The truss is twice statically indeterminate. The cross-sectional areas of the top and bottom chord members are 5.0 sq in. and the cross-sectional areas of the diagonals and verticals are 2.5 sq in. and 1.0 sq in., respectively. The truss is linearly elastic and E is the same for all members.

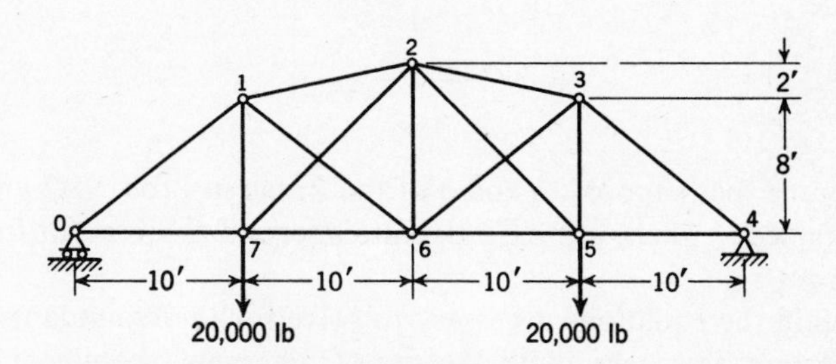

9.65–9.70. Solve Probs. 8.26–8.31 using the principle of least work instead of the unit-dummy-load method.

9.71. Compute the support reactions and construct the bending-moment diagram of the indeterminate beam shown. Use the following theorems to analyze the structure:

(*a*) Castigliano's first theorem

(*b*) Castigliano's second theorem

In part (*a*), interpret, with the aid of sketches, the stiffnesses of the structure. In part (*b*), interpret, with the aid of sketches, the flexibilities of the structure. Neglect shear deformations in the analysis.

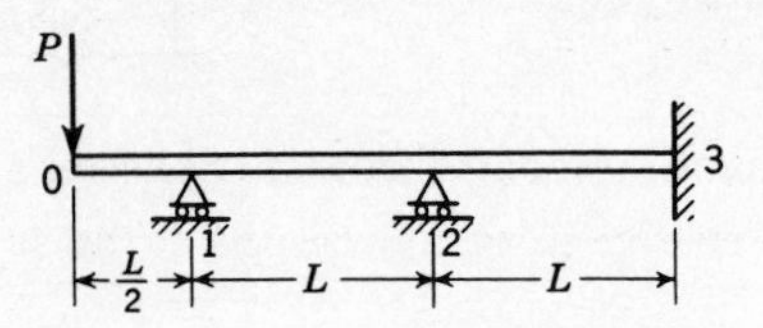

9.72. Evaluate the displacement of the load P in Prob. 9.71.

9.73. The circular arch shown is statically indeterminate to the first degree. Analyze the arch, taking into account shear and axial deformations. Compare the value of the redundant with that obtained neglecting shear and axial deformation for the following cases:

(*a*) $d/L = 1/10;\ h/L = 1/8$
(*b*) $d/L = 1/10;\ h/L = 1/100$
(*c*) $d/L = 1;\ h/L = 1/10$

Take $E/G = 5/2$.

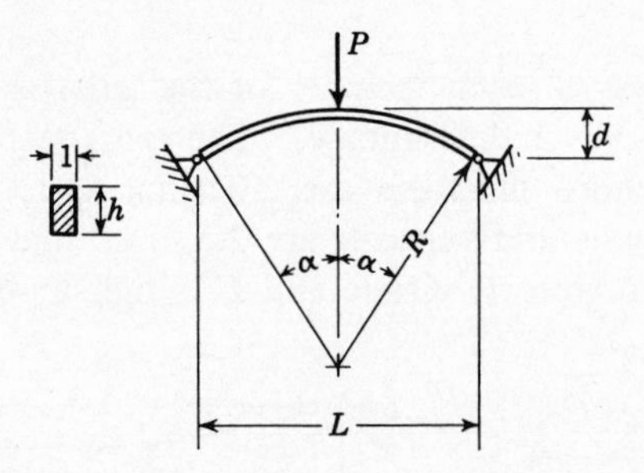

9.74. Assume that supports 0 and 3 of the frame in Prob. 9.43 are hinged instead of fixed. Then, using Castigliano's second theorem (or Engesser's first theorem),

(*a*) Obtain the equations necessary to solve for the redundants.
(*b*) Interpret physically, with the use of diagrams, the meaning of the coefficients of the redundants in one or more of the equations in part (*a*).
(*c*) Evaluate the vertical displacement of the midpoint of member 12.

REFERENCES

1. J. H. Argyris and P. C. Dunne, The General Theory of Cylindrical and Conical Tubes under Torsion and Bending Loads, *Roy. Aeron. Soc.*, pts. I–IV, February, 1947; pt. V, September and November, 1947; pt. VI, May and June, 1949.
2. J. H. Argyris and S. Kelsey, "Energy Theorems and Structural Analysis," Butterworth Scientific Publications, London, 1960.
3. F. M. Baron, Torsion of Multiconnected Thin-walled Cylinders, *J. Appl. Mech.*, vol. 9, no. 2, pp. A72–A74, June, 1942.
4. S. U. Benscoter, Numerical Transformation Procedures for Shear Flow Calculations, *J. Roy. Aeron. Soc.*, vol. 13, no. 8, August, 1946.
5. S. U. Benscoter, Secondary Stresses in Thin-walled Beams with Closed Cross Sections, *NACA-TN 2529*, Washington, D.C., 1951.
6. S. U. Benscoter, A Theory of Torsion Bending for Multicell Beams, *J. Appl. Mech.*, vol. 21, no. 1, 1954.
7. E. Betti, *Nuovo Cimento*, ser. 2, vols. 6–8, 1872.
8. F. Bleich, "Buckling Strength of Metal Structures," McGraw-Hill Book Company, New York, 1952.

9. E. H. Brown, The Energy Theorems of Structural Analysis, *Engineering*, March, 1955.

10. A. Castigliano, "Théorie de l'équilibre des systèmes élastiques," Thesis, Turin Polytechnical Institute, Turin, 1879.

11. T. M. Charlton, "Energy Principles in Applied Statics," Blackie & Son, Ltd., Glasgow, 1959.

12. A. Clebsch, "Theorie der Elasticitat fester Korper," vol. 87, B. G. Teubner, Verlagsgesellschaft, mbH, Leipzig, 1914.

13. R. Courant and D. Hilbert, "Methods of Mathematical Physics," vol. 1, Interscience Publishers, New York, 1953.

14. J. P. Den Hartog, "Advanced Strength of Materials," McGraw-Hill Book Company, New York, 1952.

15. F. L. DiMaggio, Principle of Virtual Displacements in Structural Analysis, *J. Struct. Div., Am. Soc. Civil Engrs.*, vol. 86, no. ST11, pp. 65–78, November, 1960.

16. J. Drymael, Some Theoretical Considerations on Box Beams, *J. Roy. Aeron. Soc.*, vol. 45, no. 361, January, 1941.

17. H. Ebner, Torsional Stresses in Box Beams with Cross Sections Partially Restrained against Warping, *NACA-TM 744*, 1934, translated from "Die Beanspruchung dunnwandiger Kastentrager auf Drillung bei behinderter Querschnittswolbung, *ZFM*, Dec. 14, 1933.

18. H. Ebner and H. Koller, Calculation of Load Distribution in Stiffened Cylindrical Shells, *NACA-TM 866*, 1938.

19. F. Engesser, Ueber statisch unbestimmte Träger . . . , *Z. Architekten Ing.*, vol. 35, Hannover, 1889.

20. V. M. Faires, "Thermodynamics," Macmillan Company, New York, 1957.

21. H. Ford, "Advanced Mechanics of Materials," John Wiley & Sons, Inc., New York, 1963.

22. J. E. Goldberg, The Application of Heaviside's Step Function to Beam Problems, *Proc. Am. Soc. Civil Engrs.*, no. 202, September, 1953.

23. W. J. Goodey, Stress Diffusion Problems, *Aircraft Eng.*, vol. 18, no. 213, 1946.

24. N. J. Goodier, Torsional and Flexural Buckling of Bars of Thin-walled Open Section under Compressive and Bending Loads, *J. Appl. Mech.*, vol. 9, no. 3, pp. 103–107, September, 1942.

25. Handbook of Aeronautics, No. 1, Structural Principles and Data, 4th ed., *Roy. Aeron. Soc.*, Sir Isaac Pitman & Sons, Ltd., London, 1952.

26. M. Hetenyi, "Beams on Elastic Foundations," The University of Michigan Press, Ann Arbor, Mich., 1946.

27. M. Hetenyi (ed.), "Handbook of Experimental Stress Analysis," John Wiley & Sons, Inc., New York, 1950.

28. N. J. Hoff, "The Analysis of Structures," John Wiley & Sons, Inc., New York, 1956.

29. F. E. Hohn, "Elementary Matrix Algebra," Macmillan Company, New York, 1958.

30. L. V. Kantorovich and V. I. Krylov, "Approximate Methods of Higher Analysis," 4th ed., translated from the Russian by C. D. Benster, Interscience Publishers, New York, 1958.

31. R. Kappus, Jahrbuch der deutschen Luftfahrt-Forschung, *Luftfahrt-Forsch.*, vol. 14, p. 444, 1937; also translated in *NACA-TM 851*, 1938.

32. L. Kirste, Sur le calcul des poutres en caisson, *L'Aeronautique* (*L'Aerotechnique*), vol. 19, no. 212, pp. 1–6, January, 1937; no. 216, May, 1937.

33. P. Kuhn, Stress Analysis of Beams with Shear Deformation of the Flanges, *NACA Rept. 608*, 1937.

34. P. Kuhn, "Stresses in Aircraft and Shell Structures," McGraw-Hill Book Company, New York, 1956.

35. C. Lanczos, "The Variational Principles of Mechanics," 2d ed., University of Toronto Press, Toronto, Canada, 1962.

36. H. L. Langhaar, "Energy Methods in Applied Mechanics," John Wiley & Sons, Inc., New York, 1962.

37. R. L'Hermite, "Résistance des matériaux, théorique et expérimentale," tome 1, "Theorie de l'elasticite et des structures elastiques," Dunod, Paris, 1954.

38. W. H. Macauley, A Note on the Deflection of Beams, *Messenger Math.*, vol. 48, 1919.

39. J. A. L. Matheson, "Hyperstatic Structures," vol. 1, "An Introduction to the Theory of Statically Indeterminate Structures," Butterworth Scientific Publications, London, 1959.

40. J. C. Maxwell, On the Calculation of the Equilibrium and Stiffness of Frames, *The London, Edinburgh, and Dublin Philosophical Magazine*, vol. 27, p. 294, 1864.

41. G. Murphy, "Advanced Mechanics of Materials," McGraw-Hill Book Company, New York, 1946.

42. R. E. Newton, Shear Lag and Torsion Bending of Four-element Box Beams, *J. Aeron. Sci.*, October, 1945, pp. 461–467.

43. A. S. Niles and J. S. Newell, "Airplane Structures," 3d ed., John Wiley & Sons, Inc., New York, 1943.

44. J. W. Paull, The Energy Theorems of Structural Mechanics, *Australian J. Appl. Sci.*, vol. 14, 1963.

45. D. J. Peery, "Aircraft Structures," McGraw-Hill Book Company, New York, 1950.

46. J. W. S. Rayleigh, Some General Theorems Relating to Vibrations, *Proc. London Math. Soc.*, vol. 4, pp. 357–368, 1873.

47. J. W. S. Rayleigh, "Theory of Sound," 1st ed., 1877, 2d ed. revised, Dover Publications, Inc., New York, 1945.

48. J. W. S. Rayleigh, On the Calculation of Chladni's Figures for a Square Plate, *The London, Edinburgh, and Dublin Philosophical Magazine*, vol. 22, pp. 225–229, 1911.

49. H. Reissner, Neuere Probleme aus der Flugzeugstatik, *ZFM*, September, 1926; April, 1927.

50. W. Ritz, Uber eine neue Methode zur Losung gewissen Variationsprobleme der mathematischen Physik, *J. Reine Angew. Math.*, vol. 135, 1908.

51. G. L. Rogers and M. L. Causey, "Mechanics of Engineering Structures," John Wiley & Sons, Inc., New York, 1962.

52. J. L. Sackman, Transverse Normal Stresses in Beams, *J. Eng. Mech. Div., Am. Soc. Civil Engrs.*, vol. 89, no. EM4, pp. 13–16, August, 1963.

53. D. R. Samson, The Analysis of Shear Distribution of Multi-cell Beams in Flexure by Means of Successive Numerical Approximations, *J. Roy. Aeron. Soc.*, vol. 58, no. 518, pp. 122–127, February, 1954.

54. E. E. Sechler, "Elasticity in Engineering," John Wiley & Sons, Inc., New York, 1952.

55. E. E. Sechler and L. G. Dunn, "Airplane Structural Analysis and Design," Dover Publications, Inc., New York, 1963.

56. F. B. Seely and J. O. Smith, "Advanced Mechanics of Materials," 2d ed., John Wiley & Sons, Inc., New York, 1957.

57. D. E. Smith, "History of Mathematics," vol. I, Dover Publications, Inc., New York, 1958.

58. I. S. Sokolnikoff, "Mathematical Theory of Elasticity," 2d ed., McGraw-Hill Book Company, New York, 1956.

59. I. S. Sokolnikoff and R. M. Redheffer, "Mathematics of Physics and Modern Engineering," McGraw-Hill Book Company, New York, 1958.

60. R. V. Southwell, "Theory of Elasticity," Oxford University Press, London, 1936.

61. S. P. Timoshenko and J. N. Goodier, "Theory of Elasticity," 2d ed., McGraw-Hill Book Company, New York, 1951.

62. S. P. Timoshenko, "History of Strength of Materials," McGraw-Hill Book Company, New York, 1953.

63. S. P. Timoshenko, "Strength of Materials; Part II, Advanced Theory and Problems," 3d ed., D. Van Nostrand Company, Inc., Princeton, N.J., 1956.

64. S. P. Timoshenko and J. M. Gere, "Theory of Elastic Stability," 2d ed., McGraw-Hill Book Company, New York, 1961.

65. S. P. Timoshenko and D. H. Young, "Theory of Structures," 2d ed., McGraw-Hill Book Company, New York, 1965.

66. A. A. Umanskii, "Kruchenie i iz gib tonkostennykh aviakonstruktsii," Oborongiz, Moscow, 1939.

67. J. A. Van den Broek, "The Elastic Energy Theory," John Wiley & Sons, Inc., New York, 1931.

68. V. Z. Vlasov, "Thinwalled Elastic Beams," 2d ed., translated from the Russian "Tonkostennye uprigie sterzhni" by the Israel Program for Scientific Translations for the N.S.F. and the Dept. of Commerce, U.S.A., Office of Technical Services, Washington, D.C., 1961.

69. T. Von Karman and N. B. Christensen, Methods of Analysis of Torsion with Variable Twist, *J. Aeron. Sci.*, pp. 110–124, April 1944.

70. T. Von Karman and W. Z. Chien, Torsion with Variable Twist, *J. Aeron. Sci.*, vol. 13, no. 10, pp. 503–510, October, 1946.

71. H. Wagner, Torsion and Buckling of Open Sections, *NACA-TM 807*, 1936.

72. C. T. Wang, "Applied Elasticity," McGraw-Hill Book Company, New York, 1953.

73. J. T. Weissenburger, Integration of Discontinuous Expressions Arising in Beam Theory, *AIAA J.*, vol. 2, no. 1, January, 1964.

74. H. M. Westergaard, On the Method of Complementary Energy, *Trans. Am. Soc. Civil Engrs.*, vol. 107, 1942.

75. D. Williams, The Relations between the Energy Theorems Applicable in Structural Engineering, *The London, Edinburgh, and Dublin Philosophical Magazine and Journal of Science*, ser. 7, vol. 26, no. 177, November, 1938.

76. D. Williams, "An Introduction to the Theory of Aircraft Structures," Edward Arnold (Publishers) Ltd., London, 1960.

INDEX